REACHING FOR THE
HIGH FRONTIER

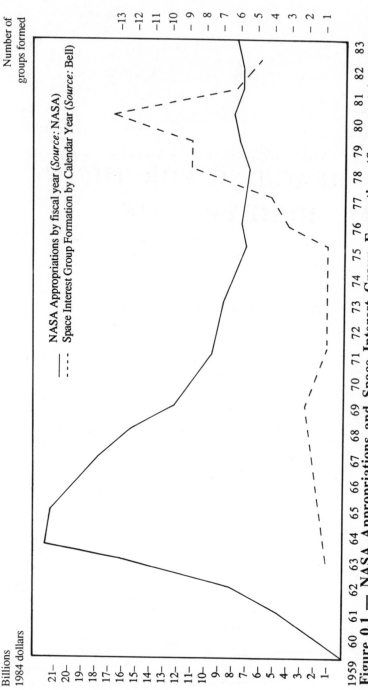

Figure 0.1 — NASA Appropriations and Space Interest Group Formation (*Source:* Adapted by the author from Trudy E. Bell, "Space Activists on Rise," *Insight* (National Space Institute), August/September 1980, 1, 3, 10 and Trudy E. Bell, "From Little Acorns . . . American Space Interest Groups, 1980-1982," unpublished paper. Material provided by NASA and Trudy E. Bell.)

REACHING FOR THE HIGH FRONTIER

The American Pro-Space Movement, 1972–84

Michael A. G. Michaud

Foreword by Joseph P. Allen

PRAEGER

New York
Westport, Connecticut
London

Library of Congress Cataloging-in-Publication Data

Michaud, Michael A. G.
    Reaching for the high frontier.

    Bibliography: p.
    Includes index.
    1. Astronautics—United States—History.
2. Astronautics and state—United States—History.
I. Title.
TL789.8.U5M53 1986   629.4'0973   86-91456
ISBN 0-275-92151-4 (alk. paper)
ISBN 0-275-92150-6 (pbk. : alk. paper)

Library of Congress Catalog Card Number: 86-91456
ISBN:0-275-92151-4 (alk. paper)
ISBN:0-275-92150-6 (pbk. : alk. paper)

First published in 1986

Praeger Publishers, 521 Fifth Avenue, New York, NY 10175
A division of Greenwood Press, Inc.

Printed in the United States of America

The paper used in this book complies with the Permanent
Paper Standard issued by the National Information Standards
Organization (Z39.48-1984).

10 9 8 7 6 5 4 3 2 1

TO THOSE

who are trying to make the pie larger

Contents

Part II: Issues

Part III: Summing Up

List of Tables

List of Figures

Foreword

As I sit at my office desk my mind sometimes wanders. I find myself looking out through the window at the flat Texas countryside stretching, seemingly forever, across the Gulf Coast toward the bustling city of Houston visible on the horizon. Were it not for this distant city and the traffic I see off to the far left, the landscape of rough grass and Texas scrub could easily be from the Wild West of the past century. But then my thoughts come back to the present and my eyes move away from the window to the model of the white Space Shuttle Orbiter poised in perpetual landing six inches above the desk, to the miniature replica of a communication satellite, to the chunk of charred heat shield suspended inside a ponderous transparent paperweight, and then to the assortment of space photos – launches, landings, and earthscapes mounted in frames around the walls – and I find myself astonished, even again, by all of this. The time cannot be the past century. We are living in the Space Age.

The newspaper on my desk today displays a headline, "Voyager Finds Two More Uranian Moons." Even as I write this paragraph, the Voyager spacecraft, eight years into its journey, is encountering Uranus, the seventh planet out from the Sun. Voyager is 2 billion miles away from us and yet communicates back in wonderful detail the secrets of this enormous planet. Perhaps I am still astonished by this because I did not grow up in the Space Age, nor did I dream as a youngster of being a space worker. In those days there were no such things as astronauts, launch directors, space suit technicians, or space workers of any kind. The only satellite that orbited the Earth was one we called the Moon.

If I dreamed then about "reaching for the high frontier," I might have been imagining myself as homesteading a plateau nestled in the Rocky Mountains, a place that *sounded* like a frontier and certainly was higher than the bluffs of the Wabash River near my home. But such dreams had nothing to do with space, and, indeed, I was in college before Sputnik made its first orbits around the Earth. Now it is hard to believe that only three decades have passed and in these years our space workers have orbited Earth, walked on the Moon, and lived in zero gravity for months at a time. In addition, we have sent our robots, spaceships which in a way are extensions of ourselves, to land on, or orbit around, or fly past every planet in our solar system but one.

Some say that out of all the generations that have preceded us on planet Earth we will be the last to live exclusively on its surface, that our children and grandchildren, some of them at least, will live elsewhere. In fact, the National Commission on Space now urges this nation to begin construction of the ultimate railroad – a space transportation system with lines to low Earth orbit, to geosynchronous Earth orbit, to the Moon, and to Mars and beyond. What just 30 years ago would have been wildly impossible is today not only possible but is even being recommended as appropriate to undertake. We are all privileged to be front-row center to these space expeditions, to watch and, to some degree, to participate in history as it is being written. But history is more than just events, it is people being motivated by convictions, beliefs, enthusiasm, and, in some cases, the sheer audacity to cause historic events. *Reaching for the High Frontier* is about such people.

Joseph P. Allen
Houston, Texas

Editor's Note: This foreword was written and mailed just before the tragic explosion of the Space Shuttle Challenger on January 28, 1986. Dr. Allen, a former Shuttle astronaut himself, was deeply moved by this event. However, he believes that the theme of his foreword should remain unchanged.

Preface

> Space is an extraordinary interplay of technology, of politics, of science, of market forces, of visionaries . . ., of backward-looking people, of strongly-held views, the rationality of which is not always obvious.
>
> Sir Herman Bondi, 1985[1]

When I was a boy, humans had not yet ventured into space. But I, and many others, knew that spaceflight was just around the corner. The 1950 film *Destination Moon* stirred our imaginations. By the early 1950s, the writings of spaceflight advocates such as Wernher von Braun, Arthur C. Clarke, and Willy Ley were spreading ideas about the excitement and promise of space.[2] Growing up in California, I was very much aware of the brave men at nearby Edwards Air Force Base, who were pushing needle-nosed aircraft through the sound barrier. By the late 1950s, pilots of the X-15 were riding that half-rocket, half-plane higher and higher into the sky, probing the edge of space. Something of historic importance was happening – something in which I believed I should be involved as more than a spectator. But how? There was no obvious means for a person without technical qualifications.

Since the last Apollo flight to the Moon in December 1972, I have been following the story of ordinary Americans, mostly of younger generations, who did get involved with the spaceflight dream. They tried to speed its coming as a real experience, not just for the few, but for the many. Some of these people have devoted vast amounts of their time, energy, and, in some cases, money to seeing the dream come true, even when they received no direct economic or professional return. As space activist David C. Webb once commented to me, "These people deserve a footnote in history."[3]

This book is not about space technology, the aerospace industry, or the National Aeronautics and Space Administration. Beyond the first chapter, it is not about the early space dreamers and space pioneers of older generations, whose achievements have been recorded elsewhere.[4] It is about more recent generations of American citizens organizing around a set of ideas and aspirations, mixing idealism and pragmatism in modern, American ways.

In this book I attempt to do two things: (1) to provide an informal, capsule history of the pro-space "movement" in the United States from 1972 through 1984 and (2) to present a simple analysis of American pro-space groups as an interest group phenomenon. I focus on the organized part of the pro-space phenomenon, the visible tip of the interest iceberg. I argue that pro-space activity in the United States has come a long way since the end of the Moon race. Something new has been added to the first rush of diffuse enthusiasm and the more prosaic concern for economic and professional advantage. A new generation of space advocates has emerged, convinced that space holds answers to such real-world problems as economic growth, environmental degradation, international tension and the threat of nuclear war, and the need for a hopeful future. Many of these people aspire to go into space themselves. By the early 1980s, the groups they had formed were showing increasing political sophistication and entrepreneurship as they pursued both direct and indirect approaches to expanding human activity in space. They were supported, if passively, by a growing pro-space constituency. The things they were doing interacted with trends that seemed to converge in the early 1980s: regular access to space through the Space Shuttle, space commercialization, heightened interest in space defense, a decision to build a space station, and revived confidence in the future of the United States.

The period January 1972 to January 1984 provides a useful frame. In 1972, the last Apollo mission to the Moon marked the end of the first, adventurous era of U.S. spaceflight. But that year also was a beginning. In January 1972, the Nixon administration announced a decision to build a Space Shuttle, the enabling vehicle for a new era of U.S. space activity. Twelve years later, with the Shuttle performing superbly, the Reagan administration announced a decision to build a permanently manned space station, the beginning of what I call "the Second Spaceflight Revolution." The years 1972 and 1984 also mark crucial turning points in the history of space-based ballistic missile defenses. In 1972, the United States and the Soviet Union signed a treaty banning space-based antiballistic missile systems. In January 1984, the Reagan administration formally set in motion the Strategic Defense Initiative, a research program that might lead to space-based defenses. Those end-pieces, those punctuation marks in time, bound an era in space advocacy, during which more Americans accepted broader uses of space, incorporating it into the U.S. cultural and intellectual heritage.

My first priority has been to tell a story. Bits and pieces of it have appeared in magazine articles and in papers presented at conferences. Science writer Trudy E. Bell has been doing biennial surveys of pro-space groups since 1980.[5] But no one, to my knowledge, has pulled the story together in a book. I believe it to be an interesting story, about highly motivated people striving to see their dreams become reality and meeting frustration in their encounters with the hard, unsympathetic world of politics and budgets. Most people do not get involved in politics or lobbying unless their interests are threatened. In the case of the pro-space phenomenon, what was threatened was not so much pragmatic interests as it was a vision of the future.

In this book I also have something to say about how ideas become interests, and how those interests are advanced and defended by people who are new at the game of interest group politics and lobbying. The story reminds us that most big technological ideas have longer tails than we think and are made acceptable only over a period of time, through repeated advocacy that makes them familiar and technological advances that make them more credible.

Members of pro-space groups that emerged after 1972 generally were rather naive about how to get things done in Washington, D.C. The learning process they went through illustrates changes in the world of interest group politics and suggests some limits to what lobbying can do.

The achievements of pro-space activity are also discussed in this book. How effective have the new pro-space activists been? Have they had any significant effect on policy, legislation, programs, and attitudes? Will their dreams become reality?

This is an eclectic book, more a survey than a thorough analysis. I could have tried to make it a more scholarly work of history, political science, or sociology, but such a book might have interested only a handful of specialists. Instead, I have sought to reach a broader audience, including those interested in space as well as those fascinated by U.S. social and political phenomena. I have avoided technical language and social science jargon, trying to make the pro-space adventure accessible to laypeople without specialized knowledge, and to emphasize readability over exactness in social science. If this "popularization" has introduced imprecision in terminology or analysis, I apologize to my more scholarly readers.

Whatever its failings may be as a work of social science, the book may provide useful raw material for professionals in those fields. Space history is becoming a legitimate field, thanks to the efforts of scholars such as Eugene M. Emme, Walter A. McDougall, and John M. Logsdon. When Joseph J. Corn wrote about a similar phenomenon in *The Winged Gospel*,[6] he had the good sense to write a generation after the events under discussion. Perhaps someone will come along 20 years from now to do a better study of the pro-space phenomenon during this period. But this is a start. The book may also be a useful reference for pro-space people and their organizations.

Some general definitions are in order, particularly for those who are not familiar with the space world. Some of the terms used in this book, such as *the space movement, pro-space,* and even the term *space* itself, are convenient shorthand rather than precise social science terminology. To a space enthusiast, *space* is not just a place, but an umbrella concept for a whole set of ideas, technologies, activities, and aspirations having to do with the exploration and utilization of that larger environment in which our planet floats, and with the often unstated hope of journeying outward to have new experiences, open up new possibilities, and escape the limitations and frustrations of life on Earth. It arches over an amazing variety of interests, political views, and personalities.

To be *pro-space* means somewhat different things to different people. It can refer to their support for larger and better government space programs; to a desire to see more activity by individuals, companies, nations, and the human species in the extraterrestrial realm; to a belief that the intelligent use of space offers solutions to national and world problems; or to a philosophical, even emotional conviction that human expansion into space is a natural and desirable next step in the evolution of life on Earth.

What about *the space movement* ? One of my tasks in this book is to examine whether there really is such a thing. Is there a social movement in the United States that supports increased activity in space? Is it coherent enough and large enough to be called a social movement? Whether the reader agrees with my conclusion or not, it is difficult to deny that we have seen a new social phenomenon in this country since the mid-1970s, an upsurge in pro-space opinion and a proliferation of pro-space organizations. One must give this phenomenon a name to make it comprehensible to the general reader, and *space movement* has proved both convenient and partly accurate.

In using the term *citizens group,* I am drawing a distinction between those organizations whose members generally do not have economic or job-related interests in space and those space-related interest groups with tangible interests in the space field, such as industry, professional, and labor organizations. Pro-space citizens groups, of course, are part of a broader phenomenon of citizen activism that became highly visible in U.S. political life in the 1970s.

When I use the term *manned* in connection with future space activities, I of course include women. "Humaned" and "crewed" sound peculiar, "piloted" is not accurate for space stations and Moon bases.

An apology to those who have been left out. I am well aware that individual executive branch officials, members of Congress, congressional staffers, aerospace executives and engineers, and space scientists have made important contributions to the pro-space effort. In most cases, however, those contributions were related in some way to their professions. They were doing their jobs, although often in an imaginative way. What is more interesting to me, and I hope to the reader, are the volunteers and enthusiasts who comprise the bulk of the pro-space movement, and who work with little prospect of reward. They are a cleaner test of the motivations that underlie the pro-space phenomenon. Even in their case, I have left out many deserving people (and many good stories) to keep the book within the limitations of length imposed by my publisher.

The same applies to many space interest groups not included here. I realize that some of the smaller groups have been left out. Instead of being comprehensive, I have tried to pick good examples that make the point. Even those included have been treated rather superficially.

The research on which this book is based was done by a variety of familiar techniques. A large number of publications, particularly those of the space interest groups themselves, were studied over a period of years. Group representatives and other space activists were observed at a variety of meetings, conventions, and congressional hearings. Statistical data concerning space interest groups were drawn on when available. A few knowledgeable academics (there are not many in this field) were asked for their thoughts.

Since there is little written history on this subject, I often have referred to interviews.[7] Oral histories are inherently treacherous, since they depend on human memories that are both fallible and selective. I can only state that I have tried to be objective and fair and to present more than one view in cases of disagreement.

The views expressed in this book are mine and do not necessarily reflect those of my present employer, the U.S. Department of State. Although I take responsibility for its contents, this book is in many ways the product of a cooperative enterprise. Special thanks go to Trudy E.

Bell, the expert science writer who did the initial research on modern U.S. space interest groups; this book would have been much more difficult without her pioneering work. She gave generously of her time and comments, reviewing drafts of all of the chapters except the last two. Marcia S. Smith, the Congressional Research Service's outstanding expert on space (and later executive director of the National Commission on Space), kindly reviewed the first two chapters. My thanks also go to others who have reviewed chapters or parts of them and who have provided comments and suggestions:

Chapter 3: Mark R. Chartrand III and Glen P. Wilson (section on the National Space Institute)

Chapter 4: T. Stephen Cheston (all) and Frederick A. Koumanoff (satellite solar power stations)

Chapter 5: Sandra Lee Adamson, David Brandt-Erichsen, Randall Clamons, Gerald W. Driggers, K. Eric Drexler, H. Keith Henson, Mark M. Hopkins, and Carolyn Meinel (all)

Chapter 7: Kathy Keeton (section on *Omni*)

Chapter 8: Tim Kyger (California space groups)

Chapter 9: Nathan C. Goldman (all), David C. Webb (Campaign for Space), and William G. Norton (American Space Foundation, Congressional Space Caucus, Congressional Staff Space Group)

Chapter 10: Noel Hinners, John E. Naugle, Clark Chapman, and Louis D. Friedman (entire chapter); John Bahcall and Richard Henry (Space Telescope); Peter Boyce (American Astronomical Society); Geraldine Shannon (Space Science Working Group); and Nathaniel Cohen (NASA Advisory Groups)

Chapter 11: Linda Billings (all), Philip Salin and Gayle Pergamit (launch vehicle entrepreneurs); Charles Chafer (Space Services Incorporated); Gary C. Hudson (GCH and Percheron); Courtney Stadd and James Bennett (Starstruck); Scott Webster (Orbital Sciences Corporation); Klaus Heiss (privately operated space shuttles); Mark Frazier (Earthport); and Charles Sheffield (LANDSAT privatization)

Chapter 12: Donald C. Hafner, Harlan G. Moen, James H. Holmes, and John Bosma (entire chapter); Maxwell Hunter II and John Rather (strategic defense); James McGovern (congressional arms control efforts); and Jim Heaphy (Progressive Space Forum)

Chapter 13: Hans Mark and David Williamson (all)

Appendix A: L. J. Carter

Special thanks also go to Michael Fulda, who kindly provided me with his files on efforts to organize the space constituency during the John Anderson presidential campaign in 1980, and to Leigh Ratiner, who allowed me to study the files of the Space Coalition. Marilyn Ehrlich, Alison Podel, and William Day, my editors at Praeger, were patient with missed deadlines and got, I hope, a better book as a result.

No one deserves appreciation more than my wife Grace and my children Jon, Cassandra, Jason, and Joshua, who tolerated my absence at night and on weekends and who put up with a clacking typewriter at strange hours.

List of Acronyms

AAS   American Astronautical Society
ABM   Anti-ballistic Missile
AEA   Aerospace Education Association of America
AIA   Aerospace Industries Association
AIAA   American Institute of Aeronautics and Astronautics
AIS   American Interplanetary Society
AMSAT   Radio Amateur Satellite Corporation
ARS   American Rocket Society
ASAP   American Society of Aerospace Pilots
ASF   American Space Foundation
ASME   American Society of Mechanical Engineers
AWA   Aviation/Space Writers Association
BIS   British Interplanetary Society
CFSD   Citizens for Space Demilitarization
CSSG   Congressional Staff Space Group
ESA   European Space Agency
FASST   Forum for the Advancement of Students in Science and Technology
IAMAW   International Association of Machinists and Aerospace Workers
IEEE   Institute of Electrical and Electronic Engineers
ISCOS   Institute for Security and Cooperation in Outer Space
ISRG   Independent Space Research Group
ISSS   Institute for Space and Security Studies
ISSSS   Institute for the Social Science Study of Space
NAR   National Association of Rocketry
NASA   National Aeronautics and Space Administration
NCSS   National Coordinating Committee for Space
NSI   National Space Institute
OASIS   Organization for the Advancement of Space Industrialization and Settlement
OSTP   Office of Science and Technology Policy
PSF   Progressive Space Forum
PSSC   Public Service Satellite Consortium

SAST	Society for the Advancement of Space Travel
SEDS	Students for the Exploration and Development of Space
SSEC	Solar System Exploration Committee
SSI	Space Studies Institute
SSP	Society of Satellite Professionals
SSWG	Space Science Working Group
USRA	Universities Space Research Association
USSEA	United States Space Education Association
USSF	United States Space Foundation
VFR	Verein fur Raumschiffahrt
WSF	World Space Foundation

The challenge of the spaces between the worlds is a stupendous one; but, if we fail to meet it, the story of our race will be drawing to a close. Humanity will have turned its back upon the still untrodden heights and will be descending the long slope that stretches, across a thousand million years of time, down to the shores of the primeval sea.

Arthur C. Clarke, 1968[1]

This is the moment of the homogenization of the world, when the diversities of societies are eroding, when a global civilization is emerging. There are no exotic places left on Earth to dream about. And for that reason there remains an even greater and more poignant need today for a vehicle, a device, to get us somewhere else.

Carl Sagan, 1973[2]

Space will forever lure us outward because it is a realm without boundaries, without limitations, without an end to the promise of understanding.

Joseph P. Allen, 1984[3]

The era of the High Frontier marks the first time Mankind has ever had a frontier that is only 200 miles away from every person on Earth.

Rockwell International Space Industrialization Study, 1976[4]

Each frontier did indeed furnish a new field of opportunity, a gate of escape from the bondage of the past; and freshness, and confidence, and scorn of older society, impatience of its restraints and its ideas, and indifference to its lessons, have accompanied the frontier.

Frederick J. Turner, 1893[5]

We're going to space if we have to walk.

Jerry E. Pournelle, ca. 1983[6]

The meek will inherit the Earth. The rest of us will go to the stars.

Pro-space slogan, early 1980s[7]

PART I
FORMATION

1

THE SPACE DREAM: REALIZED, THREATENED

> The dreamers of the day are dangerous men, for they may act their
> dream with open eyes, to make it possible.
>
> T. E. Lawrence[1]

> The politics of the moment had become linked with the dreams of
> centuries and the aspirations of the nation.
>
> John M. Logsdon[2]

In 1898, Robert H. Goddard, a boy living in Worcester, Massachusetts, read two science fiction stories in the Boston *Post?*"Fighters from Mars, or the War of the Worlds," by H. G. Wells, and "Edison's Conquest of Mars," by Garrett P. Serviss. His imagination gripped, young Goddard began to think about how to accomplish the physical marvels set forth in those stories.

On the afternoon of October 19, 1899, Goddard climbed a cherry tree to trim off some dead limbs. As he looked toward the fields to the east, he imagined how wonderful it would be to make some device that had even the possibility of ascending to Mars and how it would look on a small scale if sent up from the meadow at his feet. "I was a different boy when I descended the ladder," he wrote later. "Life now had a purpose for me."[3]

Goddard had been infected by the space dream. First through imaginative fiction, then by that moment of insight or conversion that has affected so many others, he was changed into an advocate of spaceflight.

That dream has a long history, suggesting that it reflects deep human motivations. In Western Europe, the theme of journeys to other worlds turned up periodically in works of fiction during the Renaissance and the Age of Reason, and it began to take more realistic form in works of science fiction in the latter nineteenth century, notably by Jules Verne and Kurd Lasswitz.[4] The idea was given new relevance in the public imagination in 1877 when the Italian astronomer Giovanni Schiaparelli announced that he had discovered "canali" on Mars. In the late 1890s and early 1900s, American astronomer Percival Lowell popularized the idea of an alien civilization on the Red Planet, stimulating works of fiction by Edgar Rice Burroughs and others.[5] After the first issue of *Amazing Stories* in 1926, space themes became a mainstay of the growing field of American science fiction, stirring space dreams among its readers.

Unlike most of those intrigued by the idea of space travel, Robert Goddard went on to make major contributions to spaceflight. In the best American tradition of the individual inventor helped by a few assistants, he designed and tested liquid-fueled rockets, launching one for the first time on March 16, 1926. (That date remains a major anniversary on the space interest calendar.) Goddard speculated about reaching the Moon, drawing on the resources of the Moon and the asteroids, and even about interstellar flight. However, because of criticism and ridicule about the idea of a "Moon rocket," Goddard published little, and most of his achievements were not known to the world until much later.

Other pioneering thinkers had a more direct impact within their own cultures. The deaf Russian schoolteacher Konstantin Tsiolkowsky independently conceived many of the basic ideas of spaceflight between the 1890s and the 1930s, sometimes communicating them through works of fiction, although he was virtually unknown in the West before World War II.[6] In Germany, Hermann Oberth provided a firm theoretical foundation for spaceflight in the 1920s, as well as suggesting some of the useful things we could do with space technology, and served as an adviser in the making of a 1929 film about a journey to the Moon.[7]

These seminal thinkers, plus others only slightly less famous in the space fraternity, set the intellectual context for flight into space and for travel to other worlds. They stimulated others to speculate about journeys into space, sometimes in works of science fact, at other times in science fiction. Some of the converts tried to turn the vision into reality by working on the technology and by forming organizations to advocate spaceflight.

THE EARLY SPACEFLIGHT ORGANIZATIONS

The first significant pro-spaceflight organization was the *Verein fur Raumschiffahrt* (Society for Space Travel), founded in Germany in 1927. One of its early members was the young Wernher von Braun, who had been stimulated at least in part by the Lasswitz novel *Two Planets*. An American Interplanetary Society was created by science fiction writers and readers in New York in 1930, and a British Interplanetary Society was organized in Liverpool in 1933. In the Soviet Union, rocket societies sprang up during the same period.[8]

A common theme among these societies was to advance the enabling technology of the rocket. German, American, and Russian experimenters began testing small rockets in the 1930s (such experimentation was forbidden in Britain). The ultimate goal of this work clearly was to send humans into space.

The American Interplanetary Society listed first among its purposes promotion of interest in and experimentation toward interplanetary expeditions and travel.[9] To lessen public ridicule, that group changed its name to the American Rocket Society in 1934. The one founder who remained with the organization, G. Edward Pendray, ended his 1945 book *The Coming Age of Rocket Power* with words that would ring true to later generations of space enthusiasts:

> There is something about rocket power which transcends the bleakly mechanical aspects of the subject, and changes its followers into missionaries. . . . Whatever it may be, those of us who have spent years in the study and development of rockets have acquired an emotion about them which is almost religious. We somehow feel privileged, as though we had stood in those years at some obscure crossroads in history, and seen the world change. We do not know exactly what we have loosed into the Earth, any more than Gutenberg with his movable type, or DeForest with his radio tube. But we feel in our souls that it is magnificent and wonderful, and that the human race will be richer for it in time to come.[10]

Meanwhile, a small team of enthusiasts and students at the Guggenheim Aeronautical Laboratory of the California Institute of Technology in Pasadena initiated rocket testing in 1936. This project became this nation's first officially supported rocket program and later evolved into the Jet Propulsion Laboratory, which is responsible for many of America's missions of planetary exploration.[11]

National Air and Space Museum historian Frank H. Winter, who examined the early spaceflight groups in his book *Prelude to the Space Age*, saw them as parts of an international spaceflight movement. He concluded that the greatest contribution of the early rocket societies was the subtle creation of a sort of pre-Space Age "lobby" or generation that was prepared to accept spaceflight.[12] However, the early societies did not have the resources to develop the technology of manned spaceflight themselves, and they found it difficult to sell their vision to governments, which tended to dismiss the idea of going into space as fantastic or irrelevant. Only when the rocket was linked to military need and to international political competition did the spaceflight dream become real. Primarily for those reasons, governments called on the then small cadres of experts in rocketry to develop vehicles capable of launching warheads, satellites, and later humans and formed official agencies to carry out national space programs.

THE DREAM REALIZED

University of California historian Walter A. McDougall has written the following:

Although the mathematical, chemical, and metallurgical skills necessary for practical research into rocketry were present by the 1920s, the investment required for orbital flight was so large and the immediate military or economic benefits so uncertain that the genesis of spaceflight in our time is no more self-explanatory than Iberian sponsorship of world navigation for the fifteenth century.[13]

The adoption by governments of rocketry was due in large part to dedicated advocacy by individuals, small groups, and networks of people who believed in the promise of this new technology and who often shared the spaceflight dream.

Sociologist William S. Bainbridge examined this phenomenon in his 1976 book *The Spaceflight Revolution*, chronicling how von Braun and his Russian counterpart Sergei Korolev, and their allies, persuaded their military and political masters of the military utility of rocketry, when their primary interest was going into space.[14] For them, military rockets were a means to another end. At Peenemunde during World War II, German experts took a giant stride toward space by developing a rocket (the V-2) that was essentially a form of long-range artillery but that also tested the technology for taking payloads into orbit. In the German case, the core of the advocacy was transferred to the United States after World War II, when a team of rocket scientists and engineers including von Braun went to work for the U.S. Army. That team was to be a major influence on the course of the U.S. space enterprise for its first two decades.[15]

By 1946, rocket technology had become credible enough to be considered both for longer-range missiles and for launching satellites, although the first exploratory research programs were cancelled for lack of cost-effective missions. In 1949, an upper stage launched from a captured V-2 reached an altitude of about 250 miles, well into space.[16] In the 1950s the rocket team developed the Redstone missile, an improvement on the V-2. Meanwhile, scientists proposed the use of rockets to launch satellites in connection with the 1957-58 International Geophysical Year.[17] Von Braun believed that an American satellite could have been launched as early as 1956 if the Army team had not been restrained by authorities in Washington.[18]

Despite the efforts of the small spaceflight advocacy in the United States, there was no political support for manned spaceflight. The idea of travel beyond the Earth had not been accepted by the prevailing American political culture, and such projects were regarded by many as a waste of money. Even within the technical community, there were strains between visionaries like von Braun and more conservative technologists who saw many obstacles in the way.[19]

The Classic Agenda for Manned Spaceflight

By the mid-1950s, spaceflight advocates had achieved some degree of consensus about a logical order of goals to be achieved in space. The principal stages were a rocket-powered launch vehicle (perhaps a winged rocket plane that could return to the Earth); a station in orbit around the Earth; rocket-powered vehicles to transport humans outward from the station to other bodies in the solar system; a Moon base; and a manned expedition to Mars, followed by the establishment of a base and eventual colony on that planet. There were generalized visions of travel throughout the solar system. A few visionaries predicted that someday humans would travel to other stars.

In the early 1950s, Wernher von Braun, Willy Ley, and Arthur C. Clarke – all members of early rocket societies – had particular success in communicating this vision through works of nonfiction, assisted by the visualizations of a new breed of astronomical and space illustrators. They were aided significantly by science fiction writers, notably Robert A. Heinlein and Clarke wearing his science fiction hat. One effect of their works was to spread the classic agenda for manned spaceflight.

That agenda had its roots in Europe, particularly in the German rocket team. The agenda was very similar to the early stages of what science fiction writer and publisher Donald Wollheim called "the full cosmogony of science fiction future history."[20] Long before, in the Soviet Union, Tsiolkowsky had conceived a 14-point plan for the conquest of space, which included a rocket plane with wings, space stations around the Earth, and colonies on asteroids.[21]

The classic agenda also could be seen as an upward extension of an even older dream – flight. As early as the 1930s, visionary engineer Eugen Sanger and his wife Irene Sanger-Bredt had begun designing a winged, rocket-powered craft, at one point called an "antipodal bomber," that could take humans into space and return them to Earth. A generation later it appeared that the testing of high-speed, high-altitude aircraft such as the X-15 was leading in that direction, although in fact that road into space was not taken until 1981. However, the space dream is different from the dream of flight in at least one respect: it includes the idea of finding or making new worlds, alternatives to the Earth.

The classic agenda is positive and expansionist. To the advocates of spaceflight, humanity's outward expansion was not only desirable but obviously so. Once the tools – particularly the rocket – were in hand, they thought others would see the logic of their plans. However, their agenda was altered by the realities of politics.

After Sputnik

In practical terms, the Space Age began on October 4, 1957, with the launch of the Soviet satellite Sputnik I by means of a military missile – the same approach that von Braun had proposed. This Soviet technological and political success placed the United States under strong pressure to compete. After the American rocket designed to launch the Vanguard satellite for civilian scientific purposes failed embarrassingly in December 1957, von Braun's Army team placed the first American satellite in orbit with another missile on January 31, 1958. The space race was on.

Sputnik was not necessarily an unwelcome event for American advocates of spaceflight, who knew it would accelerate the space agenda.

Gary Paiste, the Washington representative of the pro-space organization Spacepac, recalls being pleasantly surprised, although he resented the fact that Soviet secrecy "stole from us our ability to participate in the event."[22]

Later in 1958, the United States took a crucial step that significantly altered the context for spaceflight advocacy. It created the National Aeronautics and Space Administration, a government agency whose primary purposes were the development of space technology and the exploration of space. This had not been anticipated by most of the early space advocates, many of whom thought more in terms of private space ventures; the one thing that no one had envisioned, states former NASA Administrator Thomas O. Paine, was NASA.[23] This step created an institutional advocate for the space program and provided a home within the government for some members of the spaceflight advocacy. It also lead to the formation of related oversight committees of Congress that had a strong interest in the continuation of that program. In addition, the creation of NASA established in the American public mind the idea that spaceflight was essentially an open, civilian program, distinct from military activity in space.

The satellite race also led to another crucial decision: to send humans into space. The first seven American astronauts were selected and introduced to the public in 1959, in scenes dramatized in the book and the film *The Right Stuff*.[24] The spaceflight dream was becoming real.

The pace and direction of the American space program were affected profoundly when the Soviet Union put Yuri Gagarin into orbit on April 12, 1961. This was at a time when the United States was deeply concerned about the contest with the Soviet Union for world leadership, with national security in the broadest sense. An American-backed invasion of Cuba had failed disastrously; President Kennedy's legislative program was encountering difficulty in Congress; and the economy was in recession (employment in the aerospace industry had dropped significantly between 1957 and 1960).[25] Above all, the Soviets were claiming superiority in a field that symbolized technological competence and command over the future. The Kennedy administration needed a bold stroke.

Space policy scholar John M. Logsdon, in his book *The Decision to Go to the Moon*, reported that President Kennedy asked his advisers to give him options for a race the United States could win. After the first American (Alan Shepard) rode briefly into space in May 1961 atop a Redstone rocket, NASA Administrator James Webb and Secretary of Defense Robert McNamara put together a memorandum to the President that recommended landing an American on the Moon. Significantly, that memorandum included the statement that "it is man, not merely machines, in space that captures the imagination of the world." NASA, which had studied the possibility of a Moon landing for two years, assured government leaders that the project was technologically feasible.[26]

On May 25, 1961, President Kennedy went before Congress to say that, before the decade was out, the United States should land a man on the Moon and return him safely to Earth. This most dramatic of space decisions was from the top down and was not primarily the result of lobbying by economic or bureaucratic interests or of a specific demand from the broader public. However, the space advocacy had helped prepare the way, both by helping to develop the technology and by spreading the idea that spaceflight was feasible.

Because the Moon landing program was intended to win a race, time pressure led to a short-cutting of the classic agenda for manned spaceflight. That agenda suggested first building the Earth-to-orbit transport rocket that would ferry up the parts of a space station, which in turn would be the platform for assembling and launching ships to farther destinations. For a lunar landing mission, the rendezvous between the crew and the Moonship would take place in low Earth orbit. Jay Holmes, in his 1962 book *America on the Moon,* reported that NASA Administrator James Webb thought the time might be ripe by 1965 to begin work on a 10- or 12-man orbiting station.[27]

Instead, NASA accepted the arguments of engineers John Houbolt and William Michael that the mission could be done more quickly by launching from the Earth's surface directly to the vicinity of the Moon, there detaching a landing craft that later would return to rejoin the vehicle orbiting the Moon. This technique of lunar orbit rendezvous did save time and helped ensure American victory in the Moon race.[28] However, it did not leave behind the building blocks for a continuing presence in orbit. Although the Moon race may have hastened the first journey of humans to another world, it also may have postponed the realization of the classic agenda, and of a permanent human presence in space.

American Space Groups during the Moon Race

With the sudden expansion of the spaceflight field came a growth and diversification of space interest organizations. By the 1950s, the

American Rocket Society (ARS) had thousands of members, placing increasingly strong emphasis on rocketry and its missile applications rather than on space travel, and tending to be more stringent about technical credentials for membership. In the view of some who belonged to the ARS, it lost much of its adventurous quality and became relatively conservative. In 1963, the ARS merged with the Institute of the Aeronautical Sciences to form the American Institute of Aeronautics and Astronautics, which became the largest American organization of aerospace professionals (see Chapter 2).

In 1953, space enthusiasts and British Interplanetary Society members Hans J. Behm and James H. Rosenquist called the first meeting in New York City of "an American counterpart to the British Interplanetary Society." In January 1954, this became the American Astronautical Society (AAS), whose intent was to emulate some of the speculative and visionary qualities of the British organization. The AAS grew to be a respected technical society in the space field, holding numerous conferences and putting out the highly technical *Journal of the Astronautical Sciences*, as well as useful series of books on the history of astronautics and other subjects. Many individual AAS members were actively involved in turning the spaceflight revolution into space programs during the 1950s and 1960s. After the cutbacks in the American civil space program in the late 1960s and early 1970s, however, AAS membership dropped to about 500. To some, the society seemed headed in the same conservative direction as the old ARS. During a 1979 symposium on the society's history, its activist President Charles Sheffield warned that the AAS was aging and was getting out of touch with the younger generation of space enthusiasts.[29] As we will see later, the AAS has since made something of a comeback.

Sputnik prompted the formation of additional space interest groups. At the time of the launch, Erik Bergaust (later to be Wernher von Braun's biographer) founded the National Rocket Club. In 1964, this became the National Space Club, a Washington, D.C.-based organization which came to be composed largely of the Washington representatives of aerospace companies and other NASA and defense contractors. By 1983-84, the club had grown to include about 800 members. David Wilkinson, president of the club for that year, described it as "the marching and chowder society of the space community in Washington."[30] The club is best known for hosting briefings by prominent individuals in the space field, usually at business lunches on

Capitol Hill, and for its Goddard Dinner every March, the social event of the year on the Washington space interest calendar. The club also presents awards in a number of space-related fields and a scholarship to the winner of the annual Goddard essay contest.

Another product of Sputnik was the National Association of Rocketry (NAR), founded by engineer and writer G. Harry Stine in 1957. The NAR, which had no connection with the ARS, was intended to get young people involved and to promote competition and safety in the launching of rockets by hobbyists. According to Stine, the NAR in 1984 had 3,000 members all over the United States. Although it has no conscious connection with the pro-space movement, Stine notes that he finds NAR graduates in the halls of NASA.[31]

Science fiction writer Jerry E. Pournelle, a leading figure in one of the new citizens space interest groups of the 1970s, recalls a lesser-known organization of the 1950s called the Society for the Advancement of Space Travel (SAST). Consciously modeled on the *Verein fur Raumschiffahrt*, the SAST included some old VFR members such as von Braun and Willy Ley. In Pournelle's view, it provided an alternative to the ARS, which was very strict about credentials for membership. At one time, he states that the SAST had about 600 members. However, Pournelle recalls that the society simply ceased to exist after the Moon landing decision. Its members thought the job had been done.[32]

The Impact of the Moon Race

It is difficult for us to recall now the scale of Project Apollo in relation to its time. If it had been done in 1983, it would have cost an estimated $82 billion.[33] Project Apollo involved not just NASA and its employees but also hundreds of firms on contract; it was an unprecedented peacetime mobilization of skilled manpower, in a partnership among government, industry, and academia. At its peak in fiscal year 1965, NASA employed over 400,000 people directly or through contractors and absorbed an estimated 5 percent of the nation's scientific manpower.[34] New York *Times* columnist Arthur Krock called the space program the biggest public works and employment operation ever instituted in this country.[35] Such a massive effort, intended in part to upgrade American technology and American economic competitiveness, was bound to have an impact on American society; NASA official Hugh Dryden once described it as "an instrument of social change."[36] Historians once compared the space program to the railroad to examine its potential consequences for society.[37]

Federal spending on space, while spread over most states, tended to be concentrated geographically in the South and the Southwest. California became the leading "space state," in the same way it had dominated aircraft manufacture in previous years. Contract firms sprang up around all the NASA centers.

The Moon landing program also had a considerable psychological impact, magnified by television and other mass media. Millions of Americans were exposed regularly to images of real people walking in space or on the Moon. Much of the language of space entered the American vocabulary, and much of the imagery entered the American imagination. Spaceflight became something familiar, an accepted feature of American life. Younger Americans, particularly those in the "baby boom" generation (born between 1946 and 1964), grew up with space and were less likely than their elders to see spaceflight as an alien or outlandish concept.

Project Apollo raised hopes. It suggested a brilliant, positive future for America through ennobling goals, the use of cooperative social effort, and the application of high technology. "If men can visit the Moon – and now we know that they can – then there is no limit to what else we can do," said Senator Abraham Ribicoff in 1969. "Perhaps that is the real meaning of Apollo 11."[38] Space appeared to offer a new, achievement-oriented agenda, both exciting in the near term and transcendental in its implications. There also were strong transnational elements in the image of astronauts as envoys of humanity, particularly at that unique collective moment when people all over the world watched or listened as Americans touched the Moon. The space adventure seemed to many an opportunity for humans to rise above the issues that divided them, a chance for a new start in a pristine environment.

Pro-space people, believing that the tide of history was with them, made extremely optimistic predictions about achieving the goals in the classic agenda for spaceflight. As early as 1958, von Braun's team at the Army Ballistic Missile Agency put together a plan that foresaw a space station in 1962, a manned lunar expedition by mid-1965, a permanent Moon base by 1973, and a manned expedition to a planet by 1977.[39] Even after the first two steps in the classic agenda were bypassed when we went directly to the Moon with expendable rockets, space visionaries remained hopeful. In his 1970 book about the human future on the Moon, *Where the Winds Sleep*, technical publisher and space advocate

Neil Ruzic predicted a U.S. lunar base by 1975, a Russian landing on Mars in 1981, and lunar cities by 2015.[40]

Unfortunately for the space advocacy, the Moon landing program had political feet of clay. A program that was created primarily to satisfy political needs could fall when those needs were met. A program with only one specific goal, not part of a larger, agreed-on design, was vulnerable to termination when that goal was achieved. A program that reflected a new societal priority was in danger when societal priorities changed. An expensive program without strong interest groups lobbying for it was vulnerable by definition.

THE DREAM THREATENED

In many minds, the space program became synonymous with a politically motivated race to the Moon – something most space advocates had never intended and that few had foreseen. Project Apollo provided a high profile target for critics of the space program. A number of liberal intellectuals saw the Moon race as a Cold War phenomenon and a misallocation of national priorities; sociologist Amitai Etzioni called it a "Moondoggle."[41] There were criticisms directed at a "space lobby" made up of NASA and its contractors, as if it were the space analog of the "military-industrial complex." Both NASA and the aerospace industry came under harsh attack after the Apollo fire of January 1967, which killed three astronauts.[42] The leaders of the House and Senate space committees played a crucial role in preventing this from doing serious political damage to the space program, but disillusionment remained.

That disillusionment grew during the late 1960s, partly because other priorities were demanding national attention and federal government resources. The Vietnam war was draining the national treasury and dividing opinion. Riots in the cities called for more work on urban problems and minority group issues. Negative trends seemed to peak in 1968, with a psychological if not military setback in Vietnam because of the Tet offensive, the assassinations of Martin Luther King and Robert Kennedy, the riot at the Democratic Convention in Chicago, and student unrest. Meanwhile, other constituencies were pressing for a larger share of the federal budget.

Ultimately, the fundamental problem for Apollo was public opinion, as perceived by politicians. As former NASA Administrator Thomas Paine puts it, "the country had shifted under us."[43] The ordinary voter, who had never internalized the spaceflight agenda, saw no point in continuing to pour huge amounts of money into the space program once the prestige and confidence of the United States had been restored. In February 1969, five months before the first Moon landing, the Harris survey reported that 49 percent of respondents "opposed" it, while only 39 percent were in favor.[44] When, at the time of Apollo 11, Vice President Spiro T. Agnew suggested a Mars landing as a follow-on to Apollo, the idea drew public criticism. The character, Senator Grant, in James Michener's 1982 novel *Space* summed up the problem after the first Moon landing when he said: "Well, we've certainly shown the Russians. Now we can turn to other things."[45]

The late 1960s and early 1970s also saw rising criticism of major technological enterprises. NASA, much admired when support for advanced technology ran high, was much criticized when prevailing attitudes changed. There were heightened demands for democratic control over big technology and for analyses that showed that technological endeavors were worth their cost.

Underlying much of the disillusionment with the space program was a sense of frustrated hope. The idealistic vision many held of the space enterprise had been marred by politics and by public criticism of NASA and its contractors. There also was frustration with the elitist nature of spaceflight. The early astronauts were chosen from a narrow sector of society, and the number of people who actually went into space was very small. Even for space enthusiasts, the space adventure remained vicarious. Once the novelty wore off and the race to the Moon was won, the public sense of participating in a collective effort declined.

For many of those attracted to the space dream, the idea of alternate worlds had been brought into question by unmanned scientific spacecraft that showed Venus to be a furnace beneath its clouds and Mars to be a desert under its thin, cold atmosphere. For many, the vision of Mars as a smaller, exotic Earth was destroyed in 1965, when Mariner 4 sent back pictures of a barren, cratered surface looking much like the Moon.

Even within the community of people directly involved with space exploration, issues appeared that reflected fault lines persisting to this day. One of the most basic went to the heart of the spaceflight dream: should space exploration be conducted by men or by machines? Many scientists argued that human presence generally was not necessary and made missions too complex and expensive. President Kennedy's Scientific Advisory Council had not supported a Moon landing.[46] In an August 1968 report, the Space Science Board of the National Academy of

Sciences stated, "We are unable to identify a need in planetary exploration, in the foreseeable future, for the unique capabilities of Man."[47] Spaceflight advocates, by contrast, argued that humans had a versatility no machine could match and that manned missions stirred more public support for space exploration.

There may have been reasons of self-interest underlying this difference of opinion. Space scientists gained most of their public and professional recognition from unmanned missions and were upstaged (and almost excluded) by Apollo; Ralph Lapp commented in 1969 that "the heroic aspect of manned spaceflights has tended to obscure the fundamental values of space science."[48] Manned spaceflight advocates had an obvious interest in the continuation of their own programs. However, there also were hints of fundamental philosophical differences between those who saw space exploration primarily as a means of acquiring knowledge and those who saw spaceflight as an outward extension of human power and presence into a vast new environment. NASA Acting Administrator Paine, speaking after the December 1968 Apollo 8 mission around the Moon, said, "Man has started his drive out into the universe, the beginning of a movement that will never end."[49]

There also was recurrent debate over whether space should be used for activities that, to some, appeared less noble than exploration and research, particularly military and commercial uses. There was an implication that "the heavens" were different and should not be sullied by some of the more competitive aspects of human behavior. Some critics of the American space program saw it as excessively nationalistic and called for greater emphasis on cooperative international ventures in space. All these themes reappeared forcefully by the early 1980s.

The Turndown

In constant dollar terms, NASA's budgets peaked in fiscal years 1964 and 1965, when the research and development costs of the Moon landing program were at their height. After that, civil space program funding went into a prolonged downslide that lasted for ten years, finally bottoming out in fiscal year 1975 at less than one third of the peak (Figure 0.1). The cutting back of the civil space program, begun under Lyndon Johnson, was continued under Richard Nixon. The Nixon administration, which took office in January 1969, was looking for

TABLE 1.1
Comparative Program Accomplishments

	Maximum Pace	Option I	Options II, III
Manned Systems			
Space station	1975	1976	1977
Space base (50 man)	1980	1980	1984
Space base (100 man)	1985	1985	1989
Lunar orbiting station	1976	1978	1981
Lunar surface base	1978	1980	1983
Initial Mars expedition	1981	1983	1986 (II) Open (III)
Space Transportation System			
Earth to orbit	1975	1976	1977
Nuclear orbit transfer stage	1978	1978	1981
Space tug	1976	1978	1981

Source: Adapted from a chart in *The Post-Apollo Space Program – Directions for the Future: Space Task Group Report to the President, September 1969* (Washington, D.C.: National Aeronautics and Space Administration, 1969).

budget economies; the space program, closely associated with two Democratic presidents, was a tempting target.

By 1969, it was clear that NASA needed a post-Apollo agenda. In that year, James Webb was replaced as administrator of NASA by Thomas Paine, a strong advocate of a visionary future in space centered on a space station, which would make permanent the presence of Americans beyond the Earth. In February 1969, President Nixon had directed a Space Task Group of senior officials to provide him with future space program options. That September the group published a report entitled *The Post-Apollo Space Program: Directions for the Future.*[50] For manned spaceflight, this group proposed a space transportation system consisting of a "shuttle" operating between the Earth's surface and low Earth orbit, a "space tug" for moving men and equipment to other orbits, and a nuclear stage for transporting men, spacecraft, and supplies between Earth orbit and lunar orbit, or to more distant destinations. In addition, the group proposed a 6- to 12-person space station, followed by a space base of 50 to 100 people, a lunar orbiting station, and a lunar surface base. As a focus for the development of new capabilities, the group recommended that the United States accept the long-range option or goal of manned planetary exploration, with a manned Mars mission before the end of the century as the first target. The first three of the four broad options presented to the president offered this agenda on different time scales, depending on the level of funding; the fourth included no manned spaceflight.

From the point of view of many spaceflight advocates, this was a splendid plan, a formalization of the classic agenda using efficiently modular components. However, it was politically unrealistic.

The Nixon White House praised the Space Task Group's vision privately but said it was far too expensive and pared it down to one major manned spaceflight project. It came to a choice between the space station, which was the heart of NASA's near-term agenda, and the Space Shuttle. Former NASA Deputy Administrator Hans Mark recalls a crucial meeting in NASA headquarters, when Wernher von Braun said the choice must be the Shuttle, because it was technically more difficult; once the Shuttle was in operation, the station would follow naturally.[51] Space policy scholar John Logsdon adds that NASA's leadership realized that the Shuttle was the program more likely to get Presidential and congressional approval.[52] Military interest in a space transport vehicle also was a factor. Applying economic criteria to space, studies were commissioned that showed that the Shuttle would save money in the long run. One of the leaders of those studies was the Austrian-born economist Klaus Heiss, who later became a prominent advocate of commercial alternatives to government space programs.

The Space Shuttle Decision

Even after a decision was made to go ahead with the Space Shuttle program, NASA did not get what it had wanted: a two-stage vehicle, with both stages piloted and reusable. In mid-1971, the Office of Management and Budget indicated a budget level for NASA that made it impossible to proceed with the initial design. Frequent redesigns and disagreements continued until a few days before President Nixon announced his decision to proceed with a Space Shuttle on January 5, 1972. In effect, the president had chosen the minimum pace for American manned spaceflight, falling between the Space Task Group's option III and no manned spaceflight at all (Figure 1.1).

The shuttle also was threatened in Congress, where Senators Walter Mondale and William Proxmire led a campaign against it, at one point

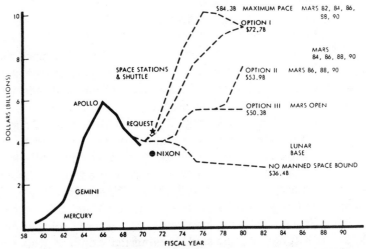

Figure 1.1 — NASA Budget Options (*Source:* Boeing Company statement on NASA authorization bill, hearings before the Committee on Science and Astronautics, U.S. House of Representatives, February 1970, vol. 1 [Washington, D.C.: U.S. Government Printing Office, 1970], p. 755).

nearly succeeding in killing the program. The space program's friends in Congress, such as Chairman Clinton Anderson of the Senate Space Committee and Chairman Olin Teague of the House Space Committee, played an important role in the Shuttle's defense.

The Lean Years

To many advocates of manned spaceflight, the year 1972 was a symbolic turning point. It was the year of the last manned mission to the Moon, Apollo 17, which flew in December of that year (three other planned Apollo missions were cancelled by the Nixon administration). Once the astronauts returned to Earth, America's outward thrust into space seemed stalled. There was no approval for a space station or a Moon base, and the idea of landing humans on Mars appeared to be dead. Although approved, the Space Shuttle was not expected to fly until the late 1970s. Wernher von Braun, who had moved from the Marshall Space Flight Center in Alabama to NASA headquarters in Washington in

1970, found his plans frustrated, and he retired in June 1972. Under pressure to be more "relevant," NASA increased its emphasis on Earth-oriented applications of space technology. For space advocates, the dream of realizing the classic agenda within their lifetimes seemed remote once again. Morale in the pro-space community sank to a low point in the early and middle 1970s, with plaintive cries, resentment, and wistful reminiscences marking much of the literature.

Today, many space activists regard the 1970s as a kind of dark age, when budget constraints and short-sighted politicians held back the manned space program, seriously delaying its next steps. "We lost ten years," says science fiction writer Ben Bova, who became President of the National Space Institute in 1983.[53] Pro-space Republican Congressman Newt Gingrich of Georgia agreed, in a March 1984 speech at the National Space Club's Goddard dinner: "I want to refer to the last decade as the lost decade," he said. In Gingrich's opinion, this was so for three reasons: (1) because of the welfare state culture, (2) because C. P. Snow's "two cultures" came to pass (Gingrich referred to the "the Ludditism of the intelligentsia"), and (3) because space interest groups failed to organize politically.[54]

In fact, there was reason for hope. Apollo systems and hardware derived from them were used in a temporary space station called Skylab in 1973 and 1974. An inspired piece of improvisation, Skylab showed that men could live and work in orbit for months at a time, in much more comfortable conditions than those suffered by the Mercury, Gemini, and Apollo astronauts. There in the mass media were images of Americans in space, saving a partly crippled Skylab with ingenuity and pluck, living in a shirt-sleeve environment, even having fun.[55] (Meanwhile, the Soviets had put up their first Salyut space station in 1971, although they encountered difficulties in making it operational and were, as usual, less generous in media coverage.) In 1975, Apollo technologies were used in the Apollo-Soyuz mission, in which an American spacecraft docked with a Soviet vehicle in orbit.[56]

American and Soviet planetary exploration reached new high points during the 1970s, the peak of an era of exploration that Carl Sagan has compared to the European voyages of discovery centuries before. Extraordinary robot spacecraft, navigated across millions of miles of interplanetary space, sent back wonderful imagery of other worlds, particularly Mars. In 1971, Mariner 9 revealed a rugged planet of huge volcanos, giant valleys, and global dust storms, far more interesting than the cratered desert implied by earlier missions. Viking spacecraft orbited and landed on Mars in 1976, showing us for the first time what it would be like to stand on the Red Planet's surface and scan its horizon. Images from the Viking orbiters suggested that water once had flowed on Mars and that it today may lie frozen in large quantities beneath the surface. Mars began to look like a friendlier place, where human habitation might be possible after all. The idea of manned expeditions to Mars, officially abandoned by NASA in 1971, was subtly revived. Soviet Venera craft landed on the surface of Venus and sent back our first images of that barren landscape.

Meanwhile, Pioneer 10 and 11 flew outward through the asteroid belt to Jupiter, giving us our first close views of that gas giant planet in 1973 and 1974; Pioneer 11 flew on past Saturn. The more capable Voyagers 1 and 2 sent back spectacular pictures of Jupiter and its moons in 1979 and of Saturn and its satellites in 1980 and 1981. The exploration of the solar system was well begun. To advocates of manned spaceflight, to those infected by the space dream, new worlds seemed tantalizingly closer.

Above all, NASA survived. Even after its budgets bottomed out in real terms in 1975, NASA still was an agency with more than 20,000 government employees and more than 100,000 others working under contracts, a budget of several billion dollars a year, and interesting programs. "We were not in an out of business phase," says astronaut Joseph P. Allen, who directed NASA's congressional relations in the late 1970s.[57] Within NASA, individuals kept spaceflight ideas alive in the hope of better times.

To some degree, the "lost decade" was a pause between generations of technology. NASA had its problems with the development of the Space Shuttle, particularly with the very advanced engines in the tail of the Shuttle orbiter. There were delays, and the Shuttle's first flight slipped into the early 1980s. These problems, and the underfunding of the project, required NASA to ask for more money, and other NASA programs were affected by transfers within the space agency's budget. Space scientists in particular were angered by delays in their projects caused by the problems of the Shuttle.[58] Yet there was an implied positive message for the advocates of spaceflight. The first step in the classic agenda was coming, a largely reusable spacecraft that would give Americans regular access to space.

2

CONVENTIONAL RESPONSES

There is no cohesive group behind our space policy comparable to such organized lobbies as the Farm Bureau, the American Medical Association, and the AFL/CIO. . . . Yet, support of, or for that matter, opposition to the space program is widely diffused. It is a personal thing.

Mary A. Holman, 1974[1]

There is a lingering image of powerful lobbying forces behind the American space program during its boom years, a kind of "space-industrial complex." "NASA became a giant under a succession of Congresses that were swayed by the arguments of the aerospace industry," wrote two authors at the time.[2] Amitai Etzioni described the space enthusiasts and "interessants" as "a formidable group."[3]

With a longer perspective, what is striking about the so-called space lobby of the 1960s is not its strength but its weakness. The picture that emerges of the aerospace industry's relationship with NASA is one of mutual dependence.[4] Even together, they were unable to prevent a spectacular downturn in the budgetary fortunes of the civil space program. "The fact that the program was cut drastically," says science fiction writer and National Space Institute President Ben Bova, "shows the absence of influential space interest groups."[5] At the time, the only significant organizations concerned outside the government were the aerospace industry, aerospace workers, and aerospace professionals, which proved unable to sustain the momentum of an Apollo-scale program. Although potentially influential organizations were involved, they were not well organized as a constituency and did not have an agreed-on agenda.

RESPONDING TO THE CUTBACKS

The drastic cutting back of the NASA budget between 1965 and 1974 was the kind of negative event that causes interest groups to mobilize. Clearly, real interests had been created by the space program during the 1950s and 1960s. NASA itself had grown to be a sizeable government agency, and congressional committees and their staffs had expanded in parallel. Large corporations and smaller companies had profited from NASA contracts. Many aerospace workers and engineers had found employment in or because of the space program. The elements of a potentially large, influential lobby with pragmatic economic interests in the space program were there. They included the components of the classic "iron triangle" of American interest group politics: a federal agency fighting to preserve and expand its budget and staffing, its oversight committees and their staffs, and groups outside the government with economic and professional interests in the continuation of the program.

The responses of these groups accord reasonably well with contemporary models of interest group behavior based on the assumption of "economic man," acting rationally in defense of his pragmatic interests.[6] NASA and the congressional authorization committees worked to ensure the continuation of the program. Aerospace industry and professional groups adapted existing associations to form governmental affairs offices and entered more systematically into that two-way exchange of information that is the essence of modern lobbying. Some also used such modern techniques as political action committees in an effort to influence political outcomes. However, this potential lobby was slow to organize. It lacked an agreed-on agenda, was organizationally fragmented, and never formed a cohesive force behind the idea of expanding American civil space activity. If the groups outside the government can be called a "lobby," it was for the aerospace industry rather than for the space program. As of the early 1970s, that lobby lacked organized support in the general population.

NASA

At the center of it all was the National Aeronautics and Space Administration. NASA went through difficult times after the end of the Apollo program. The space agency was a perennial candidate for budget-cutting because its programs were regarded by many as discretionary. There were political trial balloons suggesting that NASA be abolished, that it be merged with the Department of Defense, or that it be given a different mission more relevant to the times, such as finding solutions to the energy problem. However, few policymakers seriously thought that the United States should get out of the space business.[7] The utility of some kinds of space technology had won acceptance, and NASA had a reputation for success.

NASA had to go through a major transition, from an organization created to meet an urgent national need by performing one specific task to a permanent government agency that had to defend its interests just like any other. "At the end of the 1960s," says Klaus Heiss, "NASA had to sell itself for the first time."[8] Pragmatic visionaries in NASA's employ had to learn patience in their pursuit of a long-standing agenda. Although that transition was traumatic for many people in NASA, it was successful, particularly during the long term in office of Administrator James C. Fletcher.

NASA adapted to the times with a relatively low profile on manned spaceflight issues. The agency was under clear instructions from the White House to restrain its desire for ambitious programs, at least until the Space Shuttle was working. But NASA continued to function effectively as a technology development agency. "The 1970s were the norm," comments former NASA Administrator Thomas O. Paine; "it was the 1960s that were the aberration."[9]

Like older federal agencies, NASA had to lobby for its budgets in two stages: first with the Office of Management and Budget and the White House and then with the Congress. The agency had an Office of Legislative Affairs to represent its interests to Congress. Senior officials such as the administrator and deputy administrator also played active roles in congressional relations.

As is the case for federal agencies in general, most of NASA's congressional lobbying concerned budget requests and the provision of related information to members and committees of Congress. Unlike most agencies, however, NASA has a network of relatively autonomous centers scattered around the United States.[10] These centers have been known to lobby on their own behalf, or to encourage their contractors to do so. This gives the agency important political bases in several congressional districts, which is an advantage at budget time. NASA also has had an active public information program with a generally attractive product to sell.

NASA became a permanent player in the Washington game, an established federal agency competently advancing its interests within guidelines set down by the President and the Congress. After the early 1970s, NASA succeeded in ensuring a relatively stable funding level that allowed coherent planning for the future. However, NASA survived not only because of its own efforts and the appeal of its programs but also because it had friends.

Congress

Since its earliest days, NASA has had important allies in Congress. Senator Lyndon B. Johnson played a key role in the creation of the agency. He continued to be an active supporter as vice-president, when he was chairman of the National Aeronautics and Space Council, and later as President. Special committees of Congress with a specific interest in the space program were created in 1958 and evolved into permanent committees of the two houses in 1959.[11] These authorization committees have provided the organizational foci for congressional support of the civilian space effort.

In many cases, that support was motivated by constituency concerns or by the institutional interests oversight committees have in the agencies under their jurisdiction. However, there also were the more personal elements of respecting and admiring NASA's achievements and a conviction that these were important to the United States. Congressional interest in space has waxed and waned in cycles corresponding to the rise and decline of public interest in the program. However, key figures on the authorization committees, and individual members not on those committees, have stood by NASA in its times of travail.

In the House of Representatives, the key figure during the lean years was Olin Teague, of Texas, who provided continuity through the post-Apollo transition. Teague was chairman of the Subcommittee on Manned Spaceflight of the Committee on Science and Astronautics during the turndown in NASA's fortunes. In 1973, the Science Committee was downgraded to a non-major committee but was given important new responsibilities in non-space areas when it was renamed the Committee on Science and Technology, with Teague as its chairman. The space program was given to a Subcommittee on Space Science and Technology under Florida Congressman Don Fuqua. Together, Teague and Fuqua played a key role in maintaining support for NASA in the House, holding hearings on future programs and pushing for new American initiatives in space.[12] Fuqua became chairman of the full committee in 1979, when Teague retired, and temporarily retained his subcommittee chairmanship as well. In 1982, Harold Volkmer, of Missouri, succeeded Fuqua as chairman of the Subcommittee on Space Science and Applications.[13]

The situation in the Senate during the lean years was far more discontinuous. In the heyday of the space race, when space issues were at the center of national attention, senators sought membership on the oversight committee, and the chairmanship was held by senior Senators Robert Kerr and Clinton Anderson. Frank Moss, of Utah, took over in 1973 as support for spaceflight sagged to a low point; he recalls that many senators sought to leave the committee for those with higher political profiles and that other senators declined to join.[14] Moss worked to restore the prestige of the committee before he lost his seat in 1976. When the Senate reorganized its committees in 1977, the Committee on Aeronautical and Space Sciences was downgraded to a subcommittee of the Senate Committee on Commerce, Science, and Transportation. Senator Adlai Stevenson III, who had been actively involved in the downgrading, became the new Chairman of the Subcommittee. Stevenson followed a recommendation by Moss that he visit NASA centers and, according to Moss, came back a changed man, turning into a great enthusiast for space until his retirement. Former astronaut Harrison Schmitt replaced him in 1979 but lost his seat in 1982. In 1982, Senator Slade Gorton, of Washington, became the new chairman of the subcommittee, and by 1984 he seemed to be emerging as another friend of the space program. Another senator notable for his long-standing support of the space program is Barry Goldwater of Arizona. In 1984, NASA invited Senate appropriations subcommittee Chairman Jake Garn of Utah to fly on the Space Shuttle.

Each of the authorization committees and subcommittees has professional staffers who help to maintain continuity and who generally are sympathetic with the space program. They provide much of the day-to-day contact with NASA and other interested people and are an important source of expertise. However, the continuity of committee staffs can be interrupted when control of a house changes from one party to another and the majority brings in new staffers. For the staffs of individual members of Congress, space usually is a minor portfolio, lumped in with others. However, some of the staffers who deal with space issues are believers in the space enterprise, and several participated in the creation of a Congressional Staff Space Group in 1981 (see Chapter 9).

Appropriations and budget committees, and their subcommittees, obviously play an important role in funding the space program. However, their concerns are much broader than those of authorization committees, and their members are less likely to be advocates of a particular agency or program. It helps to place things in perspective when one notes that the NASA budget is about one tenth of the total amount dealt with by the Appropriations Subcommittee on HUD and Independent Agencies of the House Appropriations Committee. Changes at the margins of the NASA budget, which may seem important to the space authorization subcommittees and their staffs, usually are minor issues to the appropriations and budget committees, which tend to focus on "big ticket" items.

The Aerospace Industry

The aerospace industry is a descendant of the aircraft industry that grew out of World War II; many of the major aircraft companies, as well as other defense contractors, became suppliers to the space program. Despite the scale of the civil space program, however, it has been less important to the industry than defense procurement and civil aircraft. The aerospace industry is far more "aero" than "space." This is reflected in the industry's perceptions of its interests, and in its lobbying.

Figure 2.1 shows that total aerospace industry sales, which peaked in constant dollars in the late 1960s, declined to a low point in that symbolic year of 1972 before climbing back to previous levels in the early 1980s. Within that total, sales to NASA and other non-Defense Department government agencies (mostly NASA) declined from a high point in the middle and late 1960s to a low point in the early and middle 1970s but did not enjoy the same degree of recovery in the early 1980s. As a percentage of total aerospace industry sales, NASA and other agencies shrank from about 15 percent in 1968 to about 7 percent in 1982. In calendar year 1983, sales of space-related equipment led all industry gains with an increase of 30.5 percent. However, the major reason was

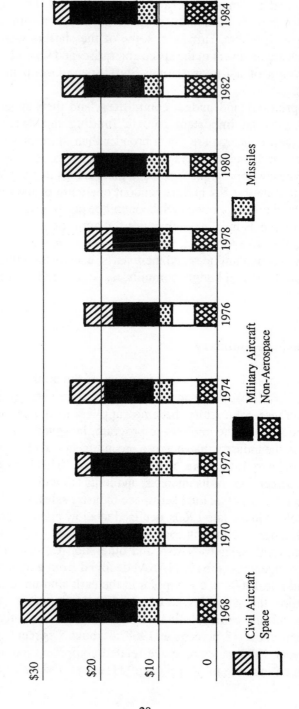

Figure 2.1 — Aerospace Industry Sales in Constant Dollars (*Source:* Adapted from data in *Aerospace Facts and Figures 1985/86* [New York: Aviation Week and Space Technology, 1985]. Used with permission.)

the significant growth in defense outlays for space systems, a trend in evidence for several years.[15] Given these facts, it is understandable that the aerospace industry would give higher priority to other categories than to the civil space program. Although there are many individuals in the industry who share and advocate the space vision, their companies generally do not; contracts, not dreams.

The cutback in space spending obviously affected the industry. But that was part of a larger decline in aerospace industry sales, including those to the Department of Defense, a cause for general concern throughout the aerospace industry.

That industry is represented as a collectivity by the Aerospace Industries Association, whose headquarters are in Washington, D.C. The AIA traces its lineage back to 1917, when the Manufacturers Aircraft Association was formed to handle aircraft production problems in World War I. In 1919, the Aeronautical Chamber of Commerce of America was established to promote aviation and advocate its acceptance as an economic force. The chamber's name was changed to Aircraft Industries Association in 1945, and that organization in turn became the Aerospace Industries Association in 1969. As of 1984, the AIA was supported by 49 large, prime-contractor member companies and worked with the National Security Industrial Association (NSIA) in the informal Council of Defense and Space Industries Associations (CODSIA) on matters of common concern.[16]

AIA has an Office of Legislative Counsel responsible for keeping Congress informed of aerospace industry views. For years, it also has had a resident space expert, James J. Haggerty, who has worked with NASA to write and edit the annual NASA publication *Spinoff*, a description of how technologies developed for the space program are applied to other needs.[17] However, the AIA did not respond to the cutbacks in the civil space program by creating a new arm focused on space interests. There is no Space Industries Association.

There is nearly universal agreement among knowledgeable people that the AIA as an institution is not a significant force in lobbying for space. The public record suggests that the AIA is more concerned with such shared interests as legal and regulatory decisions affecting procurements and exports, and with continuity and steady growth for government purchases of aerospace equipment and services, although it has testified against cuts in the NASA budget.[18] The AIA's experienced public relations officer, John (Jack) Loosbrock, says the association normally does not take positions on specific projects (an exception was the space station, which the AIA endorsed in general terms). There is little consensus within the aerospace industry as to which programs should be advocated.[19] Companies are divided by a diversity of interests, including divisions between aero and space, between defense and civil, and, according to the National Space Club's David Wilkinson, between those who favor manned spaceflight and those who are more concerned about payloads.[20] What the companies can agree on is that they want a big budget, so that everyone will get something. What the industry does not want is the boom and bust cycle of the late 1960s and early 1970s in the civil space field.

In contacting the executive and legislative branches, aerospace firms do most of their lobbying individually. In this role, they are regarded by nearly all informed observers as second only to federal government agencies as an influence on space policies and programs. The companies do most of their lobbying through briefings, providing information about the projects they are working on. "Most industry lobbying is information, not advocacy," said House Space Subcommittee Chairman Harold Volkmer in 1983.[21] Victor Reis, formerly a senior official of the White House Office of Science and Technology Policy, recalls that while aerospace companies did not lobby as a group, they came to OSTP individually "all the time."[22] As in the case of Congress, company representatives came in to tell officials what they were doing and to ask about policy developments; both sides benefited from the exchange.

Aerospace companies often work very closely with NASA, which maintains an office dedicated to liaison with industry. Although the primary purpose is to arrange and execute contracts, this close association puts company representatives in a position to do lobbying of a more informal kind, putting forward ideas and proposals. Visitors to some NASA centers find it difficult to distinguish NASA employees from the many contractors who work there. The mutual dependency relationship of the past continues, with companies careful to preserve their relations and reputation with NASA. "The aerospace companies stay pretty close to the gospel NASA is preaching," says former House Space Subcommittee Staff Director Darrell Branscome. "They respect their customer."[23]

In lobbying members of Congress, companies can appeal to constituency concerns, particularly employment. Former Senator Frank Moss recalls that the major companies had a very widespread network of subcontractors, from whom the members would hear.[24]

There was a time when companies could freely contribute money directly to political candidates. Under modern campaign financing laws, campaign contributions are funneled through political action committees (PACs) established by individual companies. *Aerospace Daily* reported in July 1984 that six aerospace PACs were among the top 12 corporate PACs in receipts and contributions.[25] As of June 30, 1984, the Rockwell International Good Government Committee was second on the list of corporate PAC spending during the 1983-84 election cycle, with contributions of $349,414.[26] Rockwell International Corporation manufactures the Space Shuttle orbiter, the centerpiece of NASA's manned spaceflight program for the near future, and as of 1984 had been NASA's largest contractor for 11 consecutive years.[27]

Aerospace companies also can have indirect influence on events by funding the aerospace-related activities of nonprofit groups. Since the beginnings of the aviation industry, companies have been aware of a need for popular acceptance of aviation and later space activities and of public support for government aerospace procurements. As one course of action, the Aerospace Industries Association (AIA) has supported a succession of organizations in the aerospace education field, the last one being the now defunct American Society for Aerospace Education (ASAE).

In 1973, the AIA hired Wayne Matson to be its director of Education Services, a new position intended to improve the industry's relations with the younger generation. For a time, Matson also was executive director of the ASAE. When that organization went out of business at the end of 1982, Matson became president of a new Aerospace Education Association (AEA), which does not receive grants from the AIA. With about 10,000 members, the AEA is composed mostly of educators interested in aviation and space. The association collects aerospace education syllabi and puts out directories useful to its members. As of late 1984, Matson was trying to make the AEA self-supporting, primarily by expanding sales of its magazine *Aviation/Space*. In December 1983, Matson became chairman of the board and president of a new International Aerospace Institute to be located in Colorado. At the end of 1984 the AEA was involved in seeking support for a National Space Council, an attempt to bring together diverse pro-space interests.

Individual companies also have been a source of financing for some of the pro-space citizens groups that began appearing in the mid-1970s. The most notable example is the National Space Institute (see Chapter 3).

Some other space interest organizations, notably the Planetary Society, chose explicitly to not solicit or accept industry contributions. Aerospace companies also have helped the educational effort in other ways; Lockheed Corporation, for example, supported the making of the film *The Dream is Alive*, about the Space Shuttle.

Another group associated with the aerospace industry is the Aviation/Space Writers Association, organized as the Aviation Writers Association by a group of writers in New York City in 1938 to "establish and maintain high standards of quality and veracity in gathering, writing, editing, and disseminating aeronautical information."[28] The AWA provides professional accreditation and opportunities for awards, holds national and regional meetings, and publishes an annual directory. With about 1,500 members in 1984, the AWA reflects the balance in the aerospace field by being much more "aero" than "space."

Aerospace Professionals

Another interest group affected by the cutback in the space program was the aerospace professionals – the engineers, scientists, administrators, and others whose major organizational expressions are the American Institute of Aeronautics and Astronautics (AIAA) and the Institute of Electrical and Electronic Engineers (IEEE). Here, too, the decline in spending on space was only part of a broader problem, which included reductions in defense research and development. As in the case of the aerospace companies, the organizational responses of the professional organizations were not directed primarily at urging a larger space program.

American Institute of Aeronautics and Astronautics

Readers will recall that the AIAA was formed by the 1963 merger of the Institute of the Aeronautical Sciences and the American Rocket Society. As long ago as 1966, then AIAA President Raymond Bisplinghoff saw a need to improve public understanding of the aerospace profession and its contributions.[29] The organization in 1967 created a Forum Committee on Aerospace Technology and Society that allowed the AIAA to engage in a dialogue with public figures. However, this did not prevent cutbacks in aerospace programs in the late 1960s and the early 1970s, when an AIAA membership of 33,000 professionals (not including students) nosedived to 25,000.

The turning point for AIAA involvement in public policy came not because of events in the space field but as a result of the 1971 congressional decision to stop funding for the development of a supersonic transport (SST). To many advocates of advanced technologies, this symbolized the negative character of the times. According to long-time AIAA official Jerry Grey, who is now publisher of the AIAA magazine *Aerospace America*, professionals were offended by the fact that the decision to end the project was based at least in part on mistaken scientific and technical information.[30] AIAA set up a Public Policy Committee, created a vice-presidency for public policy, and opened an office in Washington, D.C., under Johan Benson that was to observe what the government was doing in the aerospace field, collect information, alert headquarters to issues, identify key areas for testimony by AIAA, and maintain liaison with federal agencies and Congress. The Public Policy Committee's first project was an assessment of space transportation systems, which was provided to the relevant agencies and committees of Congress and which may have had some influence on the development of the Space Shuttle.

Since 1973, the AIAA has testified every year on NASA's proposed budget. Grey believes that testifying on the size of the budget has little effect, but the testimony provides members of the subcommittee with potentially useful arguments for internal debates within the Congress. He adds that this arrangement tends to become increasingly structured, with the pattern of testimony being developed in discussions with the subcommittee's staff before it is given.[31] This appears to be another example of the cooperative, two-way process existing between government and many interest groups.

In 1974, AIAA began holding a series of policy-oriented workshop conferences, which can influence policy through exchanges of ideas and by moving professionals within as well as outside the government toward a new consensus. As of 1984, the organization was continuing to run workshops for NASA and other agencies under contract. AIAA Executive Director James Harford describes the use of this technique:

Convene a workshop. If possible, get funds to run it right from a concerned government agency. Make sure that diverse expert views are represented . . . . work out conclusions on the scene. Circulate the draft document to the participants for final editing. Ask the AIAA Board to endorse and publish it as an Institute document – not necessarily an AIAA view but an expert view worthy of consideration. Circulate the report to key persons in government, Congress, and public. Use the conclusions as the basis for testimony before the appropriate Congressional committees. Finally, and this is the toughest job, get the AIAA members in their local sections to inform their own Congressmen and staffs on the issue.[32]

The AIAA, which has 52 technical committees, annually holds about 20 general and specialized conferences on a wide variety of subjects. Such meetings often have provided platforms for the communication of new technical ideas to larger audiences, increasing their familiarity and credibility over a period of time.

The AIAA also does assessments of technical issues for NASA and other agencies, some at its own initiative and some by request. In 1969, an AIAA team produced a document at the request of the White House called "The Post-Apollo Space Program: an AIAA View," much of which was an elaboration of the classic agenda for manned spaceflight.[33] More recently, a study for the Office of Science and Technology Policy on the commercialization of the Space Shuttle may have been influential in shaping policy.[34] In 1982, the AIAA formulated a suggested national space policy that, in the manned spaceflight field, called for improved utilization of the Space Transportation System (of which the Shuttle is the heart) and for a manned space facility in low Earth orbit, that is, a space station.[35] The AIAA magazine, now called *Aerospace America*, often has provided a medium for advocates of near-future space initiatives.

By September 1984, the AIAA had made something of a comeback in membership, climbing to 30,675 professionals and 6,960 student members; it had 63 professional sections and 134 student branches.[36] Some of these were actively involved in pro-space activity, sometimes in cooperation with other groups. However, AIAA's members still represented only 11.2 percent of the aerospace engineering and scientific community in 1983.[37]

The organization, which has an annual budget of over $10 million, decided in 1984 to move its headquarters from New York City to Washington, D.C. In that year, Jerry Grey became the new president of the International Astronautical Federation, the leading international organization in the field.[38] The AIAA has become an established and respected player in the space interest field, listened to not only because of the expertise within its membership but also because it is useful to those in government.

Institute of Electrical and Electronic Engineers

With about 250,000 members (roughly 200,000 of them American) the IEEE is the largest engineering society in the United States, and possibly in the world. One of its 27 member societies is the Aerospace and Electronic Systems Society, which has about 7,000 members.

The IEEE was stirred to action by cutbacks in funding for federal research and development, including the downturn in the space program. According to Theodore (Ted) R. Simpson, chairman of the IEEE Space Subcommittee, about one third of the scientists and engineers in the aerospace industry were laid off during the late 1960s and the early 1970s, and their employment levels have never again reached the numbers seen at the height of the space program.[39]

The IEEE's response was to set up a Research and Development Committee that analyzes the federal budget, concentrating on agencies whose budgets have a large R and D component. Since 1979, these analyses have formed the basis for testimony before the subcommittees responsible for those agencies, in which IEEE gives advice on the direction of the relatively small changes Congress normally makes in R and D budgets. Simpson emphasizes the importance of being credible and not making unrealistic demands.

Since 1974, the IEEE has had a congressional fellowship program that enables two or three of its members to spend a year working on Capitol Hill (Simpson worked for the Senate Subcommittee on Science, Technology, and Space, before which he now testifies every year). The AIAA had a comparable program until 1982.

American Society of Mechanical Engineers

Of the other technical societies with an interest in space, the most important is the 111,000 member American Society of Mechanical Engineers, which has an aerospace division with about 4,000 members. Like the AIAA, ASME opened a Washington office because it was concerned that public policy was being made without a technical input. Philip M. Hamilton, ASME's Director of Federal Government Relations, states that ASME testified on the NASA authorization for the first time in 1983, focusing on major projects and directions such as the need for a national space policy, space commercialization, and a possible fifth Space Shuttle orbiter funded by private enterprise. The society's aerospace division has an ongoing activity to review proposed

NASA budgets. ASME started a congressional fellowship program in 1973.[40]

Aerospace Workers

The key trade union in the American aerospace world is the International Association of Machinists and Aerospace Workers (IAMAW). The union's membership peaked at over 1 million in 1969 but has dropped to around 800,000. According to IAMAW leader William Winpisinger, the biggest factor in this decline was cutbacks in the space program.[41]

It was reported in the early 1970s that the IAMAW had played an important role in getting congressional approval for the Space Shuttle,[42] although Winpisinger says the union was not much involved. Since then, the IAMAW has not been particularly active in lobbying for "space." Winpisinger comments that the union had to be sympathetic with the disadvantaged community and had to give more attention to domestic problems and unrest created by the Vietnam war. Its basic posture is to support continuing U.S. efforts in space, but only as resources allow; space must be balanced against other priorities. In 1984, Winpisinger denounced President Reagan's decision to proceed with a space station as "sheer lunacy."[43] Like some aerospace companies, the IAMAW played a role in helping the National Space Institute (see Chapter 3).

The United Auto Workers has an Aerospace Department, which had about 55,000 members in December 1982.

APPLYING SPACE TECHNOLOGY

In the early 1970s, NASA put greater emphasis on the practical applications of Earth-oriented space technology. The agency, having already done a good deal of work on communications satellites, was continuing its Applications Technology Satellite series, which included a test of direct television broadcasting from orbit. NASA also was developing remote sensing satellites that observed the Earth's surface from orbit. By the mid-1970s, organized interest groups had formed around each of these technologies.

Public Service Satellite Consortium

In late 1974, former Federal Communications Commissioner H. Rex Lee and others, inspired by NASA's Applications Technology Satellite series, began discussing the use of such devices for public service functions. Public service organizations, including the Public Broadcasting Service, decided that they should band together to advance their interests in this field, since they had encountered difficulty in getting time on NASA satellites. After a year of informal meetings involving up to 27 groups, the organizers decided to form a nonprofit corporation. With the help of grants from NASA and the Department of Health, Education and Welfare, the Public Service Satellite Consortium (PSSC) was formed in 1975. The consortium represents the interests of its member organizations (97 in October 1983) to the administration and Congress, and its respected president, Elizabeth Young, regularly testifies on Capitol Hill. The PSSC was a factor in getting the Advanced Communications Technology Satellite put back into NASA's fiscal year 1985 budget after commercial interests had succeeded in getting it removed. The PSSC also coordinated experiments on two experimental communications satellites. As of 1984, it was offering a video teleconferencing service and a transportable Earth station for satellite communications. In 1982, the consortium began doing business with commercial enterprises, and in 1983 it spun off all activities in the commercial field to a profit-making subsidiary called Services by Satellite, half-owned by Fairchild Industries.[44]

The GEOSAT Committee

Another applications-oriented organization, the GEOSAT Committee, emerged from a workshop on geological remote sensing from space held in Flagstaff, Arizona, in May 1976. The new group was a response to the geological and commercial potential of the remote-sensing satellites in the LANDSAT series, the first of which had flown in 1972. A nonprofit organization based in San Francisco, GEOSAT articulates to the government the land-observation satellite interests of the exploration geology community, both in the business and the academic world (in 1983, four of the six executive committee members were from oil companies).[45] GEOSAT has recommended improvements for the geological applications of LANDSAT, is anxious to see remote-sensing satellites capable of higher resolution imagery, and was at one time pushing a STEREOSAT, which would give three-dimensionality to satellite imagery of the Earth's surface. GEOSAT's peripatetic president, Frederick B. Henderson, stays in close contact with the remote-sensing community all over the United States.[46]

A DIFFERENT KIND OF LOBBY

We have seen how NASA and the committees of Congress responded to cutbacks in the civil space program by adapting their organizations to the time and by playing accepted roles in policy, legislative, and budgetary processes. Existing aerospace interest groups, motivated by pragmatic economic and professional concerns, responded to the cutbacks in the space program and other programs involving the aerospace industry much as contemporary interest group models would predict. Seeing their interests threatened, they adapted existing organizations, increased their presence in Washington, stayed closer to the information flow about federal policies, legislation, and budgets, and applied a variety of conventional interest group techniques to increase the probability that their voices would be heard by decision makers in government. They did not form new lobbying organizations specifically intended to urge the expansion of the space program. Meanwhile, new, nonprofit associations were formed to take advantage of opportunities for the application of space technology to existing economic or public service needs.

If this were the total story, it would not be worth much more of the general reader's time. As things turned out, however, the turndown in the American space program was followed by other kinds of social phenomena as well, having some of the characteristics of a social movement. People whose economic and professional interests were not directly affected organized groups to advocate more American activity in space. In many cases, they appear to have been driven not so much by pragmatic interests as by ideas, wishes, hopes, dreams. To many of them, space symbolized a better future. In that lay the potential of a new kind of lobby for space.

3

THE BEGINNINGS OF THE
NEW SPACE MOVEMENT

Space isn't a destination. It's a way of thinking.

Stephen M. Cobaugh[1]

Humanity's first ventures into space stirred a positive emotional response in many individuals who had no direct economic or professional stake in the space program. Implicit in this was the potential for a citizens lobby for space, of the kind that was emerging around other issues in the late 1960s and early 1970s. However, this potential pro-space movement was, as science fact/science fiction writer G. Harry Stine says, "totally unorganized" in that turning point year of 1972.[2]

Homer E. Newell illuminated the problem when he wrote that "NASA had enjoyed a strong followership."[3] Many ordinary Americans were excited about space, but they were not organized to express themselves as an interest group and to make their influence felt, nor was there a consensus about what they should do. The pro-space movement of the 1970s came not from one initiative but from the intertwining of many threads.

THE SHIP OF FOOLS

Stine recalls an event that characterized the situation in the early 1970s, a gathering he calls the "Ship of Fools" expedition.[4] Noting that the Apollo 17 mission of December 1972 would be the last journey to the

39

Moon for the foreseeable future, science writer Richard Hoagland organized a voyage on the SS *Statendam* to observe the launch from offshore (Hoagland was to appear again in the role of an organizational entrepreneur in the pro-space field.) On board, invited speakers, including writers and prominent space buffs, participated in a symposium chaired by television personality Hugh Downs, who later became president of the National Space Institute. They included science fiction writers Isaac Asimov and Ben Bova. (Katherine Anne Porter, author of *The Ship of Fools*, also was present.) Writer Norman Mailer, who regarded our expansion into space as "profoundly ambiguous," nevertheless said he believed in the necessity for humans to voyage into space because, "I think it is part of our human design, part of our inner imperative."[5] While many of those present expressed concern about the future of the space program, there was no consensus as to what should be done. However, says Stine, it was recognized that "politics" had failed and that the existing organizations were not doing the job.

The "Ship of Fools" expedition reflected a broader sense that action was needed to keep the space dream alive. It was that vague, inchoate, but shared feeling that provided the constituency for the new pro-space movement.

EDUCATING THE PUBLIC

Jesco von Puttkamer, a NASA planner who has had an on-and-off relationship with pro-space organizations over the years, traces the origins of the pro-space movement back to the creation of the Alabama Space and Rocket Center (ASRC) in Huntsville, Alabama, in the late 1960s. Wernher von Braun, then based at the Marshall Space Flight Center in Huntsville, lobbied the government of Alabama successfully for funding for the ASRC, which was intended to communicate the excitement and promise of space to the public.[6] The ASRC has been a success, attracting many visitors. In 1981, as the Space Shuttle began its journeys into space, the ASRC established a Space Camp for young people that has proven to be extremely popular, attracting 2,630 boys and girls in 1984.[7]

There have been other space-oriented museums since, notably the National Air and Space Museum in Washington, D.C., which opened in July 1976. Although these may have been stimuli to some of the founders of pro-space groups, the origins of the movement appear to lie elsewhere.

LOOKING TOWARD THE FUTURE

In organizational terms, it appears that the forerunner of the new pro-space movement was a small, highly idealistic group called the Committee for the Future (CFF). Although it never had much influence, the CFF enunciated many of the themes taken up by other pro-space individuals and organizations in the middle and late 1970s.

The CFF originated from conversations in the 1960s between artist-philosopher Earl Hubbard and his wife Barbara Marx Hubbard (an heiress to the Marx toy-making fortune) and from Barbara's own search for meaning, described in her book *The Hunger of Eve*.[8] Barbara, influenced by Teilhard de Chardin's ideas about evolution and "planetisation," wanted to become an "advocate for humanity." "Just as I started that scan through literature, looking for the crucial self-image [of humanity]," she wrote, "John Glenn was fired into space from Cape Canaveral."[9] She and Earl became passionate advocates of the idea that the Space Age was the birth of a new era:

> Possibly we were witnessing the natural diversification of the species at this phase, to meet the requirement for some to be attracted to new and vital tasks of nurturing, tending, and bringing harmony to Earth – what Teilhard de Chardin calls "the agents of planetisation – the builders of the Earth," and for some to be attracted to going beyond Earth to build new worlds, and be transformed into new beings – extraterrestrials and new terrestrials – builders of new heavens and a new Earth: new worlds on Earth, new worlds in space.[10]

Barbara recalls her joy at the liftoff of Apollo 11 in July 1969: "I identified with the rocket! I felt myself rising in space, breaking through the cocoon of the sky and moving into the universe. . . . I cried uncontrollably as it rose into space, the words 'freedom, freedom, freedom' pounding in my head."[11]

July 20, 1969, the date of the first landing of men on the Moon, also was the publication date for Earl's book *The Search is On*.[12] In it he argued that it is time for men to move toward goals beyond material abundance. We must want to build a future for all mankind, he argued, by exploring the universe and developing new worlds.

In September 1969, the Hubbards discovered a fellow believer in Colonel John Whiteside, then the chief U.S. Air Force information officer in New York City. Whiteside had been a USAF Public Affairs Officer during Project Mercury. By the next year, Earl and Barbara had decided that the goal to establish the first "space community" should be announced in 1976.[13]

The Hubbards also took on the "limits to growth" school of thought, first given expression by an organization through the Club of Rome, formed in 1968. If the finite resources on this planet were being used up, the Hubbards said, greater resources could be found beyond it. Barbara quoted the response of space visionary Krafft A. Ehricke, who told the story of the erudite "fetal scientist" who is a cell in the womb of a baby in the seventh month. The scientist predicts that, from current growth rates, by the eighth month, there will be severe overcrowding and pollution; by the ninth month there will be massive starvation, suffocation, and revolution; and by the tenth month die-offs will occur in most of the poor nations of the world. "The limits to growth advocates," Barbara wrote, "are the erudite fetal scientists extrapolating from the past – with no positive vision of the future."[14]

In June 1970, the Hubbards invited a group of people to their home in Lakeville, Connecticut, to found the Committee for the Future. There they produced "The Lakeville Charter," which proclaimed the purpose of the CFF, and laid out a strategy to bring their message into the public arena in time for a presidential candidate to carry the message in 1976. In part, the charter said:

> Earth-bound history has ended. Universal history has begun. Mankind has been born into an environment of immeasurable possibilities. We, the Committee for the Future, believe that the long-range goal for Mankind should be to seek and settle new worlds. To survive and realize the common aspiration of all people for a future of unlimited opportunity, this generation must begin now to find the means of converting the planets into life support systems for the race of Man.[15]

In the charter it was declared that the development of new worlds will provide a basis for curing Earthly unemployment and offer a basis for a meaningful union of the peoples of the world. "A challenge of this magnitude," says the charter, "can emancipate the genius of Man." The CFF began putting out a publication called *New Worlds Review*.

In the fall of 1970, Los Angeles film producer George van Valkenberg pointed out to the Hubbards that two Saturn V rockets would be left over from the Apollo program. The CFF leaders realized that they could initiate the first "citizen-sponsored lunar expedition," which could pay for itself through the sale of lunar materials and through television and story rights; there could be a general subscription to let the public participate in financing the project. This came to be known as Project Harvest Moon. The CFF formed the New Worlds Company in January 1971 with the help of $25,000 from Barbara's father. The purpose was to rally support for the next great goal: a lunar community. This would help generate popular pressure for the funding of the necessary intermediate steps such as the Space Shuttle – a theme taken up by others 13 years later. Through the offering of shares in the lunar enterprise to millions of people, a constituency with a vested interest in the development of the Moon and outer space activities would be created.[16] The Hubbards clearly saw the Moon as a place where resources could be mined.

Barbara Hubbard and John Whiteside began traveling around the United States to brief space program officials on their proposal. According to Barbara, Christopher Kraft of NASA's Johnson Space Center told her, "Mrs. Hubbard, I've read your husband's book. This step into the universe is a religion and I'm a member of it."[17] In the House of Representatives, Congressman Olin Teague introduced a resolution calling for a study of the feasibility of a citizens lunar mission. However, the idea reportedly got an unfriendly reception from NASA headquarters. Hubbard and Whiteside rewrote the legislation to propose an alternative mission in near Earth orbit, called Mankind One, but that failed also. Barbara comments that "The corporate decision of NASA as a government agency was less responsive than the decision of any of its individual members."[18]

In May 1971, the CFF held a conference on "Mankind and the Universe" at Southern Illinois University, the alma mater of Colonel Whiteside. This followed up on conferences at Lakeville with students from the university. Speakers included Buckminster Fuller and other futurists. By then, Barbara, who had access to many important people, was beginning to emerge as more of a spokesperson for the CFF vision than her husband.[19]

The CFF held a series of subsequent conferences in different American cities and even in a foreign country. Colonel Whiteside came up with the idea of a large, wheel-shaped structure that housed the conference; walls between sections were removed in stages, until all participants came together in what came to be known as a "synergistic convergence," or SYNCON.[20] Two SYNCONs were held in Los Angeles; a participant recalls that one was addressed by science fiction writer Ray Bradbury, "Star Trek" creator Gene Roddenberry, and others, and that the Robert Wagner Chorale sang a "Space Madrigal." He adds that this enterprise attracted what he calls "space gypsies," young volunteers who followed SYNCON around the country.[21]

Meanwhile, things were changing in the CFF, which moved its headquarters to Philadelphia. Its emphasis began to shift away from space and toward other issues. A gap grew between Barbara and Earl, who separated. Although Earl gradually dropped out of the CFF, he went on to publish another book called *The Need for New Worlds*, which elaborated on themes in his earlier work.[22]

In September 1973, the CFF opened its new headquarters in a mansion known as "Greystone" (which was owned by Barbara's sister Jacqueline) in Washington, D.C., and called it the New Worlds Training and Education Center. At one time, about 25 volunteers got room, board, and $25 a week in a sort of commune. A media van, equipped with video cameras, frequently was parked outside.[23] Barbara's dinners and cocktail parties there provided a kind of "salon" for many people interested in the future and in space and may have facilitated some of the connections that brought about the new space movement and the elaboration of some of its ideas. World Future Society Edward S. Cornish described it as "the longest running conference on the future of Man."[24]

In July 1974, the CFF formed a "Choiceful Futures Group" at Greystone. Its meetings led to CFF support of colonies on the Moon and in space. Physicist Theodore Taylor, who was active in the group, later wrote that "among the large array of opportunities for shifting these trends is one almost totally ignored by most of the present generation of prophets of doom – the possibility for large-scale extension into space of human technology and habitats."[25]

Barbara expressed many of the seldom-articulated motivations of the emerging pro-space movement. In her book, she describes how she told Jonas Salk that she wanted to leave the Earth and to live beyond this planet, if that were possible in her lifetime. She wrote, "I've made up my mind that the future of human culture and excellence depends on the continuation of the space program."[26] Before finishing *The Hunger of Eve*, she called for a decade of open-worlds development to initiate the benign industrial revolution – "beyond the limits to growth."[27]

Although Colonel Whiteside has since died, Barbara has remained intermittently active in the pro-space cause. When the Reagan administration took office in January 1981, she wrote to Vice-President Bush, urging that the Space Council be reestablished with Bush as its head and enclosed a 20-year space plan by former NASA Administrator Thomas O. Paine.[28] An early donor to the World Future Society and later a member of its board, she also has remained a prominent figure in the futurist field, at one time heading a Futures Network. As of 1984, she still was a regular speaker at the World Future Society's conventions. In 1982, she published a book entitled *The Evolutionary Journey: A Personal Guide to a Positive Future*, in which she argued that the human species stands at the threshold of "conscious evolution," ready to make a quantum leap into a new order as it begins the great evolutionary tasks of restoring the Earth, freeing people from want, developing physical and mental potentialities, and exploring outer space.[29] As of the spring of 1984, she was the head of an organization known as the Campaign for a Positive Future and was running for the Democratic Vice-Presidential nomination.

For all its good intentions, the CFF suffered from a lack of technical expertise and credibility. Critics regarded it as "flaky," and most people in the space field did not take it seriously. It never reached the "critical mass" to become a real organization that could carry out a long-term program or lobby systematically for a cause. Alan Ladwig, a student at Southern Illinois University who was involved in the early CFF conferences, went on to head a "space-sympathetic" organization, and later to work for NASA, observes that the CFF got many people, particularly young ones, involved temporarily, but adds that there was no mechanism to keep people involved.[30]

To William S. Bainbridge, the history of the CFF suggested that there was little or no opportunity for amateurs to participate in furthering the exploration, exploitation, and colonization of the solar system.[31] Yet the CFF was a harbinger in many ways, particularly in its efforts to project a broad, positive vision of the future that fully incorporated the spaceflight revolution, and to involve ordinary citizens in the pro-space cause.

FASST

Another strain in the new pro-space movement came from young people with a serious interest in aerospace technology. In September 1970, Professor Wilbur Nelson at the University of Michigan saw that the project to build an American supersonic transport (SST) was in political trouble and brought together a group of aerospace engineering students who supported the project. A politically oriented student named David Fradin formed an organization in January 1971 called Fly America's Supersonic Transport (FASST), which shared an office with the campus Gilbert and Sullivan Society. Another active early member was the bright, aggressive Thomas A. Heppenheimer, later active in advocating space colonization. FASST became a focal point for a network of pro-technology, pro-space young people.[32]

Although clearly out of tune with the anti-big technology trend of the time, FASST collected about 3,000 signatures on a pro-SST petition at the University of Michigan. In March 1971, Heppenheimer wrote and Fradin delivered pro-SST testimony before the Subcommittee on Transportation of the House Appropriations Committee. The SST was cancelled anyway, leaving the organization without a role.

FASST regrouped and changed its name to Friends of Aerospace Supporting Science and Technology, broadening its interests as it spread to other campuses. The organization then changed its name to the Federation of Americans Supporting Science and Technology, keeping the acronym. Fradin and Heppenheimer testified before Congressman Don Fuqua's House Subcommittee on Manned Spaceflight in March 1972, with Fradin warning that "this country's respect for science and technology is changing to contempt, and, as a result, the technology required to solve global problems may never be developed."[33] FASST also supported the then embattled Space Shuttle, writing to Senator Walter Mondale in February 1972. The organization then described itself as being dedicated to the concepts of international stability, a clean environment, and social progress through science and technology.[34]

FASST moved its headquarters to Columbia, Maryland, in the fall of 1973 and succeeded in getting seed money for the Energy Youth Council from the Atomic Energy Commission and energy companies. FASST moved to downtown Washington in the spring of 1974. Fradin and Heppenheimer left the organization, and Alan Ladwig became its new leader at the end of that year. FASST also hired the young California space enthusiast Leonard W. David to help with space issues, including an assessment of student support for the space program. By the end of 1975, the organization had about 400 members and 20 active chapters on college campuses, and its staff had grown from two to seven.[35]

The organization changed its name to Forum for the Advancement of Students in Science and Technology late in 1975. Wanting to improve understanding of science and technology issues and to raise the level of debate on them, FASST made a conscious effort to make a connection between science and technology students and public policy. The new FASST got grants from NASA, the Energy Research and Development Administration, and companies to put on programs designed to involve students in the discussion of science and technology interests (as Ladwig comments, "people then took students seriously"). FASST organized two White House conferences in 1975.[36]

FASST retained a strong interest in space, helping to create the Shuttle Student Involvement Project in cooperation with NASA and the National Science Teachers Association and sponsoring or cosponsoring meetings related to space. The organization also had a contract to study student involvement in the satellite solar power station project (see Chapter 4). However, financial stringency forced a reduction in staff, eventually leaving Ladwig and David as the only officers of the organization. FASST finally folded in December 1980, a victim of changing times; the sense of emergency about youth attitudes toward science and technology had passed.

Ladwig and David were not done with space. As of 1985, Ladwig was the manager of NASA's Spaceflight Participants Program, which will choose private citizens to go into space. It is difficult to imagine a more perfect location for a member of a movement that wants to increase public access to space and broaden participation in the space enterprise. Leonard David, who remembers being called "crater eyes" by his school chums because of his interest in amateur astronomy, became the editor of the National Space Institute's newsletter *Insight*, and in 1984 he became editor of *Space World*, a previously existing magazine now published in cooperation with the National Space Institute.

THE FIRST PRO-SPACE CITIZENS GROUPS

United States Space Education Association

The early 1970s also saw the founding of the oldest still existing American "grass-roots" pro-space citizens organization with more than local membership. In 1973, two high school students in Elizabethtown, Pennsylvania – Stephen M. Cobaugh and John K. Alleman – got together to form the Lancaster County Space Education Association, largely in response to waning public interest in the space program. This became the United States Space Education Association in 1975. Describing itself as the first organization to educate the public on the benefits of space exploration, the USSEA grew to have hundreds of members, began publishing *Space Age Times*, and began putting on exhibits and other educational functions related to space.

The USSEA's basic philosophy is simple: the more people know about the space program, the more inclined they are to support it. In keeping with the spirit of the new movement, one of USSEA's objectives is to "develop the proposal for the USA to open up new worlds in space for all Mankind in a fair and equitable manner."[37] "We've been playing a vital role," says USSEA literature, "in securing a solid technological foundation for a future full of optimism."[38]

Although the USSEA remains small, it has survived because of the continuing enthusiasm of its original leaders. In its own words, the USSEA "established an early model for the pro-space movement through an integration of educational and news gathering techniques. The USSEA demonstrated the capabilities of grassroots involvement in public space policy decision-making."[39] The congressman representing that part of Pennsylvania in the early 1980s, Republican Robert Walker, is outspokenly pro-space.

The National Space Institute

The first large, national pro-space citizens organization of the new era came from different origins. As the fortunes of the civil space program declined in the late 1960s and early 1970s, some of the leading government and industry figures in Washington's space community became increasingly concerned about the lack of a broad-based, "grassroots" support organization for space. NASA officials discussed the problem with Washington representatives of interested companies who were members of the National Space Club. According to Neil Ruzic, a publisher of technical magazines who was involved in these discussions, NASA wanted something like a Navy League for space.[40] But NASA itself could not organize a citizens group to support its own program.

In May 1973 (the month when Skylab went into orbit), past National Space Club President Gene Bradley, a Boeing Company executive, recommended to the club's executive committee that the club expand its activities to the national level, encompassing tens to hundreds of thousands of members. Then Club President Tom Emerich, of Martin Marietta Corporation, appointed a committee chaired by Bradley to investigate the feasibility of this idea. Subsequently, Fairchild Industries Vice-President Tom Turner developed a program plan for the creation of a National Space Association, which he presented to the club's executive committee in May 1974. The new president of the club, Fred Everett of Westinghouse Electric, appointed Harry S. (Terry) Dawson, Jr. of System Planning Corporation to head a committee that would study the plan. The Dawson Committee reported that the club should support the creation of a National Space Association but that serious financial risks would be incurred if the activities of the two organizations were merged. As a result, an independent National Space Association was incorporated in June 1974.[41] All of those who signed the incorporation papers were past officers of the National Space Club and were employed by corporations with interests in the space field.

According to Dawson, who in 1984 was on the staff of the House Subcommittee on Space Science and Technology, the new association was to be a National Geographic type of society, complete with a quality monthly magazine. The original goal was to reach 100,000 members plus 100 corporate members within three years. The founders saw the new association as a "buffer" between government and public, which would tell citizens how space programs benefit them and get public views to the government.[42] The main purposes in its articles of incorporation were to promote U.S. space leadership, to stimulate the advancement of civilian and military applications of space and related technologies, to bring people together for the exchange of information and through them to inform the public, and to provide recognition to individuals and organizations that have contributed to the advancement of space.

The National Space Association needed a big name to attract attention and help with fund-raising. Tom Turner approached Wernher von Braun, then with Fairchild Industries, and von Braun became president of the National Space Association in August 1974. Retired rocket engineer Frederick C. Durant III, who calls von Braun "the greatest salesman for space this country has ever had," says von Braun described the new organization as his "retirement project."[43] Four months later, the association hired as its first executive director Charles Hewitt, a man with organizational and fund-raising experience but no space background.

Von Braun and Hewitt traveled around the United States to raise funds from aerospace companies. Encountering resistance to the idea of giving to "another association," they recommended that the name be changed; in April 1975, the association became the National Space Institute.

THE NSI went public at a press conference in Washington in July 1975. Von Braun picked up some of the basic themes of the new pro-space movement in his statement:

We believe that there is a need to provide the public with a voice in the direction of the space program. Properly directed and supported, the space projects planned for the future can be designed to help resolve many of our pressing energy, food, environmental, and economic problems. . . . I also believe that the much heralded idea of the Earth as a limited planet is absolutely unacceptable. . . . I am certain that future generations will say that the real significance in our space program lay in the fact that it took the lid off the limitations posed by the finite size and the finite resources of the planet Earth. For we are no longer limited to just one planet, Earth. We have all the mineral resources and all the energies of the solar system at our disposal, and maybe one day we will even go beyond that.[44]

A week later, the NSI held its first board meeting in Cocoa Beach, Florida, at the time of the Apollo-Soyuz Test Project. There was a striking emphasis on having celebrities on the board even if they had no space credentials. The initial list included comedian Bob Hope, actor Hugh O'Brian, artist James Wyeth, and Reverend Fulton J. Sheen, as well as science fiction writers Isaac Asimov and Arthur C. Clarke, Captain Jacques Cousteau, astronaut Alan Shepard, space scientist James van Allen, Senator Barry Goldwater, and the chairmen of the congressional committees dealing with space (Frank Moss in the Senate and Olin Teague in the House). Testifying before the Senate Subcommittee on Aerospace Technology and National Needs in September 1975, NSI Vice-President Hugh Downs said, "The fundamental reason for the National Space Institute's formation is to fill a communications void that exists between the national space program and the general public."[45]

Unfortunately for NSI, von Braun's declining health prevented him from devoting his full energies to the new organization; in 1976 he gave up the presidency to Hugh Downs. Von Braun's death in June 1977 was a serious blow for the institute, whose fund raising had never reached the critical mass necessary for exponential growth.

Meanwhile, astronomer Carl Sagan was approached to see if he would be willing to join the board. Sagan reportedly expressed interest but only on the conditions that more scientists be put on the board and that NSI take a broader view of space than the manned spaceflight program.[46] Sagan, with Bruce Murray, later created his own organization – the Planetary Society.

The NSI was rescued financially in 1977 by a large influx of money from the settlement of a dispute between the International Association of Machinists and Aerospace Workers and United Technologies Corporation. Union leader William Winpisinger, a member of the NSI Board, got the agreement of UTC Chairman Harry Gray that about $900,000 of the settlement could be paid to the NSI.[47] The institute used much of this money to run a direct mail campaign that increased its membership to an all-time high of 22,000. However, the ideal NSI magazine remained out of reach; members were served with a modest newsletter edited first by Julie Forbush, who went on to try something similar at the National Air and Space Museum, then by the young space activist Courtney Stadd, later active with one of the new launch vehicle companies. When Hewitt left the NSI in 1980, Stadd became acting manager of the organization and Leonard David of FASST became part-time editor.

Neil Ruzic had hoped that the institute would invest enough in a high-quality magazine from the beginning to make it self-funding, allowing NSI to use its funds for other things, even its own space projects, but this never worked out. In 1978 Ruzic suggested a linkage between NSI and the new science fact/science fiction magazine *Omni*. Under the short-lived arrangement, *Omni* conducted a direct mail campaign for a package deal that included a subscription to the magazine along with membership in NSI. According to Ruzic, this promotion brought in 10,000 to 15,000 new members.[48] However, this arrangement ended after a year, at least in part because the NSI was losing money on it. Subsequently, the NSI worked out a cooperative arrangement with the established magazine *Space World*, which included the NSI newsletter beginning in 1982. In 1984, NSI's Leonard David replaced Huntsville space journalist David Dooling as editor of that magazine.

Hewitt was replaced in 1980 by Mark R. Chartrand III, former Director of Education and later chairman of the Hayden Planetarium in New York City, and a writer of science articles for publications including *Omni*. Under Chartrand, the NSI gave greater emphasis to its educational activities, such as regional meetings and conferences intended to involve members. The NSI also has organized tours to space launches from Kennedy Space Center in Florida. Perhaps the NSI's best-known program was the "Dial-a-Shuttle" arrangement launched in 1983 under the direction of Bonny Lee Michaelson, through which people can listen in on communications between astronauts and mission control by dialing a special number. Over 800,000 calls were placed to that number during one of the early Shuttle flights.[49]

Since 1983, the NSI has testified each year before the House Subcommittee on Space Science and Applications, the only pro-space citizens group that is regularly invited to do so. Ben Bova, who became president of NSI in 1983 after publishing the pro-space and pro-technology book *The High Road*, believes that one of the NSI's major accomplishments has been to act as the voice of the people in these hearings. "It is the only broad, pro-space group to tell Congress what its members are thinking," he says.[50] Both NSI officers and congressional staffers agree that the NSI does no real lobbying on Capitol Hill. Chartrand says that the NSI's contacts are with congressional staffers, not members, and that its highest level contact in the administration is the President's science advisor.[51]

The NSI did show some sign of activism in 1981 when it appeared that the Reagan administration's budget cuts might threaten the space program. A letter from Chartrand asked members to call the White House opinion number and to write President Reagan and suggested some arguments to use.[52]

As of 1984, the NSI was just holding its own in terms of members, having about 11,000 at the end of 1983. It has continued to rely on periodic appeals to corporations for donations. At the end of 1983, a reorganization moved new names onto the board, including former astronaut Frank Borman (Chairman of Eastern Airlines), House Science and Technology Committee Chairman Don Fuqua, actress Nichelle Nichols (who played Lieutenant Uhura in *Star Trek*), and space artist Robert McCall, who painted a huge space mural in the Smithsonian Institution's National Air and Space Museum. Mark Chartrand resigned as of March 1984. Glen Wilson, long-time staffer of the old Senate Committee on Aeronautical and Space Sciences and later director of NASA's academic programs, became executive director later that year.

The NSI achieved a milestone in 1984 when Space Shuttle payload specialist Charles Walker became the first NSI member to go into space. In June 1984, the NSI was given a tremendous boost when President Reagan announced that the institute would work with NASA to develop the Young Astronaut program, which is part of "Operation Liftoff," an effort to encourage young people to study science and technology.[53] However, by the end of 1984 it looked as if the Young Astronaut Program was entirely independent of NSI.

As the first major example of the new pro-space citizens groups, the NSI was setting precedents. Under the scrutiny of hopeful pro-space people, it had to deal with a series of issues that faced later groups as well. Given its status as a nonprofit educational institution under section 501(c)3 of the Internal Revenue Act, how aggressive could it be in lobbying? Should a citizens organization depend on donations from industry? Should a pro-space organization follow the staff model (as the NSI basically did) or the grass-roots model? Can one find an executive director who combines the talents of administrator, fund raiser, expert on space, and communicator?

The high expectations many attached to the NSI left them disappointed by its performance. There is a widespread feeling in the pro-space community that more could have been done with the unique opportunity the NSI had to mobilize pro-space opinion, possibly turning itself into the single, powerful, pro-space organization some space advocates dream about.

Some critics see the NSI as an organization that was created from the top down, rather than from the grass roots, and one designed primarily to serve those already involved with the space program rather than the average space enthusiast. "NSI is a NASA fan club," says space activist James Muncy, who founded his own small pro-space organization.[54] "NSI was seen as representing the aerospace industry," comments Planetary Society Executive Director Louis D. Friedman.[55] "They got paid bureaucrats instead of enthusiastic volunteers," says Hughes Research Laboratory physicist Robert L. Forward, who in 1975 had sent out a circular letter inviting others to join NSI.[56]

Ben Bova, who is known to favor a politically more active NSI, recognizes that the institute is seen partly as a monument to von Braun and partly as an industry lap dog. However, Bova points out that "if you support the civil space program, you must support NASA. They have the responsibility of making things work."[57] It presumably was no coincidence that the NSI, which does support NASA, was chosen to help the Young Astronaut Program. In any case, Bova believes that it never was possible to create a single, umbrella group for pro-space Americans; there is too much diversity in their interests, and there are too many interest group functions to be performed.

If that ever were possible, it now appears to be too late. At least three other pro-space citizens groups formed since NSI was created have grown to a size larger than or comparable to that of the institute. NSI's inability to mobilize or gather together all pro-space interests was a factor in the proliferation of space interest groups that took place in the late 1970s. Dennis Stone, president of Spaceweek puts it this way:

> If NSI had been a grass-roots organization, there would have been no L-5 Society or Spaceweek. If NSI had included planetary exploration, the Planetary Society might not exist. If NSI had done lobbying, there might be no American Space Foundation. Other groups have arisen to fill perceived needs.[58]

The NSI may not have lived up to everyone's expectations, but it did provide the first gathering place at the national level for citizens interested in space.

SUBVERSIVE IDEAS

During the early 1970s, the pro-space constituency was still in the early stages of getting organized and had no influence in reversing the downward trend in NASA budgets and aerospace procurement. Despite the gloom among enthusiasts of manned spaceflight, however, inventive people continued to think about how this new environment could be used. Out of this speculation emerged two old but refined concepts that were to prove important in motivating much of the new space movement in the United States: space industrialization and the use of extraterrestrial resources.

These ideas can be traced back well before our own time, particularly in science fiction. The man who first put them together in systematic, credible ways, who was the father of both space industrialization and the mining of other celestial bodies, was Krafft A. Ehricke. A member of the German rocket team that included von Braun, Ehricke had even more sweeping visions of the human future in the universe, proposing (among other things) orbiting mirrors to reflect solar energy where it is most needed, the use of nuclear explosives to mine the Moon and the asteroids, lunar industries, and mobile space colonies.

In 1957, Ehricke published "The Anthropology of Astronautics," laying out three "laws" of astronautics that are an extreme statement of the space dream:

I. Nobody and nothing under the natural laws of this universe imposes any limitations on Man except Man himself.
II. Not only the Earth, but the entire solar system, and as much of the universe as he can reach under the laws of nature, are Man's rightful field of activity.
III. By expanding through the universe, Man fulfills his destiny as an element of life, endowed with the power of reason and the wisdom of the moral law within himself.[59]

This paper set the stage for Ehricke's concepts of space industrialization, which he came to call "exoindustrialization." In 1971, he began spelling out in publications his "Extraterrestrial Imperative," which argued that human expansion into space was both inevitable and desirable, challenging the limits to growth.[60] Specifically, he developed plans for the industrialization and urbanization of the Moon. Sadly for the pro-space cause, Ehricke's impact on the general public was limited by his heavy, Teutonic style. "A Krafft Ehricke speech is a data dump," said aerospace engineer Gerald W. Driggers, an admirer of Ehricke's work who became president of a pro-space organization.[61] Ehricke died in December 1984.

The other major figure, also little known outside space circles, was the late Dandridge M. Cole. He, too, was advocating the mining of the asteroids and suggested that some might be hollowed out to become mobile space arks. His 1964 book *Islands in Space* , written with Donald W. Cox, was the principal statement on space colonization of its time.[62] Meanwhile, other thinkers also contributed to the foundations of space industrialization and the use of extraterrestrial resources.

The first practical application foreseen by most advocates was materials processing in space (MPS). American experiments on MPS began in a primitive way in 1968, on the Apollo 8 mission, and were continued on Skylab. Space-processing experiments were flown on rockets beginning with SPAR-1 in 1976.[63]

A group of experts at the Marshall Space Flight Center in Huntsville, Alabama, where the presence and influence of the German rocket team were strongest, began developing more specific ideas about space manufacturing. G. Harry Stine, stumbling on these developments in 1971 while researching an article on energy in space, realized that the implications could be revolutionary. In January and February 1973, he published his article "The Third Industrial Revolution" in *Analog*.[64] This was elaborated into a book, published in 1975.[65] In language that nonexperts could understand, Stine explained that many of the constraints on industrial processes would be removed if factories were located in space, where gravity would be a negligible factor, hard vacuum would be easily available, and waste heat could be dumped into the infinite sink of the Cosmos. There was an environmentalist message as well; by placing our most polluting industries in space, we could protect the Earth while continuing economic growth.

Jesco von Puttkamer, who came to Washington in 1974 to help work on a long-term plan for NASA, recalls that the space industrialization concept was developed within NASA during 1975 as an input into the study called *Outlook for Space*, which was published in 1976.[66] With the drive to realize the classic spaceflight agenda blunted by political realities, NASA planners were looking for an idea that allowed multiple options, and space industrialization fit the bill.[67]

The first conference on space industrialization was held in Huntsville in May 1976 under the cosponsorship of the Marshall Space Flight Center, the local AIAA chapter, and the University of Alabama in Huntsville. During that year, NASA commissioned Rockwell International Corporation and Science Applications Incorporated (SAI) to do studies of the prospects for space industrialization. Some of those who worked on those studies were believers in the space dream and were to have strong connections with the citizens space movement; consultants to the Rockwell study included Krafft Ehricke and Barbara Marx Hubbard, and those to the SAI study were almost a "Who's Who" of American space advocates.[68]

To their advocates, these concepts seemed to respond to the worries of the 1970s and to open up new possibilities for the human future. The problem was to sell the new ideas and get support for them. Few of these concepts attracted much public attention. However, they were being circulated within the pro-space community, discussed, reformulated, and made more acceptable. In these ideas lay the potential of earning public and political support for new ventures in space.

Most of the people generating these ideas were unknown outside the ranks of space experts and enthusiasts. However one man, Gerard K. O'Neill, created a new synthesis that caught the imagination of a broader section of the public, opening up a major new opportunity for pro-space activity, and sending the pro-space movement off in a new direction.

4

THE BIG IDEA

Utopias seldom begin to be written in any society until after its
members have lost the expectation and ambition of making
further progress.

Arnold J. Toynbee, 1934[1]

A space program directed toward exhibiting that there are no
visible limits to man's future in the universe could be a most
important help in reviving faith in the idea of progress.

Arthur Kantrowitz, 1971[2]

We need utopias. Without utopias the world would not change.

Thornton Wilder[3]

In the summer of 1970, a year after the first Moon landing, a group
of researchers at the Massachusetts Institute of Technology began a study
of the implications of continued worldwide growth. Drawing on the
interactive world economic models of economist Jay Forrester, they
examined the interactions of five basic factors: (1) population increase,
(2) agricultural production, (3) the depletion of nonrenewable resources,
(4) industrial output, and (5) pollution. Feeding these into a global
computer model, the team got results showing that the Earth cannot
support present rates of economic and population growth much beyond
the year 2100, if that long. Humans could live indefinitely on Earth only

57

if they impose limits on themselves and their production of material goods, seeking a state of global equilibrium.

The results of this study were published in nontechnical form in 1972 as *The Limits to Growth*.[4] This much-discussed book was a direct challenge to Western industrial and technological optimism, raising doubts about the continuing expansion of the capitalist system. The conclusions fit in with the cultural pessimism so widespread among Western intellectuals at the time, once described as "bleak chic." One of the authors, Dennis L. Meadows, reportedly said, "What we need is a Copernican revolution, and we don't have one."[5]

The limits to growth thesis was given a mighty boost by the oil crisis of 1973-74, when many oil-producing states cut off exports to the United States and some other Western nations. At the United Nations General Assembly in the autumn of 1974, many less developed countries banded together to demand a New International Economic Order, involving a massive transfer of resources and technology from richer to poorer nations, a redistribution of the world's wealth.

Meanwhile, the environmental movement had been gaining strength. An Earth Day was organized in April 1970, and a United Nations Conference on the Human Environment was held in Stockholm in June 1972. Seeking an end to the abuse and pollution of our natural surroundings, the movement showed an antigrowth bias. It also demonstrated in its early days a considerable hostility to large-scale, bureaucratized technological efforts, particularly nuclear power.

An important challenge to an expansive view of the future came from the perception that the Earth is finite, with a thin, fragile biosphere. Ironically, this may have been triggered in part by images from spacecraft that showed a cloud-flecked blue and brown ball floating against the blackness of space.

Other events seemed to accelerate a loss of American confidence in the future. The war in Vietnam, which may have contributed to antitechnology attitudes, led to a final American withdrawal in 1975. The Watergate political crisis of 1973-74, which led to the resignation of President Nixon, brought government into disrepute, heightening suspicion of public policy at the federal level. Throughout the late 1960s and early 1970s, the first cohorts of the post-World War II "baby boom" generation were entering their twenties. The emergence of this population bulge into political awareness seems to have had much to do with the climate of protest that characterized those times. Many young people challenged existing ways of doing things, were cynical about government, and looked for hidden agendas behind "establishment" policies. Related protest movements (such as women's liberation and minority causes and the anti-war movement) tended to see the space program as a white male-dominated, establishment project irrelevant to the social issues of the day and to the needs of ordinary people. The intellectual and political climate for major new ventures in space was poor.

Yet many people were looking for a more positive vision of the future, and some thought of space. "The realization that the Earth was not all we had available was stirring in many minds at that time," says H. Keith Henson, later to be the first president of the pro-space L-5 Society.[6]

What was needed was a mobilizing idea. In his excellent 1978 book *Spaceships of the Mind*, British science writer Nigel Calder wrote that the prerequisite for a leap forward is the "Big Idea," which he defined as a concept that "captures the enthusiasm of people who will struggle against great difficulties to make it happen."[7] By its nature, the spaceflight field tends to generate daring, sometimes sweeping concepts, but most of them never move the public. For example, proposals for a technologically fascinating orbiting "sky hook" or "space elevator" were circulating in the technical community in the 1960s and 1970s but generated little interest outside it.[8] Something more appealing and relevant to the times was required.

OUT OF FRUSTRATION

In 1967, when the Moon race was still on, two bright, energetic scientists who were competing for selection into the astronaut corps found themselves roommates at Brooks Air Force Base in Texas, where detailed medical tests were done. One was the sunny, curly-haired young Brian T. O'Leary, who had just received his doctorate in astronomy. The other was Gerard K. O'Neill, a slender, intense assistant professor of physics at Princeton University. O'Neill, a bright, hard-working experimental scientist who had invented a storage-ring technique for colliding particle beams, was then unknown outside his profession. Both men were highly motivated to go into space, and both were finalists in the competition.

Brian O'Leary was chosen for astronaut training. However, he had entered the astronaut corps at a time when the prospects for future manned spaceflights were dimming; his class became known as the "Excess Eleven." After a year, O'Leary left the program and later wrote a charming, if disillusioned, book called *The Making of an Ex-Astronaut*. [9] O'Leary's life since then has been an odyssey of personal experiences, often connected with the spaceflight dream. At different times he was a liberal critic of the Space Shuttle, then a supporter of space colonization, the teacher of a course called "Physics for Poets," an advocate of asteroid mining, and most recently a staff scientist with Science Applications, Incorporated, where he has worked on space station studies. O'Leary has authored two more books, suffused with technological optimism: *The Fertile Stars* and *Project Space Station*.[10] As of late 1984, O'Leary was chairman of the National Space Council, a new effort to bring the pro-space community together.

Gerard O'Neill was not selected for the astronaut corps. Crushed, he went back to Princeton and pursued his interest in space another way. From his efforts sprang the single most interesting phenomenon of the new space movement, a "big idea" of potentially historic proportions.

O'NEILL'S VISION

In the fall of 1969, after the first Moon landing, O'Neill recalls that he was concerned that his students were growing cynical about the potential human benefits of science. "Everything scientific or technological was under suspicion," he says. O'Neill, who regarded technology as a tool, wanted to provide an ethical connection, showing that it could be used for good.[11] The device he used was to turn physics into engineering solutions to the world's problems. Leading a group of selected students from his freshman physics seminar, he posed a question: "Is a planetary surface the right place for an expanding technological civilization?"[12]

O'Neill led his students through calculations of how big a rotating pressure vessel in space could be, examining the basic shapes – the sphere, the cylinder, and the torus (doughnut). The answers that came back from the students looked intriguing because it appeared that human colonies in free space were technologically feasible. Calculations showed that a steel shell, rotating to provide Earth-normal gravity and loaded with soil and atmosphere, could be built in sizes as large as several miles in diameter – a miniature, inside-out planet. O'Neill himself put together a model of a cylindrical colony, showing how its surface might be divided into long panels, some of which would open to bring sunlight into the interior through huge windows.

This began the first five-year period of the O'Neill enterprise, during which he did most of his work alone, without outside support, in the best individualist American tradition. After the seminar ended in 1970, O'Neill continued his calculations in occasional spare hours on weekends or late at night. As the numbers continued to come out right, he became interested in communicating his work to others. There followed years of frustrating, unsuccessful attempts to publish an article in a reviewed journal on the concept, which was seen by most editors and reviewers as too "far out."

In 1972, O'Leary arranged for O'Neill to lecture at Hampshire College in Amherst, Massachusetts. This was the first public presentation of O'Neill's ideas. O'Leary recalls that the lecture went over well. O'Leary, who describes himself as having a "left-wing patina" at the time, encouraged O'Neill to "take his ideas to the people."[13] O'Neill began lecturing at colleges and universities at every opportunity.

O'Neill consulted with experts in other fields as he refined his ideas, sometimes sending them papers for comment. Perhaps the most prominent was the distinguished physicist and mathematician Freeman Dyson, at Princeton's Institute for Advanced Study, who pointed out to O'Neill the earlier work of other thinkers in this field.

National Science Foundation policy analyst George Hazelrigg, then in the Aerospace and Mechanical Engineering Department at Princeton, recalls that O'Neill telephoned him in 1973 to say "I've got this idea about colonies," and to ask about the cost of putting payloads into space. The engineers invited O'Neill to give a seminar on the subject, where they raised problems he had not considered. Hazelrigg pointed out that the L-4 and L-5 gravitational libration points, 60 degrees ahead of and 60 degrees behind the Moon in its orbit around the Earth, would be stable locations for space colonies (Figure 4.1).[14] This became part of the emerging O'Neill synthesis and inspired the name of an organization dedicated to propagating O'Neill's ideas: the L-5 Society.

In the early spring of 1974, O'Neill came up with the "Mass Driver," a linear accelerator that could be used either as a space propulsion system or as a means of throwing lunar material to high orbital manufacturing facilities. O'Neill says this was inspired by the work on magnetic levitation of the Massachusetts Institute of Technology's Henry Kolm and by an article in *Scientific American* on the same subject, although the idea had appeared much earlier.[15]

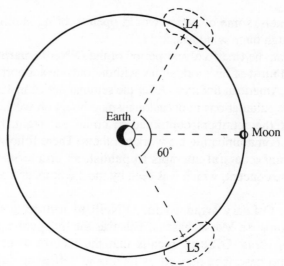

Figure 4.1 — Diagram of Earth-Moon Libration Points
(*Source: L-5 News.*)

By early 1974, a number of technical problems had been worked out in principle. With the support of the Point Foundation of San Francisco (associated with Stewart Brand and the *Whole Earth Catalog*), O'Neill arranged for a conference at Princeton in May 1974 "to discuss the colonization of space as a serious possibility."[16] The first day consisted of a closed session for invited guests only, including three NASA officials. O'Neill provided a detailed description of his plans, which included a linear accelerator on the Moon's surface to launch materials to be used in colony construction. He visualized a Model One colony of about 10,000 people, which he estimated could be built for $31 billion. That colony could build larger successors. "None of the technical tricks which are needed for this task is beyond 1970s technology," said O'Neill. "If these ideas are followed, people could be living and working in space within the next fifteen to twenty years, if they want to."[17]

Topics discussed at that meeting were presented in lectures at an open meeting the next day. New York *Times* Science Editor Walter Sullivan, who covered the conference, wrote a front-page report in the May 13 edition entitled "Proposal for Human Colonies in Space is Hailed by Scientists as Feasible Now."[18] An interview with O'Neill was published in *Mercury*, the magazine of the Astronomical Society of the Pacific, in July/August 1974.[19] The news began to spread. O'Neill later wrote that, while he had shunned publicity before, this experience taught him that publicity is a powerful force.

One of the first nonexperts to seize on O'Neill's ideas was Barbara Marx Hubbard, who called him after seeing the New York *Times* article. "The amorphous hunger of Eve," she wrote later, "suddenly found itself in the presence of a scientist providing a way to develop universal life that corresponded precisely with my intuitive sense of the future." Her Choiceful Futures group accepted O'Neill's concept that we can build new communities in space and provide inexpensive solar energy for them and for Earth. When O'Neill mentioned the need for a $1,000 research grant to his assistant K. Eric Drexler, she gave it immediately.[20]

O'Neill's article "The Colonization of Space" finally appeared in *Physics Today*, the nonspecialized publication of the American Institute of Physics, in September 1974 (his letter on the same subject had appeared in the British science journal *Nature* in August).[21] In this article, O'Neill challenged the orthodox view that the Earth is the only practical habitat for Man or that the planets are the only sites for extraterrestrial colonies, calling this the "planetary hangup." According to O'Neill, we can colonize space, move nearly all our industrial activity away from Earth's fragile biosphere within a century, and encourage self-sufficiency, small-scale governmental units, cultural diversity, and a high degree of independence. He also took on the limits to growth thesis, saying the ultimate size limit for the human race on the newly available frontier is at least 20,000 times its present value if we make use of asteroidal resources.

THE NEW SYNTHESIS

The idea of human colonies in space was not new. The pioneering Russian space thinker Konstantin Tsiolkowsky had written about space habitats early in this century, providing a fictional vision that would ring true to many space enthusiasts today:

> You . . . know how vast and free is the space that surrounds the Earth; you know that it is filled with light; you know that it is empty. It's a sad thought that we are crowded on Earth, treasuring every sunny corner where we can raise crops and build our homes and live in peace and tranquility. While I was wandering in the emptiness about our rocket, it was the vastness, the freedom and lightness of movement that impressed me – that tremendous quantity of solar energy going to waste, uselessly. Who is there to stop Men from

building their greenhouses and their palaces here and living in peace and plenty? . . . This will give us the possibility of operating various kinds of solar engines, welding metals and performing a great many manufacturing operations without the use of fuel.[22]

Tsiolkowsky foresaw that our successors would someday occupy "this space, or ring," around the Earth, adding "we could even get much more space and solar light by forming a ring round the Sun with our new dwellings, outside the Earth's orbit." "Dwelling houses," he wrote, "will be built for millions of people."[23] Clearly stating the utopian intent of space colonies, he listed as the final step in his 14-point plan: "Human society and its individual members become perfect."[24]

In a book published in 1929, British scientist J. D. Bernal also envisioned inhabited spheres orbiting the Sun.[25] The basic concept was well established in the space advocacy community by the 1950s, when J. N. Leonard wrote of it in his book *Flight into Space*:

True space habitations would be space ships large enough to contain all the necessary components of human culture. They would start in a modest way as satellite stations staying near the earth to draw on its resources. Eventually they would become self-sufficient, taking energy from sunlight and setting up carbon cycles to supply their inhabitants with food and oxygen. They would have rocket motors of some sort to give them maneuverability, and they would cruise at will all around the solar system. . . . The asteroids would be convenient sources of solid raw materials. . . . Feeding on sunlight, asteroids, and planetary gases, the space habitations could grow to any desirable size. They would be, in fact, mobile planets, custom-built to fit human needs. So, say the space planners, humans could thrive upon them as they never have thriven on the earth, and better kinds of men should develop quickly. . . . They can multiply almost without limit.[26]

In his 1954 novel *Islands in the Sky*, Arthur C. Clarke detailed many features of living in a large station in space, made primarily of materials launched electromagnetically from the Moon.[27] In 1956, aerospace engineer Darryl Romick advanced a proposal for a giant space station that was, in effect, a space colony; Jerry Grey called this a clear precursor to O'Neill's space settlements.[28] By 1963, Dandridge M. Cole had described habitats in hollowed-out asteroids and in metal-skinned colonies made from them, and had outlined the concept he called "macrolife."[29] Cole's vision of a cylindrical space colony is quite similar to O'Neill's early designs, and space writer James E. Oberg still calls space colonization "the Tsiolkowsky-Cole-O'Neill concept."[30] Krafft A.

Ehricke proposed mobile colonies he called "androcells," and a Lockheed Corporation design for a "space city" had appeared in the early 1970s.[31]

What O'Neill did, apparently without realizing that similar proposals had been made before, was to revive the idea of space colonies in more sophisticated form, with many of the numbers worked out. He brought the idea forward in time by showing that space colonies could be built with familiar materials and techniques (Tsiolkowsky had predicted the date of the first flight into space as 2017.) He avoided the problem of lifting the entire mass of material from the Earth into orbit by proposing that most come from the Moon, which is much more accessible in gravitational and energy terms for colonies in high orbit. Unlike earlier visionaries, O'Neill got involved with the technology and spelled out a step-by-step approach to building the colonies. As he developed his ideas further, he put together a grand synthesis of several ideas, proposing a complete system rather than a single large technological project.

O'Neill also linked space colonization to solutions to problems that preoccupied many people in the mid-1970s: raising global living standards, protecting the biosphere, finding high-quality living space for a growing population, developing clean, practical energy sources, and preventing overload of the Earth's heat balance. He even suggested that space colonization could help prevent territorial wars.[32] Here was the idealistic vision of new worlds for humanity, made far more credible and relevant. Here, too, was a breakout from the limits to growth.

O'Neill's *Physics Today* article may have been one of the most photocopied science articles in history. Keith Henson recalls making at least 500 copies himself.[33] Word of the concept, which Henson describes as a perfect example of a "self-replicating idea," spread rapidly among those interested in space and in the use of technology to solve global problems. O'Neill found himself flooded with correspondence. In January 1975, he began sending out informal newsletters to people on his growing mailing list. As late as 1977, he was typing these himself.

In late 1974, O'Neill began briefing government officials in Washington, including people working on NASA's *Outlook for Space* study. NASA planner Jesco von Puttkamer recalls that "O'Neill's ideas were just what I was looking for."[34] O'Neill persuaded Captain Robert Freitag, then in charge of NASA's advanced programs, to provide the first small grant of government funds ($25,000 for six months) to enable him to elaborate his ideas. According to O'Neill, he got backing from senior NASA officials James Fletcher, John Yardley, and Hans Mark.[35]

This initiated the second five-year phase of the O'Neill enterprise, during which he received government support for his research.

O'Neill had consciously related his proposals to the concerns of the time, and a major American concern in 1975 was energy. In 1968, engineer Peter Glaser had suggested that a large collector in orbit could gather solar energy uninterrupted by night or weather and beam it to the Earth.[36] O'Neill believes that he first heard about this idea when Glaser briefed the *Outlook for Space* group, although von Puttkamer believes he planted the idea by suggesting to O'Neill that the space colonists build something other than colonies from lunar resources, such as satellite solar power stations.[37]

O'Neill used the idea to make his synthesis even grander. In a lead article in *Science* magazine, published in December 1975, he linked space colonization to a solution to the energy problem, arguing that manufacturing facilities in high Earth orbit could build satellite solar power stations from lunar materials.[38] Once again, he presented numbers that made the concept look credible to many readers. Brian O'Leary recalls that it was the linkage of space colonization to satellite solar power stations that won him over to O'Neill's cause; he took a position on the research faculty at Princeton in 1975 to help O'Neill, replacing student Eric Hannah.[39]

One of those who had heard of O'Neill's ideas was Konrad Dannenberg, another member of the German rocket team that came to the United States in 1945. "O'Neill's ideas came in a dry zone, when nothing was happening," says Dannenberg, who has retired from the Marshall Space Flight Center in Huntsville, Alabama. "As a space enthusiast," he adds, "I was anxious that something get started." Dannenberg, who arranged for O'Neill to come to Huntsville to speak to a combined meeting of the local AIAA chapter and the World Future Society, recalls that most NASA officials at the Marshall Space Flight Center kept O'Neill at arm's length.[40]

One of those who did meet O'Neill in Huntsville was Gerald W. Driggers, who was later to become president of the citizens organization inspired by O'Neill's ideas (he also was co-manager of the Space Applications Incorporated space industrialization study). Driggers found O'Neill naive about the way government and the aerospace industry worked and feared a conflict between the visionary O'Neill and more conservative professionals in the aerospace business. Space industrialization, he believed, was a way to bridge the gap.[41]

In May 1975, O'Neill hosted the second Princeton Conference on Space Manufacturing Facilities (Space Colonization), the first to be attended by large numbers of interested outsiders.[42] Gerald Driggers, who gave a paper on the "Construction Shack" that would house people building the first colony, recalls that the atmosphere was one of enthusiasm and camaraderie. Richard Hoagland, of the *Statendam* voyage, was active as a master of ceremonies.[43]

In July 1975, O'Neill was invited to testify before the Subcommittee on Space Science and Applications of the House Committee on Science and Technology, which then was holding hearings on future space programs. This was arranged largely through the efforts of committee staffer Darrell Branscome.[44] O'Neill first testified before the Senate Subcommittee on Aerospace Technology and National Needs in January 1976.[45] Both occasions provided rallying points for his small but growing band of citizen enthusiasts. Meanwhile, T. Stephen Cheston, then an associate dean of Georgetown University's Graduate School, arranged for O'Neill to meet with White House and other senior administration officials.

From June to August 1975, a group of scientists, engineers, sociologists, and economists met at NASA's Ames Research Center in California for a summer study of O'Neill's concept. Their report, issued in August, did not find any insurmountable problems that would prevent humans from living in space, although the practical engineering and social problems were seen to be quite difficult. Noting the large estimated price tag of $100 billion, the study commented, "In contrast to Apollo, it appears that space colonization may be a paying proposition." Significantly, the authors also said, "Space colonization appears to offer a way out from the sense of closure and of limits which is now oppressive to many people on Earth." The group recommended that the United States, possibly in cooperation with other nations, take specific steps toward the goal of space colonization. The report's concluding words suggest the idealism attached to the concept:

The possibility of cooperation among nations, in an enterprise which can yield new wealth for all rather than a conflict over the remaining resources of the Earth, may be far more important in the long run than the immediate return of energy to the Earth. So, too, may be the sense of hope and of new options and opportunities which space colonization can bring to a world which has lost its frontiers.[46]

O'Neill called a press conference to attract attention to the report, generating further press stories.[47] To O'Neill's annoyance, NASA was slow to publish the complete results of this and other summer studies.

Despite this, O'Neill's ideas received generous publicity in the media during 1975 and 1976. Former NASA Administrator Thomas O. Paine wrote a supporting column in *Newsweek* in August 1975.[48] O'Neill's concepts also were described in the *National Geographic*, the *New York Times Magazine*, *Saturday Review*, and *Harpers* (in an article attractively entitled "The Garden of Feasibility").[49] In February 1976, *Smithsonian* magazine carried an article entitled "Colonies in Space Might Turn Out to be Nice Places to Live."[50] O'Neill also began appearing on television talk shows, while continuing to lecture. Meanwhile, O'Neill's ideas received a huge, if subtle, boost from the widely reproduced paintings of Don Davis, who visualized the interiors of space colonies to be attractive places with many of the qualities of familiar Earthly environments. One painting even depicted a suspension bridge across a body of water, reminding the viewer of San Francisco Bay. "Suburbia was utopia to people coming home from World War Two," observes G. Harry Stein. "Space colonies were the new utopia."[51]

Another NASA-sponsored summer study took place at Ames Research Center in 1976.[52] Meanwhile, O'Neill and a small number of helpers were working on the actual technology needed to open the High Frontier. During the first half of 1977, a volunteer group at the Massachusetts Institute of Technology under Henry Kolm and O'Neill (who was there for a sabbatical year) put together the first working model of a mass driver. Brian O'Leary focused his research on the mining of asteroids and published an important paper in *Science* in July 1977.[53] O'Neill and others, particularly Harvard economics student Mark M. Hopkins, also were developing economic arguments for what came to be called "high orbital manufacturing."[54]

In January 1977 O'Neill published his book *The High Frontier: Human Colonies in Space*, which provided a popularized version of his ideas. Describing what day-to-day life would be like in a colony, O'Neill said that by the year 2150 more people may be living in space than on the Earth. Reprinted several times, the book has been translated into eight languages. *The High Frontier* concluded with this paragraph:

> More important than material issues, I think there is reason to hope that the opening of a new, high frontier will challenge the best that is in us, that the new lands waiting to be built in space will give us new freedom to search for

better governments, social systems, and ways of life, and that our children may thereby find a world richer in opportunity by our efforts during the decades ahead.[55]

In May 1977, O'Neill hosted the third Princeton conference on space manufacturing, cosponsored by the American Institute of Aeronautics and Astronautics and NASA. This has become a biennial event, with conferences held in 1979, 1981, and 1983. Although many of the papers presented are technical, listeners are reminded of deeper motivations. At the 1981 conference, Henry Kolm reportedly said, "The survival of life depends on the success of this movement." Professor James Arnold of the University of California at San Diego, an advocate of asteroid mining, reportedly agreed, saying, "Humanity is taking its final exam, on survival."[56]

With the help of Stephen Cheston and Brian O'Leary, O'Neill organized an advisory panel on large space structures to the Universities Space Research Association, a consortium of 52 universities, as part of his continuing effort to improve his concept's standing among his fellow scientists. He also continued his efforts to persuade NASA that space industrialization based on extraterrestrial resources deserved serious consideration.

In the summer of 1977, a third summer study was held at Ames.[57] Brian O'Leary believes that this was the high point of the O'Neill enterprise, producing the most definitive and mature version of the concept.[58] After that, O'Neill began to move away from colony design because he had concluded that "we knew how to make space colonies. The next problem was to find out how to get from here to there."[59]

THE IMPACT OF O'NEILL

One of the most striking things about the High Frontier concept is that it excited many people who had not been space enthusiasts and who had no connection with NASA, the aerospace industry, or space science. Many saw it as a breakout from the limits to growth and accepted the idea that space was an essential part of an optimistic scenario for the future. Even the distinguished foreign policy journal *Foreign Affairs* carried an article by Louis J. Halle entitled "A Hopeful Future for Mankind," half of which was about O'Neill's vision.[60] O'Neill's synthesis suggested that the imaginative use of technology could provide solutions to a host of interlocking problems, including energy shortages and the protection of the environment.

Many of those in the environmental movement of that time were unsympathetic to the space program, but O'Neill turned some into supporters of the new vision. Stewart Brand, publisher of *The Whole Earth Catalog* and editor of *Coevolution Quarterly*, took up the O'Neill cause in print in the fall of 1975 with an editorial entitled "Apocalypse Juggernaut, goodbye?" and later produced a book on space colonies.[61] Brand says O'Neill's appeal lay in the prospect of alternative societies and of breaking the limits to growth. "It was the first new ball game since the American continent filled up," he adds. Brand himself became a supporter of space colonization because it offered a constructive alternative to the use of high technology for war,[62] a reminder of the older "moral equivalent to war" argument.

Thomas A. Heppenheimer, one of FASST's "Michigan Mafia," believes that the appeal of O'Neill was that we could have livable, attractive colonies in space within our time. There was a sense of feasibility because O'Neill had worked out the engineering figures. And the idea had a hopeful potential for the world.[63]

Carl Sagan, who testified to the House after O'Neill in 1975, said:

Our technology is capable of extraordinary new ventures in space, one of which is the space city idea, which Gerard O'Neill has described to you. That is an extremely expensive undertaking, but it seems to me historically to be of the greatest significance. The engineering aspects of it as far as I can tell are perfectly well worked out by O'Neill's study group. It is practical.[64]

Another striking aspect of the response to O'Neill's ideas is that they attracted interest and support from a broad range of professions and academic disciplines. His ideas were inherently interdisciplinary and not the exclusive province of space scientists and engineers; his concept gave more people an opportunity to identify with, and contribute to, the space enterprise. Unlike most spaceflight ideas, colonization drew some support from social scientists. Stewart Brand notes that anthropologist Margaret Mead was attracted to the idea because of her concern about the "monoculturization" of Earth; the colonies would be islands that would be less like other islands.[65] Brian O'Leary believes that some anthropologists saw in the colonies opportunities for social experiments.[66] Magoroh Maruyama, an anthropologist then with Portland State University, published papers describing the opportunities for social diversity and tolerant participatory societies that such colonies would provide.[67]

Early converts helped the cause with their own written works. In October 1975, scientist J. Peter Vajk, then with the Livermore National Laboratory and later active with the L-5 Society published a computerized economic analysis entitled "The Impact of Space Colonization on World Dynamics," which included the idea of the solar power satellite.[68] In 1978 he turned this into a book entitled *Doomsday Has Been Cancelled*, arguing that O'Neill's concept was a breakout from the limits to growth.[69]

The appeal clearly included a utopian element. Earlier examples, such as steam power, electricity, and atomic energy had suggested that a new technology can imply utopia to some of its advocates. However, O'Neill is quick to point out that what he advocates is the opposite of a utopia in the sense that it does not tell people how they should live.[70]

O'Neill also provoked opposition. Many argued against space colonization on the basis of cost, and the economic analyses done by O'Neill's team were seen by most outside observers as optimistic. There were doubts among technical people. There also was an ideological, sometimes emotional quality to some of the criticism, as if people found the idea of space colonization offensive. Lewis Mumford was quoted as saying that "I regard space colonies as another pathological manifestation of the culture that has spent all of its resources on expanding the nuclear means for exterminating the human race. Such proposals are only technological disguises for infantile fantasies."[71] There also was the argument that humans should not take their evil ways into the heavens. In a 1977 article in the *The Futurist* entitled "Space Colonization: An Invitation to Disaster?" physicist Paul L. Csonka argued that space communities would be undemocratic and would fight each other; without effective control from Earth, "most colonies could easily turn into space variants of the prison island concept." Csonka, who believed global government must come first, called for a moratorium on large-scale space colonization.[72]

The old elitist argument used against the Apollo program resurfaced. *Newsweek* reported in November 1978 that "To diehard Earthlings O'Neill's space colonies remain hopelessly utopian and even dangerous," and described author-philosopher William Irwin Thompson as fearing that space colonization might bring about an elitist world government of scientists and bureaucrats.[73] An even more drastic version of this argument was implied by the James Bond film *Moonraker*.

LEGISLATING THE HIGH FRONTIER

Despite his critics, O'Neill had succeeded brilliantly in developing and publicizing an exciting new idea. Where he failed was in his encounter with the unfamiliar world of Washington policymaking and congressional politics.

O'Neill's troubles began after the arrival of the Carter administration in January 1977 (ironically, one day before O'Neill's book came out). Space clearly was low on the new administration's list of priorities, and the political climate for a major new space enterprise was poor. NASA was placed under further constraints in planning for advanced missions, and funding cutbacks led to reductions in NASA's financial support for O'Neill's research. "After 1977," says Brian O'Leary, "the bottom began to fall out."[74]

O'Neill and trusted supporters like O'Leary and Cheston were continuing the long, tiresome job of calling on people in the administration and the Congress that is the essence of lobbying. Cheston recalls that O'Neill had a long meeting with Presidential Science Adviser Frank Press in January 1978 in which Press asked "quality questions."[75]

O'Neill's relationship with NASA had not always been an easy one. Impatient with bureaucracy, he sometimes irritated NASA officials. Stewart Nozette, the young executive director of the California Space Institute, argues that O'Neill did not "work the system" to get a major program going; he did not find a way to make it happen within the existing infrastructure. "You can't do it just because it is scientifically provable," observes Nozette.[76] O'Neill was dealing with a gunshy NASA, which had been the target of many critics, which was having trouble with the Space Shuttle, and which had uncertain backing from the White House. An added blow was an attack by Senator William Proxmire in 1977 after the "60 Minutes" television program did a report on O'Neill and his followers. Proxmire said of O'Neill's ideas: "Not one cent for this nutty fantasy." According to Thomas Heppenheimer, this hurt O'Neill's prospects for getting more NASA funding.[77]

The next major blow from the administration came when the White House released a fact sheet on U.S. civil space policy in October 1978. That document stated, "It is neither feasible nor necessary at this time to commit the United States to a high-challenge space engineering initiative comparable to Apollo."[78] In his newsletter, O'Neill called this "myopic." O'Leary, who recalls making frequent trips to Washington in O'Neill's private plane, describes him as being very frustrated at this point.[79]

O'Neill and his allies sought action through the legislative branch. In October 1977, Barbara Marx Hubbard's Committee for the Future organized a seminar on the High Frontier on Capitol Hill. Representatives Olin Teague, Edward Pattison, and Barbara Mikulski invited their colleagues to attend. This event provided another rallying point for the loose but growing constituency for O'Neill's ideas among young people. Young space activist Harrell Graham reported later that the Committee for the Future had called space-sympathetic people around the country, asking them to urge their local congressman's office to see that a Washington staffer was sent to the seminar. "This last minute, quite unorganized and spontaneous long distance phone campaign," he wrote later, "resulted in not only increased congressional staffer attendance at the seminar but laid down the beginnings of a network of people and organizations around the country who can be relied upon in the future."[80]

Two months later, Teague introduced a pro-High Frontier resolution (H.R. 451), which included the following:

As longer-range, high priority national goals, it is anticipated that by the year 2000 these explorations will have opened the resources and environment of extraterrestrial space to an as yet incalculable range of other positive uses, including, but not limited to, international cooperation for the maintenance of peace, the discovery and development of new sources of energy and materials, industrial processing and manufacturing, food and chemical production, health benefits, recreation, and conceivably, the establishment of self-sustaining communities in space.

In a circular letter Barbara Hubbard wrote, "the resolution represents a first effort of citizens to take leadership in calling for long-range decisions on space."[81] A companion pro-High Frontier resolution (H.R. 447) was introduced by Barbara Mikulski, Lindy Boggs, and David Stockman, who later was to be budget director for the Reagan administration. An ad hoc committee to support the two House resolutions was formed by Brian Duff, who had been NASA press spokesman until 1971, and Carol Rosin, who later became an outer space arms control activist.

Teague held hearings on future space programs, including discussion of his own resolution, in January 1978. O'Neill and Hubbard were invited to testify, along with former astronaut Neil Armstrong, engineer and third industrial revolution advocate G. Harry Stine, scientist-entrepreneur-science fiction writer Charles Sheffield, Presidential Science Adviser Frank Press, and NASA Administrator Robert Frosch. However, Teague's resolution never left the committee. Meanwhile,

Senator Harrison Williams of New Jersey (where Princeton is located) introduced legislation that would have funded a National Science Foundation study of High Frontier concepts. Although the bill passed the authorization committees, the appropriations committees declined to provide the funds.

More general space policy bills were introduced at the next session, but none passed. In 1979, Don Fuqua sponsored a "Space Industrialization Act," which would have created a public corporation to finance promising space industrialization projects, and held hearings in May and June, but that bill died, too. In the Senate, Adlai Stevenson III introduced a space policy act in 1978, apparently in reaction to the Carter administration's space policy document, and reintroduced it in the following year. Senator (and ex-astronaut) Harrison Schmitt introduced his own space policy act in 1979 and 1982, advocating a series of steps toward an "orbital civilization." Both senators included references to space industrialization; Schmitt's bill, the bolder (and politically less realistic) of the two, included High Frontier concepts. Congressman George E. Brown of California introduced a similar space policy bill in the House. None of these bills got out of committee.

THE SOLAR POWER SATELLITE FIGHT

Meanwhile, the focus for the hopes of the High Frontier advocates had become the satellite solar power station (SSPS), which provided the concept with an economic rationale. Peter Glaser had testified on the concept in 1973. After the oil shortages of 1973-74 had spurred interest in alternative energy systems, the Energy Research and Development Administration (ERDA) established a Task Force on Satellite Power Stations in 1976. NASA, which wanted to get into the energy field, negotiated an arrangement for a joint study with the Department of Energy (the successor to ERDA), which ran from 1977 to 1980. NASA conducted the systems definition of the SSPS, while the Department of Energy worked on the environmental, health, and safety factors, as well as economic, international, and institutional issues. O'Neill briefed the study group, arguing for the use of nonterrestrial materials for building these satellites.

In 1978, Peter Glaser and his allies formed their own pro-SSPS organization called the Sunsat Energy Council to push the idea, using this vehicle to attract attention to the SSPS concept on Sun Day, May 3, 1978. Sunsat had an informal alliance with O'Neill and his colleagues, who saw the SSPS as the way to get space colonization. Other players included people at the Marshall and Johnson Space Flight Centers and at the Boeing and Rockwell International Corporations. One of the Boeing engineers involved in this effort was Gordon Woodcock, who became president of the pro-O'Neill L-5 Society in 1984.

The SSPS study, which included inputs from a variety of experts all over the United States, enabled a number of young space enthusiasts to get involved in this public policy issue. The study concluded that there was no major scientific or technological barrier to building the system, although it would be expensive. However, people who favored decentralized solar power sharply criticized SSPS, which they saw as another centralized, bureaucratic system. Some critics saw SSPS as a major aerospace project masquerading as an energy option. Others used environmental arguments, claiming that the microwave beam from the SSPS would disturb the ionosphere and be dangerous to life.[82] Washington lawyer-lobbyist Leigh Ratiner, who later worked closely with the L-5 Society, recalls that the satellites were called "bird-fryers" on Capitol Hill.[83] Some opponents even claimed that these large objects would interfere with astronomy.[84]

Leonard W. David, who worked on a related study, comments that SSPS was a litmus test of attitudes toward big technology.[85] One critic called SSPS supporters "techno-messianic zealots."[86]

There were even more powerful forces in opposition. According to George Hazelrigg, who did extensive analyses of SSPS, it ran into the combined power of the fusion energy lobby in the Department of Energy and the major energy companies, which did not want it.[87]

In 1978, Representative Ronnie Flippo of Alabama (his district included the Marshall Space Flight Center) introduced a Solar Power Satellite Research Development and Demonstration Act, which would have provided $25 million for further SSPS studies. The Flippo bill passed the House but was killed in the Senate, largely through the efforts of a single professional lobbyist acting in partnership with one Senate staff member and a retiring senator.[88]

The crisis came in January 1980, when the Carter administration "zeroed" SSPS in its new budget. Supporters of SSPS and other High Frontier ideas launched a desperation campaign in Congress to get funding restored. Flippo got $5.5 million authorized, but this was killed by the House Appropriations Committee.

The SSPS campaign effectively died in the summer of 1980, when the Department of Energy study was completed. That study concluded that the SSPS had the potential to serve as a future (post-2000) electrical power option but that the present portfolio of electrical power options appeared adequate.[89] The National Academy of Sciences published its own study of the SSPS in 1981, concluding that the concept of an SSPS is presently faced with sufficiently serious difficulties that no funds should be committed during the next decade to pursue development of the system.[90] The Congressional Office of Technology Assessment also published a study that year, concluding that too little is currently known about the technical, economic, and environmental aspects of SSPS to make a sound decision on whether or not to proceed with its development and deployment.[91]

The solar power satellite was a fine example of space advocates seizing the theme of the time, in this case energy. But it was premature and did not have enough friends. Frederick Koumanoff, who directed the Department of Energy effort, states that "the bottom line on the SSPS was that there was and still is a need for significant improvement in some basic technologies before an SSPS concept could be fully considered and the economics of it be determined with some precision."[92] The legislative effort in support of it was fragmented and uncoordinated.

The space revolutionaries had pushed too far, too soon. With the SSPS died any early realization of O'Neill's dream with government funding. "O'Neill hung himself on the SSPS," says Hazelrigg.[93] "SSPS queered the deal," agrees Ratiner.[94] The attempt to legislate the High Frontier had failed and was not revived seriously again. Todd Hawley, a young space activist who later led the student pro-space organization Students for the Exploration and Development of Space, calls it "the death of utopia."[95]

THE SPACE STUDIES INSTITUTE

During the mid-1970s, O'Neill was doing his research on a shoestring. His first research assistant, Eric Drexler, was funded in large part by a grant from Barbara Marx Hubbard. O'Neill's wife Tasha offered to provide room and board for young researchers at the O'Neill home and to allow experiments in the basement. O'Neill also faced uncertainty in funding from NASA, which was vulnerable to budget cuts and political criticism.

Seeing a need for independent, private financing for his research, O'Neill, in 1977, founded the Space Studies Institute (SSI) in Princeton, with himself as its unpaid president and Tasha as its unpaid secretary-treasurer. The SSI provides a mechanism through which donors can fund what O'Neill calls "critical-path" research. Early major contributors included William O'Boyle and David Hannah, who later was to found the private launch vehicle company Space Services Incorporated. By the summer of 1978, O'Neill had formed a Senior Advisory Board for the institute, which included physical scientists Freeman Dyson and Hannes Alfven, anthropologist Carleton Coon, designer Buckminster Fuller, former NASA administrators Thomas Paine and James Fletcher, Chairman of the United Nations Committee on the Peaceful Uses of Outer Space Peter Jankowitsch, and old allies Barbara Marx Hubbard and Stephen Cheston. This began the third phase of the O'Neill enterprise.

Starting with a mailing list of several thousand, O'Neill quickly attracted support; there were over 1,000 paying subscribers by 1979 and about 10,000 by the end of 1983. In addition to regular members, who contribute $15 a year, the SSI by 1984 had nearly a thousand senior associates who had pledged $100 a year for five years, despite the fact that few of them will realize any tangible benefit from their contributions. The SSI has a small but highly computerized office with a lively and dedicated staff and plenty of enthusiastic young volunteers (for example, members of the Students for the Exploration and Development of Space have stuffed envelopes for mailings of the institute's newsletter). SSI also has local support teams, which are like small chapters of the organization.

The SSI's research has been concentrated on the development of a mass-driver and on techniques for processing lunar and asteroidal materials, with some work on the mathematical and telescopic search for asteroidal materials near the Earth. In 1984, the SSI awarded a grant for the design of a satellite solar power station.

O'Neill's decision to set up the SSI proved to be a wise one. As of September 30, 1979, all NASA support of private space manufacturing research, initiated in 1975, was ended.[96] This was the main factor leading to the departure of Brian O'Leary, who had worked on High Frontier research since 1975.

Writing in the spring of 1984, O'Neill said SSI's current research program was designed to do the following:

> to bring us well along the way to SSI's central goal: to establish at the earliest possible time a productive, peaceful industrial facility and associated human colonies in space. All that we have learned convinces us that when that goal is reached, humanity will never turn back. It will be our first step into the high frontier, the principal domain of humanity from now on.[97]

THE INSTITUTE FOR THE SOCIAL SCIENCE STUDY OF SPACE

Meanwhile, Stephen Cheston, desiring to get more social scientists involved with the space enterprise and to defuse potential criticism, formed a Space Utilization Team at Georgetown University in the spring of 1978 and got a NASA grant to study social science interest in space development. Cheston was assisted by Charles Chafer, who later became the Washington representative of the private launch vehicle company Space Services Incorporated, and by Chafer's wife Sallie. Together, the three formed the Institute for the Social Science Study of Space, affiliated with the Universities Space Research Association, to develop "an objective, scholarly community of researchers and analysts concerned with the systematic investigation of the human factors involved in the exploration, industrialization, and utilization of outer space."[98] In May 1979, the ISSSS put out volume 1 of the *Space Humanization Series*, containing papers on several space-related subjects (the editors were Cheston and space activist David C. Webb, and the managing editor was the National Space Institute's Courtney Stadd).[99] Volume 2 had not been published as of 1985. The institute also began putting out a newsletter called *Space and Society* in the spring of 1980. This effort eventually led to a NASA publication entitled *Social Sciences and Space Exploration* to aid college faculty in teaching space subjects in social science and humanities courses (several colleges and universities have given courses on space industrialization and colonization).[100] However, Cheston discontinued the ISSSS in 1984 because both he and the Chafers had become involved in commercial space activity.

GEOSTAR

Although he has focused the SSI on "critical path research," O'Neill realizes that a modestly funded SSI will not bring his vision to early realization. In his April 1979 newsletter, O'Neill announced his decision that any potentially lucrative patents of his own would be registered in the name of the SSI and that all income from such patents would go to the SSI. This set the stage for O'Neill's newest venture.

An active private pilot, O'Neill was disturbed by inadequate navigational and positioning systems for aircraft, particularly after a 1978 mid-air collision between a private plane and an airliner over San Diego that killed 150 people. Later that year O'Neill began developing ideas for a satellite positioning system, later adding to it a digital communications system. An article on the subject in *AOPA Pilot* in July 1982 was well received; O'Neill obtained patents in November and set up the GEOSTAR corporation in February 1983. By the end of 1983, successful ground proof tests of the system had been conducted. GEOSTAR hoped to launch its first satellite in 1987.

O'Neill structured the new company so that the SSI is a major, although nonvoting, shareholder. When GEOSTAR begins generating revenues (perhaps in the late 1980s) a significant portion will go to SSI. If GEOSTAR proves to be a commercial success, SSI could find itself generously funded for the first time. "GEOSTAR money," says Stephen Cheston, now a vice-president of the firm, "could be critical in moving up the whole High Frontier idea by one to three decades."[101] But the SSI must hold on until the 1990s.

SUMMING UP

Has O'Neill made a difference? At the very minimum, he revised and broadened the classic agenda for spaceflight, opening up for public discussion potential new options for human activity in space. His was the first widely known proposal for an economic return from a massive investment in space since the communications satellite. He took a giant step toward making space industrialization and the use of extraterrestrial resources appealing to people outside as well as inside the aerospace community. O'Neill's colonies and the satellite solar power stations they were to build may have been an overreach at the time, but they stretched the limits of the conceivable. They changed the intellectual culture for space, making it possible for such ideas to appear regularly in mainstream space literature.

O'Neill also changed the rationale for sending humans into space, from exploration and adventure to economically useful permanent residence. In tune with the ethos of the emerging new pro-space movement, part of which he inspired, he democratized the idea of spaceflight, making it an aspiration for the many rather than for a select few. Irwin Pikus of the National Science Foundation, who addresses space law issues at the Princeton conferences, believes that O'Neill's tireless campaigning for humans in space also helped revive the permanently manned space station, which became U.S. national policy in 1984.[102]

O'Neill showed that a dedicated individual can have an impact even in a field as bureaucratized as space. But the pursuit of this dream has been personally demanding. O'Neill sometimes seems a driven man, seeking the realization of the High Frontier within his lifetime. Absorbed with GEOSTAR, O'Neill in 1984 could not even give much time to his first love, the SSI. He has even less time for people in the pro-space movement, many of whom see him as remote and reclusive. He has clashed with other bright people such as Heppenheimer and Hazelrigg and has been accused of being less than generous in giving others full credit for their ideas.

O'Neill personally experienced what was to be the full cycle of the new pro-space movement. First, enthusiasm about the big idea and a desire to communicate it with others. Second, the encounter with the realities of government and politics, leading to frustration. Third, the scaling down of near-term goals and their pursuit by indirect or private means. However, O'Neill has not surrendered his dream. In his 1981 book *2081: A Hopeful View of the Human Future*, O'Neill wrote that "the fundamental transformation that space colonies will bring about is from an economics of scarcity – the zero-sum game that we are forced to play on Earth – to an economics of abundance. . . . By 2081 there may be more Americans in space colonies than are in the United States."[103]

G. Harry Stine believes that O'Neill stepped into the vacuum Wernher von Braun's departure left.[104] If von Braun was a leader of the first spaceflight revolution, O'Neill deserves credit for at least outlining and trying to launch the second: space industrialization and permanent human residence beyond the Earth. Pikus believes that O'Neill "has had real influence, creating a new option after the limits to growth."[105] Space activist Courtney Stadd (formerly with the National Space Institute and later the Washington representative of the private launch vehicle company Starstruck) says, "O'Neill may die embittered, but he provided a conceptual revolution, a paradigm shift."[106] This may turn out to have been the Copernican revolution Dennis Meadows reportedly said we did not have. Concludes Stephen Cheston, a trained historian, "1975 was a turning point in U.S. cultural history. The idea will continue to wedge out."[107]

5

O'NEILL'S CHILDREN

We must help bring the Copernican Revolution to politics, to
business, and to the people.

K. Eric Drexler[1]

In the fall of 1974, young aerospace engineer Thomas A.
Heppenheimer was a passenger on a bus traveling across the deserts of
northern Mexico. A few days earlier, this veteran of FASST had visited
the Pasadena office of Jet Propulsion Laboratory engineer Louis D.
Friedman (who later became the executive director of the Planetary
Society). There Heppenheimer had picked up a copy of the September
1974 issue of *Physics Today*. As the bus traversed the sere Mexican
landscape, Heppenheimer began to read Gerard K. O'Neill's article.

By the time he finished, he was "entranced." "I knew I had to get into
it," he says. What particularly struck him was that he could make a
personal contribution and could get in on the ground floor. When O'Neill
lectured at the California Institute of Technology in January 1975,
Heppenheimer introduced himself and got on O'Neill's mailing list. He
gave a paper at the May 1975 Princeton conference and participated in the
1975 summer study. There he was stimulated by an inspirational talk
given by Richard Hoagland, who spoke of space colonization as "The
High Frontier." Heppenheimer knew an old Kingston Trio song, "The
New Frontier," which had been written to honor President John F.
Kennedy. Together with its composer, John Stewart, Heppenheimer
revised some of its lyrics and called the result "The High Frontier." The
words say much about those who responded to O'Neill's ideas:

Some to the rivers and some to the sea,
Some to the soil that our fathers made free,
Then on to the stars in the heavens for to see,
This is the High Frontier, this is the High Frontier.
Let the word go forth, from this day on
A new age of Mankind has begun.
Hope will grow for the human race!
We're building a colony deep in space!
This is the High Frontier, this is the High Frontier.
Let us begin, for it shall take long,
Let everyone sing a freedom song.
Not for ourselves that we take this stand,
Now it's the world and the future of Man.
This is the High Frontier, this is the High Frontier.
The day will come, it's going to be,
a day that we will someday see
when all Mankind is reaching out
without a limit, without a doubt!
This is the High Frontier, this is the High Frontier.[2]

Heppenheimer went on to make real contributions to the space colonization idea. Through a difficult computer analysis, he showed that the L-4 and L-5 locations for colonies might be less desirable than the "2:1 resonant orbit," taking 14 days to traverse through the Earth-Moon system. With the help of Barbara Marx Hubbard as a go-between, in 1977 he published a book entitled *Colonies in Space*, which became a Book of the Month Club selection. Partly because of friction between himself and O'Neill, Heppenheimer last participated in a Princeton conference in 1981. As of 1984, he was a full-time writer, publishing books of science fact and science speculation like *Toward Distant Suns* and *The Man-Made Sun*.[3]

Heppenheimer was only one of many examples of the strong effect on some younger people of O'Neill's ideas, which offered the prospect of personal participation in the space dream. Mark M. Hopkins recalls trying as early as his junior year in high school to accelerate the space program by finding a way to travel faster than light. He spent nearly all of his time studying, and his only recreation was watching "Star Trek" once a week. By his junior year at the California Institute of Technology, he decided that if anyone was going to discover a way of traveling faster than light, it was not going to be him. He switched his major to economics and entered Harvard, where in 1972 he cofounded the Harvard-Radcliffe Committee for a Space Economy. Members of the committee believed that once space activity reached a certain level, due to dramatic economies of scale and other factors, the economic development of space would "take off" in a self-sustaining way, feeding on itself rather than on noneconomic factors such as exploration, military advantage, and national prestige. In that context, writes Hopkins, O'Neill's September 1974 article "hit like a bomb."

Hopkins recalls calling up O'Neill and asking, "How are you going to pay for this?" O'Neill put him on the mailing list for the May 1975 conference, where Hopkins filled a gap in economic analysis by doing "back of the envelope" calculations. Impressed, O'Neill invited Hopkins to the 1975 summer study, where he did most of the economic work. "This changed my life," says Hopkins. With Heppenheimer, he published an article on "Initial R and D Requirements for Space Colonization" in *Astronautics and Aeronautics* in 1976. Hopkins, who has devoted thousands of volunteer hours to activism in support of O'Neill's ideas, became vice-president of the L-5 Society, a citizens group inspired by those concepts. He also played the major role in creating Spacepac, the largest space political action committee. As of 1984, Hopkins was an analyst with the RAND Corporation in Santa Monica, California.[4]

K. Eric Drexler, then a student at the Massachusetts Institute of Technology, had been thinking independently about the potential of large-scale space activities even before O'Neill's article. Asking Professor Philip Morrison for advice on whom to contact, he was referred to O'Neill. He was one of those who critiqued a draft of O'Neill's *Physics Today* article, and he helped plan the 1974 Princeton conference. Drexler initiated a student seminar on space colonization in January 1975 and established the M.I.T. Space Habitat Study Group. He also became O'Neill's first research assistant on space colonization. Remaining closely involved for the next decade, he has been one of the L-5 Society's most active board members. As of early 1986, he was about to publish his own pro-technology book, entitled *Engines of Creation*.[5]

OUT OF THE DESERT

The first large organization stimulated by O'Neill's vision appeared not at Princeton or M.I.T., but in Tucson, Arizona. There two bright, technologically literate young people named Keith and Carolyn Henson were running a small high-tech firm called Analog Precision, which they had started from virtually nothing. "They lived like church mice," says Mark Hopkins, "and saved every penny to put into the business."[6] Their dedication and entrepreneurial skills were to prove invaluable in creating from scratch the most daring of the major new citizens pro-space organizations.

The blunt, energetic Keith, who has a degree in electrical engineering, says he has had for many years an interest in "high tech space futurist views."[7] Carolyn, a tall, slender, dynamic woman with a presence some find charismatic, is the daughter of astronomers Aden and Marjorie Meinel (he was the first director of nearby Kitt Peak National Observatory). Carolyn recalls being an advocate of asteroid mining even before discovering space colonization.[8]

Like others, Keith and Carolyn learned about O'Neill through the *Physics Today* article and immediately became excited by the concept and the roles they could play. "I wanted to be one of the people who change history," says Keith.[9] Carolyn was to write in 1979 of "a chance to make history, to make a real difference in when, and how, we reach into space."[10]

The Hensons contacted O'Neill. When a speaker on space colony agriculture dropped out of the May 1975 Princeton conference, O'Neill asked them to step in. There followed a crash research program assisted by experts in closed-environment (greenhouse) agriculture at the University of Arizona. The Hensons' paper was a success. At the post-conference banquet, Carolyn gave a rousing pro-colonization speech.[11]

At that conference, the Hensons met Gerald W. Driggers, an aerospace engineer then with the Southern Research Institute in Birmingham, Alabama. Driggers, although born into a poor farm family, had begun building telescopes at a young age to look at the stars. He was like many others at the conference in that he had discovered O'Neill through the *Physics Today* article, but he was unlike most of them in having considerable background and experience in the aerospace business. He had been invited to fill another vacant slot, this one to give a paper on the "construction shack" space station that would build the first colony.[12] O'Neill later selected him for the 1976 and 1977 summer studies on space colonization.

At the conference, Driggers expressed his belief that the space colonization idea needed a newsletter. He passed around a sign-up sheet and got a copy of the existing mailing list compiled by student Eric Hannah, then an assistant to O'Neill. When Driggers was unable to follow through on this, the names eventually were turned over to the

Hensons. Carolyn obtained O'Neill's own mailing list and used the combined mailing lists in creating the first public group dedicated to the colonization of space.[13]

THE L-5 SOCIETY

Richard Hoagland, then a proselytizer for O'Neill, volunteered to come to Tucson, where Carolyn set up interviews and public appearances for him in June 1975. In discussions among Hoagland, the Hensons, and a few other enthusiasts, the idea emerged of founding a High Frontier Society, which was incorporated quickly. Hoagland then went off on a promotional and fund-raising tour for the new venture.[14]

Meanwhile, the new group was making the first of its many forays into politics. Carolyn, who had been active politically in Arizona, used her contacts to arrange a meeting between O'Neill and Congressman Morris Udall, then a serious contender for the Presidency. Udall reportedly was fascinated by the idea and agreed to support the concept, asking only that his support be recognized publicly. Brian O'Leary, who then was on the staff of Udall's Interior Committee, wrote a pro-O'Neill letter for Udall's signature to Robert Seamans, then head of the Energy Research and Development Administration.[15]

By August 1975, some of the founders of the High Frontier Society had concluded that a break had to be made with Hoagland and that the existing name for the organization should be buried. A new organization evolved out of the structure of the old. Keith, with O'Neill's blessing, decided that it should be called the L-5 Society, after one of the gravitational libration points where colonies might be located. He also invented a striking goal for the society: to disband itself at a mass meeting in a space colony.[16] The L-5 Society was incorporated in August 1975, with Keith as its first president. Even then, he recalls, Carolyn did most of the work.[17]

In contrast to the National Space Institute, the L-5 Society began as a shoestring operation, run out of a corner of Analog Precision. The Hensons and a small band of helpers had to borrow money, labor, and equipment to set up an office.

Some potential donors said they would contribute money if there were a newsletter. The first issue of the *L-5 News*, featuring Udall's support of O'Neill, appeared in September 1975. According to the new publication, the society was formed to "educate the public about the benefits of space communities and manufacturing facilities, to serve as a clearinghouse for information and news in this fast developing area, and to raise funds to support work on those concepts where public money is not available or is inappropriate." Its short-term projects (dependent on financial response) were to aid O'Neill in completing his book on the L-5 concept and to provide part of the paid staff to help him.[18] Here were the patterns of earlier pro-space organizations, seeking to provide a home for enthusiasts and to educate the public while supporting, if not actually doing, work on the technology.

Sending membership forms to the names on its mailing lists, the L-5 Society quickly attracted a hard core of enthusiasts for the High Frontier, plus a variety of other pro-space people. However, there also was something about the society that appealed to a broader spectrum of opinion than one found in most pro-space organizations. Many who had shown no previous interest in space, including a number of environmentalists and advocates of alternative or utopian societies, joined the new group. Randall Clamons, who served as administrator of the L-5 Society from 1978 to 1984, recalls that the early members tended to be younger and more idealistic and that some were escapist.[19]

Space colonization also attracted some people from the fringes of American culture. Keith Henson wrote in the May 1976 *L-5 News*:

> A major purpose of the L-5 Society is to arouse public enthusiasm for space colonization. In this, we have been quite successful. However, one unfortunate result is that we attract "groupies" whose thoughtless enthusiasm (fortunately, short-lived) is often destructive. When these individuals come in contact with the media, politicians, or government officials, they give our goals a bad name.[20]

Carolyn recalls "space groupies" turning up at the L-5 office in Tucson with their life's possessions. In her opinion, some needed psychiatric help ("No," she recalls telling one, "I can't turn your pool table into a starship").[21] As of late 1979, she still was writing of "harassment by kooks and con artists."[22]

In the meantime, professor and counter-culture leader Timothy Leary also had come to the conclusion that migration into space was the right direction for humanity. Perhaps as early as 1973, Leary had created a group called "Starseed." After completing a prison term for drug offenses, Leary did proselytizing tours for a concept he called SMI²LE: Space Migration, Intelligence Increase, Life Extension. He also became a supporter of the L-5 Society, bringing in some of its early members.[23] In the August 1976 *L-5 News*, Leary wrote:

> Since Peenemunde and Sputnik it has been obvious that the most intellectually and survivally interesting issue on this planet concerns Humanity's transition to extraterrestrial life. . . . It is both the self-appointed liberal humanists and the civil service engineers who threaten space migration with their clashing doctrinaire opinions – the former peevishly suspecting that any escape from Earth gravity is an elitist liberation from the compulsory egalitarian plans of limited growth and intraplanetary bussing.[24]

The small initial band of L-5 activists, including the Hensons, Thomas Heppenheimer, Eric Drexler, Mark Hopkins, and J. Peter Vajk, showed great energy in making appearances wherever possible. At a "Limits to Growth" conference near Houston in October 1975, some of them made contact with the futurist Herman Kahn, who later endorsed the space colony concept. Peter Vajk, who gave a paper on the world dynamics implications of space colonization at what Keith Henson calls "the gloom and doom" conference, comments that he was "naive enough to think that the people there wanted a solution."[25] In May and June 1976, L-5 activists including Vajk, Magoroh Maruyama, anti-war activist Norie Huddle, and futurist/science fiction writer Robert Anton Wilson went on the road to the Habitat conference in Vancouver, British Columbia, where they participated in the Non-Government Organizations meeting. Enthusiasts in New York made an unsuccessful effort to put space colonization on the agenda of the United Nations Committee on the Peaceful Uses of Outer Space. In the spring of 1976, the L-5 Society considered moving its headquarters to New York, but it never did so.

Advocating what it called "the new space program," the L-5 Society soon acquired the reputation of being the most extreme of the major pro-space citizens groups, with the grandest dreams and the least modesty in proclaiming them. Some L-5 activists also had a style radically different from that of established space interest groups, being pushy, utopian, and even messianic in the view of some critics. Both the Hensons were considered by many who encountered them to be aggressive to the point of being abrasive.

In part, this style may have been due to the nature of the society's appeal, which offered a special kind of life goal. It also may have been because some of the group's leading personalities had backgrounds in protest movements, notably those connected with the Vietnam war, the environment, and women's rights. Carolyn Henson in particular fell in this category. Norie Huddle, who was associated with Leary as well as L-5, had written a book on the pollution crisis in Japan, and was a coordinator of Mobilization for Survival and the protest against the Seabrook nuclear reactor (in 1984 she published a book entitled *Surviving: The Best Game on Earth*).[26]

This style, and doubts about the way the organization was being run, led to a growing alienation from the man whose ideas had inspired the society: Gerard O'Neill. "The O'Neill-L-5 rift was deep by 1976," says Gerald Driggers.[27] Concerned about adverse publicity, O'Neill kept the Hensons at arm's length. In a 1984 interview, he referred to them as "those people in Arizona."[28] The other side of this argument should be noted, however. In Keith Henson's view, O'Neill wanted more control of the idea than was possible.[29]

The L-5 Society, which had little money, relied heavily on volunteers from the beginning. Although the society got financial help from Barbara Marx Hubbard, George Koopman, William O'Boyle, and others, money to pay the staff was sometimes scarce. However, the idea was a motivating one, and chapters began to spring up in other locations; there were seven in 1976, including one in the United Kingdom. The society also had organizational allies such as FASST (FASST leader David Fradin was on the L-5 board).

At first, L-5 was a very small society, having only 126 members at the end of November 1975. In September 1976, L-5 established an important link to the science fiction community when writer Jerry E. Pournelle persuaded Keith Henson to come to the MIDAMERICON science fiction convention in Kansas City, where Keith spoke to an estimated 1,500 people. L-5 significantly increased its membership at that meeting. "This was the start of L-5 as a viable organization," claims Pournelle.[30] Keith Henson met the prominent science fiction writer Robert A. Heinlein, who later was generous to the L-5 Society with his name, his letters of support, and his money. Pournelle later helped L-5 by putting its address in his column and in some of his books.[31]

Despite its activism, the early L-5 Society was cautious about direct lobbying. "In order to maintain our tax-exempt status, no one in the organization may represent the society in any activity that could be construed as lobbying," Keith Henson wrote in the May 1976 *L-5 News*.[32]

L-5 activism was given a new stimulus in the summer of 1977. Carolyn Henson, who had left Analog Precision in 1976 to finish her degree at the University of Arizona, quit school and became the salaried second president of the society, quickly putting on it the stamp of her dynamic, if controversial, personality. She may have been the first pro-space leader to also be an ardent feminist; in editing articles for the *L-5 News*, she would change "manned" to "piloted." She livened the magazine, becoming what sociologist B. J. Bluth calls "the gossip columnist of the space movement."[33] Working with very limited resources, she was effective in keeping members informed about events in Washington, using a sense for politics and an irreverent style to good advantage. One regular feature of the *L-5 News* was a column of rebuttals to those who claimed the space program was a waste of money.

The Hensons and other space activists, notably Jon Alexandr, proved adept at inventing slogans, including some with a strong environmentalist flavor. "Declare the Earth a wilderness area," said one; "If you love it, leave it," said another, across a picture of the full Earth. In the "small is beautiful" vein, one slogan said "decentralize – get off the planet."

During Carolyn's two-year presidency, the society grew and turned increasingly toward political activism. Noting that Internal Revenue Service regulations had changed to allow educational organizations to spend up to 20 percent of their income on lobbying, Carolyn began moving the society in that direction. In July 1977, she and other L-5 activists campaigned (mostly by telephone from Tucson) to support efforts by space scientists to save the Jupiter Orbiter Probe (JOP) project. By then, Tucson had become a major center for space science; one of the leaders of the floor fight in the Congress to save JOP was John Rhodes of Arizona. The society's leaders believe that it played an important, if supporting, role in getting funds restored. However, congressional staffers and administration officials interviewed on this point generally think the society's influence was marginal at best.

In 1977 advent of the Carter administration, with its negative attitude toward major new space ventures, was a challenge to the growing L-5 Society. Carolyn Henson launched a Legislative Information Service to bring fast-breaking news to "those of you with a special interest in space politics."[34] This included a space politics "hot line" staffed by Marc Boone. This service, funded by separate donations, was a shoestring operation, with a budget of about $500 a month. L-5 used some of its funds to support the activities of young space activist Ken McCormick, who with Eldon James had launched a new organization in Washington, D.C., called the National Action Committee for Space by the spring of 1978. This small and underfunded operation was the closest thing to a permanent presence in the national capital that L-5 then had, and it was a forerunner of other political action arms of the citizens pro-space movement. McCormick, Boone, and others periodically wrote articles for the *L-5 News* on the basics of lobbying.[35] At the time, the only specific reference material available was a book by Robert A. Freitas, Jr., called *Lobbying for Space*.[36]

In 1978 the L-5 Society got involved with the satellite solar power station issue, specifically the Department of Energy's outreach effort; L-5 got a contract to poll members and other pro-space people about SSPS. In the end, the society lost money on the contract, which proved to be a drain on the organization's slender resources. Randall Clamons believes this effort interfered with the normal growth of the society, which then had only abut 3,000 members.[37] However, it was typical of the risk-taking L-5 tendency to overreach, taking on tasks other small organizations would have avoided.

In the summer of 1979, the time was approaching for the transition to a new president of the society to replace Carolyn Henson, who then was bedridden and pregnant with her fourth child (born in December 1979, she was named Virginia Heinlein Henson, after the wife of the famous science fiction writer). This and other events led to some infighting within L-5's board, which by then included people representing a wide range of philosophical, ideological, and political views.[38] However, all that was overshadowed when L-5 took on the biggest political fight of its short life, and won.

The issue was the Moon treaty, a proposed international agreement that had been negotiated during the 1970s within the United Nations Committee on the Peaceful Uses of Outer Space. The full official title of the draft treaty was "Agreement Governing the Activities of States on the Moon and other Celestial Bodies." The L-5 Society's leaders had learned of the treaty during the spring of 1978, when members of the U.N. Committee reached tentative agreement on the complete text. Eric Drexler and others quickly realized that the treaty had potentially ominous implications for the society's goals.[39] The Moon treaty would have established a legal regime for space far more detailed than that in the primary existing agreement on international space law, the 1967 Outer Space treaty. In the opinion of many of the people who favored the industrialization and colonization of space, the draft treaty's provisions on the exploitation of natural resources would discourage private investment and inhibit the mining of extraterrestrial materials, slowing the growth of economic activity in space and delaying the time when large numbers of people would live there. There also was a fear that the treaty would limit individual and group freedom in space, an important issue for many L-5 members. Writing in the October 1979 *L-5 News*, Houston patent lawyer and space enthusiast Arthur Dula described the Moon treaty as "the most far-reaching international agreement ever written."[40]

The U.N. Committee reported the agreed treaty to the General Assembly in July 1979, with the next step to be its adoption by the Special Political Committee of the General Assembly before it was opened for signature by governments. Some L-5 activists believed that the society should mobilize against the treaty. Keith Henson later wrote, with some poetic license:

> On the Fourth of July 1979 the space colonists went to war with the United Nations of Earth. . . . The treaty makes no provisions for the civil rights of those who go into space. In fact, it authorizes warrantless searches. . . . The treaty makes about as much sense as fish setting the conditions under which amphibians could colonize the land.[41]

L-5 Society leaders quickly realized that they did not have the political clout, the lobbying expertise, or the presence in Washington to stop the treaty. They needed professional help. Keith Henson had read an article about Washington lawyer-lobbyist Leigh Ratiner, who was then engaged in lobbying against some of the seabed mining provisions of the draft Law of the Sea treaty, still under negotiation. Ratiner had considerable expertise in oceans law as it affected mining interests and had directed the Interior Department's Office of Ocean Affairs under the Ford administration. He later became a principal architect of the Reagan administration's position on this issue.

When Carolyn Henson contacted him in August 1979, Ratiner told her that the treaty could be defeated for about $100,000 (it actually took less).[42] Since the natural resource exploitation provisions that either were in the draft Moon treaty or were likely to be considered in its further elaboration were analogous to the seabed mining provisions in the Law of the Sea, Ratiner found it both important and congenial to fight the Moon treaty as well. It was a natural political alliance, based on shared interests.

Carolyn Henson took the Moon treaty issue to L-5's board in September 1979, where it became entangled with the issue of who should succeed Carolyn as president. Jerry Pournelle, a leading candidate, reportedly thought the Moon treaty was a lost cause, and argued that the society should not waste money on fighting it. International lawyer Edward Finch, another board member, favored the treaty.[43] In the end, the board voted to oppose it, and Ratiner was hired to represent L-5. There began a sustained campaign of telephone calls and letters to Congress and attempts to get the issue into the media. L-5 put together a packet of materials on the question. Although Keith Henson was the titular head of the effort, Carolyn was the engine.

Ratiner played the key role in the lobbying effort, although he had energetic help from L-5 activists, notably Eric Drexler and Chris Peterson. After a briefing by Ratiner, the young lobbyists fanned out across Capitol Hill, briefing staffers on why the treaty should be opposed. Ratiner comments that the L-5 connection added to his credibility; he could say he represented citizens concerned with principle rather than with economic advantage.[44]

Ratiner contacted his friend Congressman John Breaux of Louisiana, chairman of the House Fisheries, Wildlife, and Environment Subcommittee, and was invited to testify against the treaty. After stirring up concern about the treaty in the House, Ratiner then went to the staff of the Senate Foreign Relations Committee, where he was able to show that "it was not just me plus 3,000 kooks."[45] He provided a draft letter from the committee to Secretary of State Cyrus Vance. In November 1979, Senators Frank Church and Jacob Javits of that committee sent a letter to Secretary Vance, urging that the United States not sign the treaty.[46] This occurred just as the Special Political Committee of the U.N. General Assembly passed the treaty, with the U.S. delegate voting for passage. Congressman Breaux, an opponent of some provisions of the Law of the Sea treaty, sent a similar letter. The result was that the Department of State suspended action on the signing of the treaty. Subsequently, an interagency group was formed to study the matter, and the issue was shelved. As of this writing, the United States still has not signed the Moon treaty.

L-5 leaders point proudly to the defeat of the Moon treaty as the first major political victory won by one of the new pro-space citizens groups. Some space activists consider this event as a turning point for their cause. Space commercialization advocate Gayle Pergamit was quoted as saying in 1984 that "the initial defeat of the Moon Treaty had preserved the right for private entities to act in space."[47]

It seems clear that members of Congress (and Ratiner) opposed the treaty primarily for Law of the Sea reasons and not because they were advocates of space development. Their major concern was to prevent restrictions on seabed mining. Support for the treaty was weak and unorganized outside the small group of international lawyers involved in its negotiation and some sympathetic academics; it was relatively easy to kill. What L-5 should have learned from this experience is the value of weak interest groups having more powerful allies. "L-5 was the tail on a very large dog," comments Thomas Heppenheimer.[48] The Moon treaty fight also illustrated the fact that it is easier to organize people against something – a threat to their interests and their dreams – than for something as generalized and far off as living and working in space.

Meanwhile, the infighting within L-5 had led to the selection of Gerald Driggers as the person most acceptable to all factions, and he became president of the society in November 1979. Barbara Marx Hubbard, again acting as an "angel" of the emerging pro-space movement, provided funding to allow Driggers to leave his job and also become a full-time executive director for L-5.

Under Driggers, L-5 in early 1980 shifted its focus to nearer term goals, particularly a manned space station. "We wanted to turn the Society into a respectable public interest organization," says Driggers, "and a space station focus was a way of doing this."[49] The new emphasis not only was more pragmatic but also was a return to the classic spaceflight agenda. In a July 1980 column, Driggers wrote that the society's goals had been reformulated to make them realizable.[50]

Superficially, at least, this seems an almost classic case of an early radical leadership being displaced by a more conservative one as the organization achieved success and greater stability and sought broader acceptance. "You need fanatics to get started," says Pournelle, "but you need other kinds of people to operate the organization."[51] "The talents needed to start an organization are different from those needed to manage it," agrees space writer James Oberg, a former L-5 board member.[52] But Pournelle is quick to add, "Without the Hensons, there would be no L-5 Society."[53]

In early 1980 Driggers and Ratiner went on a tour to raise funds, primarily from aerospace companies. This was largely unsuccessful. Ratiner then came up with the idea of a separate pro-space lobbying organization called the Space Coalition that would be financed largely by aerospace companies while drawing on the grass-roots base of pro-space citizens groups, beginning with L-5.[54] This suggestion led to friction with many members of the L-5 board, who saw the Space Coalition as a rival organization – particularly Jerry Pournelle and Mark Hopkins, who had become increasingly influential within the society's leadership. Forced to draw on his own financial resources to remain a full-time worker for the society, Driggers eventually had to resign.

Meanwhile, a second wave of members was beginning to appear in L-5. Many had been stimulated by reading O'Neill's *The High Frontier* and Heppenheimer's *Colonies in Space*, which came out in paperback during 1978. Two good examples were David Brandt-Erichsen and Sandra Adamson, who both discovered O'Neill through Heppenheimer's book. Brandt-Erichsen says he realized that this was a real possibility for his generation; they could do it now, not 200 years from now. He started an L-5 chapter in the San Francisco Bay area and another at Oregon State University, and he became friends with Adamson. Moving to Tucson in August 1982, they became full-time space activists. They note that both came from a background of "trying to make the world better." Brandt-Erichsen was an active worker for Zero Population Growth but "that was negative"; he was stimulated by O'Neill's ideas, which gave him "something to rally around." Adamson, who is writing a book on the space movement as a sociological phenomenon, argues that the younger members in L-5 were not activists from the 1960s who had been disillusioned before finding O'Neill; they grew up with a positive viewpoint without going through an alienated phase.[55] By the end of 1984, Adamson was secretary of the L-5 Society and Brandt-Erichsen was its treasurer.

Just as L-5 was recovering from internal strife and the Moon treaty effort, the Satellite Solar Power Station issue came to a head. When the administration's budget proposals announced in January 1980 left out $5.5 million for further SSPS studies, L-5 leaders contacted Ratiner, who then was the society's representative in Washington, and got advice on tactics. The society's leaders then asked members for money for a lobbying campaign to save the SSPS.

In April 1980, a large meeting was held in Lincoln, Nebraska, to review the results of studies of the SSPS concept. L-5 activists were among those who presented technical papers. In the face of strong criticism from opponents of the SSPS, Carolyn Henson made an impassioned plea for the system.

At an L-5 board meeting held in Lincoln, Mark Hopkins called for a major political effort to save the SSPS program and formed a "phone tree" of 500 to 700 volunteers.[56] At a signal from L-5 headquarters, members of the phone tree contacted others on their lists, who then contacted others in a spreading net. All then were to call or write appropriate persons in Washington to express L-5 opinions on the issue.

Driggers, the "dyed in the wool technologist," says he refused to stake the society's future on the SSPS.[57] He was quoted in 1980 as saying that the space migration movement did not need SSPS to justify off-planet expenditures.[58] However, Hopkins and others put on an energetic lobbying effort, with telephone calls, letter writing, and personal visits to congressional staffers by Ratiner and a team of L-5 people who included Eric Drexler, Robert Lovell, and David C. Webb, who was to wear many hats in the pro-space movement. They got help from Jerry Grey of the American Institute of Aeronautics and, reportedly, from aerospace workers leader William Winpisinger. One of those involved was Stewart Nozette, who was later to become executive director of the California Space Institute.[59]

After winning in the two authorization committees, SSPS supporters lost when the House Appropriations Subcommittee on Energy killed the bill for new funding for SSPS studies. According to Hopkins, the main result of this effort was the phone tree, which has grown since then. Getting permission to use the mailing lists of some other pro-space organizations, L-5 activists were able to organize the phone tree from a list of about 18,000 people.[60]

The collapse of the SSPS dream in 1980 took away much of the society's confidence in seeing its dreams realized soon. Kenneth McCormick had written in the June 1979 *L-5 News*: "No other project on the horizon even comes close to offering the spur to space colonization that SPS offers."[61] Subsequently, the society became more of a mainstream pro-space organization, although one strongly advocating space industrialization and the use of extraterrestrial resources; colonization was mentioned with declining frequency in L-5 literature.

After Driggers left office as its president, the society was run *de facto* by Mark Hopkins and Jerry Pournelle, with its new president, ex-astronaut Philip K. Chapman, an arbitrator of disputes. A direct mail campaign, urged by Mark Hopkins and Jerry Pournelle, was particularly successful with science fiction readers and increased the society's membership significantly.[62]

Meanwhile, 1981 had seen the divorce of Keith and Carolyn Henson; she resumed her maiden name of Meinel. For a time, Keith moved around the United States in a variety of jobs in the computer industry before returning to Tucson in 1984 to resume the presidency of Analog Precision. He had become a vocal advocate of space-based defenses against ballistic missiles, suggesting that such a program might speed space industrialization and settlement.

Carolyn Meinel, although no longer a leading figure in the society, continued to advocate space development and to publish articles on space and technology subjects. In 1983 she married conservative activist John Bosma, who had been a defense analyst at the Boeing Company and who later was a member of the Reagan administration's transition team, an aide to Colorado Congressman Ken Kramer, vice-president of the Congressional Staff Space Group (see Chapter 9), and, as of 1985, editor of the Washington-based newsletter *Military Space*. By 1984, Carolyn was working on advanced space system analyses in one of the capital's many "think tank" contracting firms. Like Keith, both Carolyn and her new husband are advocates of a space-based missile defense.

L-5 in 1984

As of 1984 the L-5 Society seemed to have become a relatively permanent part of the space interest group scene, with about 9,500 members. Its new president was Gordon Woodcock of Boeing, and the Chairman of its Board of Directors was physicist and former Avco-Everett Research Laboratory executive Arthur Kantrowitz; both were respected figures from an older generation of space advocates. L-5 remained a highly decentralized organization, with its key figures scattered all over the United States. "L-5 runs by telephone," says Mark Hopkins.[63] The society, which relies almost entirely on membership dues, had a budget of about a third of a million dollars. On a day-to-day basis, it was still run from Tucson by its administrator Gregory Barr. Surveys showed that L-5's membership still included an unusually high percentage of libertarians. Like the National Space Institute, L-5 achieved a symbolic success in 1984 when member Charles Walker flew on the Space Shuttle.

One of the most striking features of L-5 is its national and international network of over 70 chapters, which gives it the strongest grass-roots base of any pro-space citizens group. Largely autonomous, sometimes fractious, these chapters often work closely with the local branches of other space interest organizations such as the AIAA in local pro-space activities. L-5 leaders believe that their organization is more in touch with grass-roots opinion on space than any other. This strong tradition of decentralized activism suggests that the L-5 idea has a firm foundation and will survive changes in its national leadership or crises in funding. O'Neill associate Stephen Cheston, a student of Russian history, notes that "the cell structure makes revolutionary organizations hard to kill."[64]

L-5 still relies heavily on volunteers. "Without them, you're dead," says Hopkins, who estimates that 300 members put in 20 or more unpaid hours a week in chapter work or in the phone tree. Only about 20 of these are at the national or international level.[65] Observes Sandra Adamson, "L-5 allows individual participation,"[66] one of the most important elements in its attraction and staying power.

L-5 has remained highly active in promoting manned space activity. It has spun off related organizations described elsewhere in this book, such as Spacepac and the Citizens Advisory Committee on National Space Policy. L-5 activists campaigned in support of the permanent manned space station proposed by President Reagan in January 1984, although none testified before Congress until March 1984.[67] Recalling the Society for the Advancement of Space Travel, Jerry Pournelle says "This time we will stay with it until the job gets done."[68]

The L-5 Society held its first Conference on Space Development in Los Angeles in 1982. The second, organized in Houston by Arthur Dula, had as its theme "Doing Business in Space," reflecting a growing emphasis on space commercialization rather than the earlier idealistic visions of utopian communities. About 700 people attended the third Conference on Space Development, held in San Francisco in April 1984. The fourth was held in April 1985 in Washington, D.C.

L-5 remains cheeky and anti-authoritarian in style. "Kiss my asteroids," announces an L-5 T-shirt made for a science fiction convention. Addressing the San Francisco conference, Congressman George E. Brown said he had decided that L-5 members were "5 percent Democrat, 5 percent Republican, and 90 percent anarchist." This drew a rousing cheer. Certainly, liberty remains a powerful refrain. Jerry Pournelle ended his toast to the society in April 1984 with, "Ladies and Gentlemen, I give you freedom and the stars."

Assessing L-5

Many L-5 members have a high opinion of their society's role, seeing it as the cutting edge of space advocacy. "Most of the space movement is L-5," says Mark Hopkins. "It has 80 percent of the activists."[69] The society certainly remains the radical wing of the pro-space movement. "L-5 is the credible far-out organization in the space field," says Jerry Pournelle.[70]

L-5 activists also believe that their organization has stretched the limits of the possible in the space field by making it easier for other groups to advocate bold space goals. "L-5 exists to make organizations like the National Space Institute look respectable," says Pournelle.[71] At the first L-5 conference, he reportedly described L-5 as the communists and NSI as the social democrats.[72] L-5 also pushes other groups to be more daring. "L-5 is the burr under the saddle of the space movement," observes James Oberg.[73]

L-5's goals and style have been the subject of considerable criticism. The early L-5 was outside the political culture of existing pro-space activity, and its stridency was seen by many other pro-space people as counterproductive; it has taken years for the society to live down its early reputation.

Those who do not share the desire to live and work in space have been particularly critical. University of Oklahoma anthropologist Stephen I. Thompson has suggested somewhat mockingly that the followers of O'Neillism constitute a revitalization movement (those in the past commonly have had a religious character) and that O'Neill is its "prophet."[74] However, many L-5 members would not object to the definition of a revitalization movement used by Thompson: "deliberate, conscious effort(s) by members of a society to construct a more satisfying culture." Sandra Adamson, a student of anthropology, believes that what L-5 is doing will result in the most radical change in social structure humanity has ever seen.[75]

One of L-5's most striking achievements has been to bring together believers in the space idea who come from very divergent ideological and political backgrounds, at one time or another including as supporters individuals ranging from Timothy Leary to Senator Barry Goldwater. These differences in political origins were a major cause of the tensions sometimes visible within the organization. It is the shared commitment to space that allowed the alliance and gathering place known as the L-5 Society to survive.

In recent years, L-5 has become somewhat more respectable in style, with its lobbyists dressing in "straighter clothes," notes Sandra Adamson, who refers to "radicals in three-piece suits."[76] But its original purposes have not been forgotten, only expressed differently. "The goal of L-5," says David Brandt-Erichsen, "is to bring space manufacturing from nonterrestrial materials to the point of economic payback."[77] There are still more sweeping visions, more in the tradition of the early L-5. At the society's April 1984 conference, Pournelle told the receptive audience that the L-5 Society was "the advanced planning department of the human race," adding, "This bunch of 'kooks' and 'flakes' just might save the world."

OTHER MEMBERS OF THE FAMILY

The L-5 Society was not the only space advocacy organization stimulated by O'Neill, who was a significant factor in the launching of the new pro-space movement. Several of the groups included in L-5's April 1976 listing of other space colonization organizations (all on the West Coast) give some of the flavor of early responses:[78]

Space Colonization Now of Walnut Creek, California, which claimed 400 members

Phoenix Foundation of Klamath Falls, Oregon, a religious group that believed in the basic goodness of man

International Society of Free Space Colonizers of Seattle, Washington, "the first wave of an emerging hard-core capitalist revolutionary movement which will bring forth the triumphant rise of the Creator class on and off this planet," dedicated to defeating "altruist-collectivism"

Earth/Space of Palo Alto, California, dedicated to free space enterprise

The Network of Berkeley, California, "interested in a gestalt of starflight, immortality research, and higher intelligence," which said it was a clearinghouse for persons interested in Leary's SMI^2LE concept

United For Our Expanding Space Programs (UFOESP) of San Diego, California, "dedicated to propaganda and education for the increased exploration and exploitation of the space environment."

Most of these early organizations went out of business or remained small and local. But many other groups sprang up around the United States, and some achieved enough momentum to continue on despite the failed attempt to legislate the High Frontier into existence. Some became L-5 chapters; others have remained independent, in some cases because of disagreements with the L-5 leadership, a distaste for the early L-5 style, or a desire for autonomy. At one time, says Randall Clamons, there were "hundreds" of pro-space organizations that did not want to be L-5 chapters.[79] This proliferation also reflected organizational entrepreneurship, the desire to found one's own group rather than be part of a larger organization.

One of the oldest of these other groups to survive was the Maryland Alliance for Space Colonization, which started in 1977 as an organization affiliated with L-5 at the University of Maryland, near NASA's Goddard

Space Flight Center. In the fall of 1978, Paul Werbos, Gary Barnhard, and Ray Hoover founded the MASC, which later changed its name to the Maryland Space Futures Association because they thought the term *colonization* reduced its credibility. (Werbos had been a co-founder, with Mark Hopkins, of the Harvard-Radcliffe Committee for a Space Economy.) The MSFA claims to be the oldest student pro-space organization in the United States and had about 800 members on its mailing list in 1984.[80] Each year, it sponsors a Space Futures Day on campus, with lectures, films, and exhibits. As of 1984, it was completely independent of the L-5 Society.

A related phenomenon occurred in Boston, where Massachusetts Institute of Technology student Peter Diamandis was involved with L-5 in the late 1970s. Feeling that existing space groups did not fully satisfy the needs of the college audience, he founded Students for the Exploration and Development of Space in 1980. SEDS grew to be by far the largest pro-space student organization in the United States, mushrooming to about 120 chapters before the application of stricter criteria reduced the number to about 25. Since 1982, SEDS has held a conference each summer, at which it presents the Arthur C. Clarke award for space education. The group places particular emphasis on space careers and planned a space careers conference in Tucson for March 1985. In 1983, it became affiliated with the American Astronautical Society, and in 1984 it became the student auxiliary of the Space Studies Institute.

The U.S. chairman for SEDS in 1984, Todd Hawley, is a true child of the Space Age. He was born on April 13, 1961, the day after Yuri Gagarin became the first human to orbit the Earth. SEDS leaders Diamandis, Hawley, and Robert Richards, the founder of SEDS Canada, traveled to the UNISPACE conference in Vienna in 1982. There they became friends with Arthur C. Clarke, referring to him as "Uncle Arthur."[81] Clarke has been quoted as saying, "It seems very appropriate that at this moment in time the cycle should begin again, so you [SEDS members] can regard yourselves as the reincarnations of Wernher [von Braun] and his colleagues."[82]

Another example of the centrifugal/centripetal forces within the O'Neill family occurred among students at the University of Arizona in Tucson, home of the L-5 Society. Deciding that becoming an L-5 chapter would make them a little fish in a big ocean, they became the Students of Space and joined SEDS instead.[83] In West Virginia, space enthusiast Cynthia Riedhead wanted to start a pro-space organization but did not want it to be just another L-5 chapter; the result was the Piedmont Advocacy for Space, still active in 1984.

At Niagara University in upstate New York, O'Neill's work inspired a Space Settlement Studies Project, begun in 1978. The project studies the societal and cultural aspects of human communities in extraterrestrial habitats and publishes a quarterly newsletter called *Extraterrestrial Society*. Co-directors Stewart B. Whitney and William R. McDaniel, both in the university's Department of Sociology, say their intention is to provide stimulation to students, sensitize space planners to the necessity for detailed social planning, provide a focal point for social science knowledge related to space utilization and humanization, and develop a cadre of scholars and scientists in this field.[84]

In Chicago, the bright, aggressive young lawyer Gregg Maryniak founded the Chicago Society for Space Settlement in 1977. As the High Frontier idea suffered reverses, the name was changed to the Chicago Society for Space Studies. The CSSS, which had about 200 members in 1984, has monthly meetings, gives courses on space, and puts out a newsletter called *Spacewatch*. Maryniak himself became vice-president of Gerard O'Neill's Space Studies Institute.

In June 1978, southern California space enthusiast Terry Savage founded an L-5 chapter of about 30 people in Los Angeles, which became the Organization for the Advancement of Space Industrialization and Settlement (OASIS). Thomas Heppenheimer joined in 1979. OASIS, which grew steadily during the next few years, included conservative aerospace engineers who did not like the "flaky" L-5 style and distanced itself from the "mother" organization. Since then, changes in L-5's leadership have encouraged OASIS to reinvolve itself, and it looked in 1984 as if it might combine with other L-5 members in southern California to become by far the largest L-5 chapter, with 1,200 to 1,500 members.

Two of the people who met through the social networking provided by OASIS were Howard Gluckman and Janelle Dykes, who later married. Howard remembers being interested in space since the 1961 launch of Mercury 3 when Alan Shepard became the first American to go into space. In 1976 he turned to the National Space Institute but found that it did not provide a way of getting involved. After reading *The High Frontier*, he joined L-5 instead and then OASIS. Janelle Dykes had her interest in space rekindled by reading Tom Wolfe's book *The Right Stuff*, but then found it "tough" to locate pro-space organizations. "They did not do good marketing," she says. Each of the Gluckmans went on to serve as president of OASIS, and Janelle applied her professional skills in designing L-5's first major direct mail membership drive. Although they want to devote more time to their own careers and claim to be "burning out," the Gluckmans remain enthusiastic. As of 1984, they still were traveling to sites near Vandenberg Air Force Base on the California coast to watch space launches. "It's the pyromanic aspect," says Howard, only half facetiously. More seriously, he comments, "A lot of us want to go into space ourselves."[85]

6

THE NEW DEMOGRAPHICS
OF SPACE

Your generation stands on the verge of greater advances than
Humankind has ever known. Nowhere is this more true than
America's next frontier – the frontier of space.

President Ronald Reagan, 1984[1]

I want to GO.

World Space Foundation Slogan, 1984[2]

The O'Neill/L-5 phenomenon may have been the precursor of a broader
movement of American opinion, the leading edge in a new surge of pro-
space sentiment. After hitting a low point in the early 1970s, the
percentage of Americans favorable to doing more in space began to grow
in the middle of that decade. That shift was particularly noticeable among
younger people. Unlike the sharp ups and downs of attitudes toward
space during the 1960s, this upward trend seems more gradual and long-
term, implying that it may be less tied to specific events.

TRENDS

One of the most useful presentations of long-term trends in American
opinion about space activity was prepared by Professor John M.
Logsdon of George Washington University for the Office of Technology

Assessment's 1982 report entitled *Civilian Space Policy and Applications*. Figure 6.1 shows the results of asking the question: should the United States do more in space or less?

The trends seem clear. Until 1980, those who favored doing less consistently outnumbered those who wanted to do more. There was a short-lived peak of pro-space opinion in the latter part of 1968, when Apollo 8 flew around the Moon. After that, however, "do more" responses declined to around the 20 percent mark until the mid-1970s. (This surviving "do more" group may have reflected the existence of a core pro-space constituency that survived hard times.)

The "do less" school of thought grew rapidly during 1966 and 1967, a period marked by the end of the Gemini program, the Apollo fire, the emergence of other pressing national issues such as Vietnam, and cutbacks in NASA's budget. This sentiment declined during the 1968 boom in pro-space opinion. However, it rose again in 1969 and 1970 to about half the population, a level that was sustained until the mid-1970s.

There was a clear change of trends from 1975 on. "Do more" opinion grew all through the late 1970s, approaching half the total responses for the first time ever when the first Space Shuttle flew in April 1981. "Do less" opinion declined during the late 1970s, dropping sharply in 1980, *before* the first Shuttle mission.

Logsdon does not speculate about the causes for these changes. However, he does comment that "more recently, public understanding of the space program, and a supportive public attitude toward that program, have increased to the point where they may have political impact."[3]

Similar trends appear in surveys reported by Robert D. McWilliams of Virginia Polytechnic Institute and State University in papers presented to the Princeton/AIAA conferences on space manufacturing. In 1981, McWilliams reported that the percentage of people believing that too much was being spent on the space program rose in the early 1970s to a high of 61 percent in 1974, while those believing too little was being spent scored as low as 7.4 percent in 1974 (those who thought spending was about right fluctuated between 27 and 30 percent during this period).

A change began in 1976 and 1977. Those who wanted less spending declined to 39.1 percent in 1980, while the percentage of those who wanted more spending rose to 18.0 percent that year. Meanwhile, the percentage of people who thought space spending was about right also was growing. By 1980, those who wanted space spending either increased or kept at the same level exceeded those who wanted it cut for the first time during the years the survey was conducted.[4]

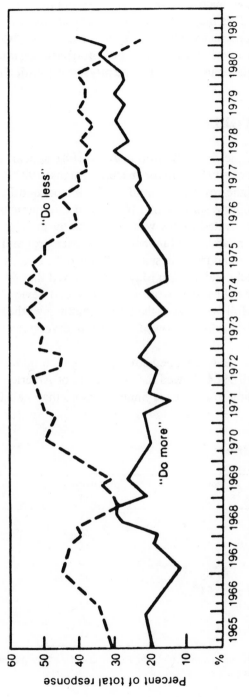

Figure 6.1 – Long-Term Trend Polling Results of U.S. Public Opinion on the Federal Space Effort (*Source: Civilian Space Policy and Applications* [Washington, D.C.: Office of Technology Assessment, 1982], p. 136.)

105

There seems to have been a crossover point in opinion about space around the year 1980. This may turn out to have been an historic moment in the history of pro-space activity in the United States. As we will see in Chapter 7, it coincided with a proliferation of pro-space organizations, and with a growing self-awareness within the pro-space constituency.

POST-SHUTTLE POLLS

The first flight of the Space Shuttle in April 1981 appears to have generated a new peak of enthusiasm for space. In August 1981, the New York *Times* reported the results of a poll by the Associated Press and NBC News. This sampling showed that 60 percent of American adults said they thought the United States was not spending enough or was spending about the right amount of money on the space program. Forty-two percent said they would travel in space if they had the chance.[5]

A Louis Harris poll conducted in May 1981 showed that 63 percent of those interviewed believed it worthwhile to spend several billion dollars to develop the full potential of the Space Shuttle over the next ten years. However, the breakdown showed interesting divergences (Table 6.1).[6]

Support appears to have remained strong through the mid-1980s. A poll conducted in April 1984 showed that 45 percent of American adults were interested in riding in the Space Shuttle – more than half the men

TABLE 6.1
Poll on Space Shuttle

	Worth it (%)	Not worth it (%)	Not sure (%)
Total	63	33	4
College educated	71	26	3
Men	76	21	3
Women	52	43	5
Blacks	45	53	2
Republicans	71	26	3
Democrats	57	39	4
Conservatives	66	30	4
Liberals	57	41	2

Source: Louis Harris. *The Harris Survey*. (Orlando, FL: Tribune Media Services, Inc., June 1, 1981). Reprinted with permission.

and 36 percent of the women. Younger people were the most eager to go, and the West had a higher percentage expressing interest than any other region.[7]

CATEGORIZING THE PUBLIC

Jon D. Miller, director of the Public Opinion Laboratory at Northern Illinois University, did a major opinion study for NASA in 1982 entitled "The Information Needs of the Public Concerning Space Exploration."[8] Miller's analysis identified four segments of the public on this question:

1. The attentive public, which was both interested in and knowledgeable about space exploration. This constituted about 10 percent of the total public in 1981, or about 15 million adults. Miller found these people to be highly supportive of the space program. They also were high-volume consumers of news and science information and were committed to the idea that interested and informed citizens can have an impact on the formulation of public policy.
2. The interested public, which had a high level of interest but believed itself inadequately informed. This amounted to just over 15 percent of the total, or about 24 million adults. Despite the information deficit, the interested public held very positive attitudes toward space exploration.
3. The residual science-oriented public, which had a high level of interest in science and technology, but a relatively low level of interest in space exploration. A majority of this group, which included over a fourth of the population, had positive attitudes toward space exploration.
4. The nonattentive public, which reported a low level of interest in and little knowledge about space exploration. Even this group, which made up almost half the population, favored a continuation of the space program. However, these people placed a low priority on federal spending on space relative to health care, education, and human service programs.

When the attentive public and the interested public are added together, we find that about a fourth of American adults were very positive about space in 1981. This is about 39 million people – a large potential constituency by any standard.

In October 1983, the National Space Institute and *Science Digest* magazine conducted a poll on attitudes toward space. The results tended to support Miller's findings and showed the science-oriented public (that is, the readers of *Science Digest*) to be nearly as space positive as the attentive public (that is, members of NSI). Asked to rate their level of agreement with various statements, respondents gave the highest marks to the statement that spending on the civilian space program has a beneficial effect on the nation technologically and economically. Close behind were strong agreement on the importance of a manned space station by about 1990 and the necessity of increasing the planetary exploration budget. Respondents strongly agreed that they would be willing to pay about $50 a year in taxes to support the overall civilian space program. About 23 percent of the NSI members and 15 percent of the *Science Digest* readers said they would be willing to spend $10,000 for a two-day ride into orbit aboard the Space Shuttle.[9]

PRIORITIES

Robert McWilliams looked at the question of where space ranked in relative priority among selected federal programs, finding that it ranked low on the list, just above foreign aid. However, the percentage of people favoring increased spending on space rose from 10.7 percent in 1977 to 19.6 percent in 1980. Only the military, armaments, and defense category showed a more rapid increase.[10]

SPACE, DEFENSE, AND SCIENCE

Is the pro-space phenomenon simply riding the coattails of rising support for the military and for science and technology? McWilliams looked at this question in his 1981 paper and found that attitudes toward defense and the space program are only mildly associated.[11] He also found that favorable disposition toward science in general has not kept pace with that toward space exploration for several years.[12] McWilliams concludes:

> On the basis of such evidence, it appears that the increasing approval of space activities among Americans over the past several years is not a trend that is riding mainly on the coat tails of militarism or growing faith in science and

technology. Rather, it seems that Americans may be coming to view the space program as being conducive to the achievement of other types of goals of which they are in favor.[13]

CORRELATING KNOWLEDGE AND ATTITUDES

McWilliams found a strong correlation between attitudes toward federal spending on space and levels of accurate information about the program, especially its cost. "The results indicated strongly that those who know more about the space program tend to be decidedly more in favor of increasing its funding," he wrote in 1981.[14] McWilliams speculates "It may be that the improvement in NASA's popularity over recent years is an indication that such misconceptions are eroding, and that space-positive attitudes will continue to increase as such fiscal misinformation is cleared up."[15]

Johan Benson, the Washington representative of the American Institute of Aeronautics and Astronautics, puts it more informally. "To know space is to love space," he observes.[16]

WHAT DO PEOPLE LIKE ABOUT SPACE?

The Office of Technology Assessment reported surveys that suggest reasons why people are interested in or support the space program. Table 6.2 lists perceived benefits from space exploration.

The perception that space spending will improve other technologies is not surprising, since it has been a major selling point for the space program for years. What is more intriguing is the public interest in more expansive ideas, such as finding new sources of minerals and energy and new areas for future habitation, both favorite themes of the O'Neill/L-5 school of space industrialization. Also notable is the level of interest in contacting other civilizations. One may speculate that much of the credit should go to the educational efforts of Gerard K. O'Neill (space colonization and extraterrestrial resources) and Carl Sagan (planetary exploration and the search for extraterrestrial life and intelligence).

Quoting a study by William S. Bainbridge and Richard Wyckoff, McWilliams reports these results on the popularity of various justifications for spaceflight, estimating percentages for the total U.S. adult population based on a survey in Seattle (Table 6.3).

TABLE 6.2
Perceived Benefits from Space Exploration

	First or Second Mention
Improve other technologies	272
Find mineral or other wealth, other resources, sources of energy	200
Increase knowledge of universe and man's origins	190
Find new areas for future habitation	134
Contact other civilizations, other forms of life	107
Improve rocketry and missile technology	43
Find industrial uses of space	27
Find new kinds of food, places to raise more food products	26
Create jobs and other economic benefits	16
Learn about weather and how to control it	13

Source: Institute of Survey Research, Temple University, National Survey of the Attitudes of the U.S. Public Toward Science and Technology, submitted to the National Science Foundation, May 1980, p. 164. Reprinted with permission.

TABLE 6.3
Popularity of Justifications for Spaceflight

	Percent
Information	69.8
Economic-industrial	41.9
Military	40.7
Educational-idealistic	27.9
Colonization	24.7
9 Unfactored justifications	39.4
All 49 justifications	38.4

Source: Robert D. McWilliams, "The Improving Socio-Political Situation of the American Space Program in the Early 1980s," in Jerry Grey and Lawrence A. Hamdan, eds., *Space Manufacturing – Proceedings of the Fifth Princeton/AIAA Conference, May 18-21, 1981* (New York: American Institution of Aeronautics and Astronautics, 1981), p. 256. Reprinted with the permission of William S. Bainbridge.

PROFILES OF THE PRO-SPACE PERSON

Studies show a significant degree of agreement as to the profile of the pro-space American. Surveys reported by the Office of Technology Assessment show that the people who are the most pro-space tend to share certain characteristics: male, between 25 and 34, college educated, professional or technical employment, working for the government, income over $25,000 a year, living in the West, and slightly more likely to vote. Those most opposed to the space program tend to be female, over 65, with less than a high school diploma, laborers and service workers, and with an income under $5,000. The report comments that while pro-space attitudes have increased substantially among whites, they increased only negligibly among blacks.[17]

Jon Miller reports that surveys done in 1979 and 1981 showed that the attentive public for space exploration was the youngest and best educated of the four segments, with the interested public ranking second in both youthfulness and level of formal education. The nonattentive public tended to be the oldest and least well-educated segment of the population.

Miller also found the following:

The attentive public for space exploration was predominantly male, by about two to one, while the interested public had a male majority (however, the residual science-oriented public included a majority of females).

The attentive public included a higher level of professional and technical workers.

A college level science course was strongly and positively associated with attentiveness to space exploration.[18]

McWilliams reported similar results, finding that males are somewhat more likely than females to express opinions in support of the space program, that whites tend to be more pro-space than blacks, and that individuals between 22 and 35 years old show the greatest propensity toward pro-space attitudes. Among the most significant trends in evidence, writes McWilliams, was a constant tendency for space exploration supporters to score notably higher than their space-negative counterparts on all usual measures of socioeconomic status, such as class and education. In fact, the level of formal education showed the most marked association with space-positive attitudes. In his 1981 paper,

McWilliams reported that differentiation of space-spending opinion along educational lines had increased since 1977.[19] He also reported an overrepresentation of "space positives" among persons who expressed no formal religious preference and among persons who had never been married.

Although females have tended to be far less enthusiastic about space than males, in 1981 McWilliams reported that "the relative differentiation along sexual lines is now slightly less marked than it was in 1977 due to a greater proportional increase in pro-space sentiment among females than among males." However, "the notable schism in space spending attitudes along racial lines has become more marked since 1977," reported McWilliams.[20]

The class association also has changed in recent years. Formerly, McWilliams notes "lower" and "working" classes were more anti-space than were "middle" and "upper" classes. Recently, however, the "middle" and "working" classes have become more space positive than either "upper" or "lower" class respondents. McWilliams comments that "space exploration antagonists can no longer legitimately claim that, from a public opinion viewpoint, the American space program is an essentially elitist endeavor."[21]

Summing up, McWilliams said the American who favors increased government investment in space exploration tends to be white, male, and between the ages of 22 and 35. He is 50 percent more likely than the average American to have had some college education and nearly twice as likely to have earned a college degree. In addition, he shows a markedly greater than average tendency to have a high socioeconomic status and is notably more likely than his "space-negative" counterparts to exhibit a less than average concern for formal religious affiliation.[22]

SPACE INTEREST GROUP PROFILES

Available statistical profiles of members of pro-space citizens groups tend to confirm the general patterns found by Miller and McWilliams. In a survey of its members, released in 1983, the National Space Institute came up with this profile:[23]

Average age is 33.
Average household income is $20,000 to $30,000.
Fifty-eight percent have college degrees.

Most are not in aerospace-related fields, but 98 percent are members
    because of a lifelong interest in space exploration.
Members range across the political spectrum, but are one and a half times
    more likely than the average person to vote and to contact their elected
    representative.

In a 1984 survey of its members, the National Space Institute found
that respondents had an average age of 34 and an average annual income
of over $32,000. Eighty-four percent had some college education, and 30
percent had done postgraduate work. Forty-four percent considered
themselves professional, and 28 percent, technical. Although only 15
percent of the respondents were directly involved in the space program,
98 percent said they had a lifelong interest in space.[24]
    The L-5 Society, considered by some to be the "hard" wing of the
pro-space movement, surveyed its members in 1983, finding that
members were:[25]

86 percent male
61 percent single
71 percent with college degrees
18 percent in the computer profession (the next largest category was 13
    percent for students)

L-5 members responding to the survey fell into these age groupings:

Age	Percent (rounded)
25-35	52
35-50	24
18-25	14
50-65	5
Under 18	2
Over 65	1
No response	1

Gerard O'Neill's Space Studies Institute also conducted a
demographic survey of its members.[26] The results showed SSI
supporters to be 87 percent male, with 84 percent with college degrees,
54.5 percent in professional employment, and 54 percent single. The age
distribution by percentage was:

30 to 45	43
18 to 30	37
45 to 65	10
Less than 18	6
Over 65	4

In 1985, the Planetary Society reported the results of a survey of its members, which showed that they were 85 percent male, that 88 percent had some college education (including 24 percent with advanced degrees), and that 64 percent were in the 26 to 49-year age bracket. The largest income bracket was between $20,000 and $30,000.[27]

A smaller organization, the United States Space Education Association, reported that a demographic survey in early 1983 showed that the "average" USSEA member was a single white male between the ages of 21 and 30, a Protestant who had completed 3.52 years of college, and someone who was making between $11,000 and $20,000 annually in a field other than the aerospace industry.[28]

PRO-SPACE GEOGRAPHY

Statistics on the geographic distribution of pro-space opinion are difficult to find. One of the most useful presentations was made by Hugh Millward in his article "Where is the Interest in Space Settlement?" published in the January 1980 *L-5 News* (Figure 6.2). Millward's map showing the density of L-5 Society membership in 1979 showed a strong tilt toward the West and Southwest, with pockets in New England, the Mid-Atlantic states, Michigan, and Wisconsin. As we will see in Chapter 8, this roughly coincides with patterns in the formation of new pro-space groups (Figure 6.2).

Millward also rated the major factors influencing L-5 Society membership rates, in descending order of importance:

December 1976	*March 1979*
College students per thousand of population	Same as in 1976
Per capita income	Same as in 1976
Presence of NASA facility or major NASA contractor	Same as in 1976

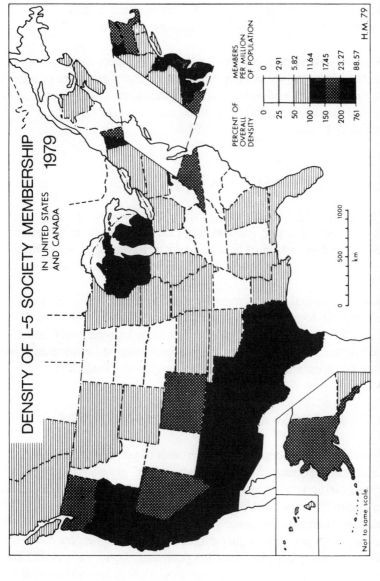

Figure 6.2 – Density of L-5 Society Membership in the United States and Canada – 1979 (*Source:* Hugh Millward, "Where is the Interest in Space Settlement?" *L-5 News*, January 1980, pp. 8-10.)

Percent increase in state income, 1970-76	Percent of population in metropolitan areas
Percent of population in metropolitan areas	Percent change in state population, 1970-76
Percent change in state population	Percent increase in state income, 1970-76

What is striking here is that all the factors (except the presence of a NASA facility or contractor) have to do with general measures of wealth, education, growth, and urbanization. Interest in space seems to be correlated positively with common socioeconomic indicators of prosperity and knowledge.

THE NASA CONTRACT FACTOR

The geographic pattern suggesting that pro-space sentiment is stronger in the West and Southwest appears to be at least superficially related to the distribution of NASA's contract awards. Prime contractor awards from 1979 to 1982 were distributed by region as shown in Table 6.4.[29]

NASA's Annual Procurement Report for Fiscal Year 1983 listed the top ten states in terms of prime contract awards (Table 6.5).

Cutbacks in NASA contractor employment between 1966 and 1971 showed patterns that may be related to later interest in space activism, with the Far West suffering by far the worst blows (Table 6.6).[30]

THE BABY BOOM AND PRO-SPACE ATTITUDES

There is some evidence that the growth of pro-space attitudes in the United States is correlated with the "baby boom" generation (people born between 1946 and 1964), particularly its later cohorts. Space interest and space group profiles tend to be tilted toward the younger end of the population spectrum. The space group boom that began in the late 1970s (described in Chapter 8) roughly correlates with the emergence into their twenties of people born between the peak baby boom year of 1957 and the last baby boom year of 1964. Randall Clamons, then administrator of the L-5 Society, wrote that "the membership of the L-5 Society is made up of the products of the baby boom." According to him, people reaching

TABLE 6.4
NASA Prime Contractor Awards by Region

Region	Amount (millions of dollars)
Far West	9,402
Southeast	5,190
Northeast	2,767
Southwest	2,257
Rocky Mountain	1,080
New England	918
Great Lakes	641
Plains	113

Source: NASA Pocket Statistics – January 1984 (Washington, D.C.: National Aeronautics and Space Administration), p. C-2. (It should be noted that these figures are not necessarily final.)

TABLE 6.5
NASA Prime Contractor Awards by State

State*	Percent
California	37.6
Florida	13.7
Texas	9.3
Maryland	7.2
Louisiana	6.1
Utah	4.5
Connecticut	3.0
Virginia	2.8
Colorado	2.7
Alabama	2.1

*All but Utah, Connecticut, and Colorado contain NASA centers or major NASA installations.

Source: Annual Procurement Report – Fiscal Year 1983 (Washington, D.C.: National Aeronautics and Space Administration, 1983), p. 38.

TABLE 6.6
Geographic Distribution of NASA Contractor Employment Reductions

Region	Fiscal 1966	June 1971	Change
Far West	167,000	30,000	−136,700
Southeast	70,900	26,100	−44,800
Mideast	64,000	19,900	−44,100
Southwest	26,200	14,100	−12,100
New England	25,200	6,900	−18,300
Plains	17,500	2,800	−14,700
Great Lakes	13,700	2,800	−10,900
Rocky Mountains	1,500 [sic]	5,500	−4,000

Source: Testimony of George M. Low, Acting NASA Administrator, in *1972 NASA Authorization,* Hearings before the Committee on Science and Astronautics, U.S. House of Representatives, March 2, 1971 (Washington, D.C.: U.S. Government Printing Office), p. 76.

age 25 between 1970 and 1985 accounted for 91 percent of the society's membership.[31]

OTHER EVIDENCE

There have been other signs of increased public support for space. One measure is the size of the crowds that continue to turn out to see the launches of manned spacecraft. An estimated 600,000 people observed the Space Shuttle as it lifted off on its first mission in April 1981, and 400,000 to 450,000 drove out to Edwards Air Force Base in California to watch it return.[32] Hundreds of thousands more came to see the Shuttle orbiter as it visited various cities on the back of its Boeing 747 carrier. These crowds were comparable to those that turned out at the height of the space race in the 1960s. The fact that there is no comparable sense of national emergency today makes this demonstration of popular interest all the more impressive. It also is striking that the Space Shuttle drew large crowds when it toured Western Europe in 1983; NASA Administrator James Beggs estimated that the total was about 5 million people.[33] Interest in space is not just an American phenomenon.

Another measure is the popularity of the "Dial-a-Shuttle" telephone service sponsored by the National Space Institute. Under this arrangement, a caller who dials the correct number can listen in to radio conversations between Shuttle crews and ground controllers. Despite the rather vicarious nature of this experience, hundreds of thousands of calls have been placed during each Shuttle flight. According to Bonny Lee Michaelson of the National Space Institute, who coordinates this program, 861,000 calls came in during Shuttle mission STS-9 in December 1983 (the one that carried the first Spacelab). About one third of the calls were international, probably because the mission carried the first Western European to ride into space on an American spacecraft.[34]

Another sign is attendance at the National Air and Space Museum in Washington, D.C., which has become the most popular museum in the world. The museum drew 4 million visitors in the first five months after it opened in July 1976, at a time when no Americans were going into space.[35] On May 24, 1984, the museum welcomed its 75 millionth visitor.[36] During that calendar year, the museum had 14,438,799 visitors.[37]

NASA continues to be deluged with applicants for its astronaut, mission specialist, and payload specialist programs. In announcing a new class of 17 astronaut candidates in May 1984, NASA reported that it had received 4,934 applications.[38] Hughes Aircraft Company received nearly 600 applications from Hughes employees to fill two payload specialist positions available on Space Shuttle missions to fly in 1985.[39] Clearly, the desire to go into space and to be associated with the space enterprise is widespread.

POLITICAL CHARACTERISTICS

Jon Miller found that the attentive public for space exploration was more politically aware and active than the other strata he described. Furthermore, the level of contacting of public officials was significantly higher for the attentive and interested publics than for the other segments of the population. Not surprisingly, a higher proportion of attentives indicated that they perceived contacting officials to be efficacious in influencing policy. Two thirds of the attentive public, when faced with a hypothetical policy issue, indicated that they definitely or probably would participate personally in the dispute.[40] The National Space Institute's 1984 survey showed that 86 percent had voted in the last presidential election, almost twice the national average.

Miller found that the attentive public for space exploration is not composed of a single slice of the ideological spectrum but is broadly representative of a wide range of political philosophies. However, a higher proportion of the attentive public labeled themselves "liberal."[41] (The National Space Institute's 1984 survey showed the reverse, with the balance being slightly conservative, although less so than the public at large.)

Miller also found that the attentive public for space exploration was bipartisan, with 39 percent expressing a preference for each of the two major parties. The nonattentive public was predominantly Democratic.[42]

The May 1981 Harris survey quoted in Table 6.1 showed that Republicans and conservatives tended to have more favorable attitudes toward the Space Shuttle than Democrats and liberals. In a survey of L-5 Society members in 1983, 31 percent identified themselves as Republicans, 25 percent as Democrats, 8 percent as Libertarians, and 33 percent as none of the above.[43] All this suggests that, as of the early 1980s, support for space activity in the general public was not closely tied to partisan views, although Democrats were more likely to be unenthusiastic.

POLITICAL IMPLICATIONS

There is political potential in a large number of Americans who are well above average in socioeconomic status, level of information, and willingness to vote or to approach their representatives. Candidates from both major parties might find it useful to seek the support of pro-space people, not because of that characteristic but because of their socioeconomic profile and their tendency toward activism. The space theme also might be one way of attracting younger voters if pro-space attitudes are indeed more common in the baby boom generation. As of 1984, that generation constituted over 40 percent of the voting age population, and politicians seemed to be showing greater interest in the views of its members.[44]

Some evidence suggests that pro-space opinion is particularly strong in the West and Southwest. Census results show that those areas have been growing more rapidly than other regions of the country.[45] "Sunbelt" states will continue to gain members in the House of Representatives, and their influence on national decisions seems likely to increase. One study predicts that nearly 40 percent of the 270 electoral votes needed to elect a President in 1992 will be in three states: California, Texas, and Florida.[46]

We also should consider the other side of this coin: the strength, or weakness, of anti-space opinion. Miller concludes that negative opinions toward the space program are rarely based on substantive information, suggesting that for segments other than the attentive public these views may be classified as "opinions" that are not deeply rooted or strongly held.[47] Furthermore, anti-space opinion is unorganized, while pro-space attitudes are reflected in organizations that are active and numerous, even if they are not very effective in direct lobbying.

THE POTENTIAL PRO-SPACE CONSTITUENCY

McWilliams concluded in his 1981 paper that the base of support for increasing federal allocations for space exploration is broadening rapidly.[48] There is an implied political opportunity here for pro-space people. However, Aerospace Industries Association officials John Loosbrock and James J. Haggerty are correct in saying, "watching a launch is not a constituency."[49] The question for pro-space activists is whether they can turn this attitudinal constituency into a political one.

7

SCIENCE FICTION, SCIENCE FACT, AND SPACE

> The ability to extrapolate . . . came to be only when men realized that their lives were changing, and could be changed, through their own efforts, and that Man's logic could determine the course of this change and possibly influence it.
>
> James Gunn[1]

> Culture preserved in a redoubt may later emerge into the larger society. Spaceflight itself might be the best example of an idea protected and developed to some extent within science fiction and later brought to realization and full acceptance in the larger society.
>
> William S. Bainbridge[2]

The changing demographics of interest in space may have been related to another phenomenon of the late 1970s and early 1980s: a surge of public interest in mass media science fiction and science fact, with a notable emphasis on space themes. Again, this was particularly visible among younger people. The timing of this surge of interest appears to be roughly simultaneous with the boom in space interest groups described in Chapter 8.

SCIENCE FICTION: THE ENDLESS JOURNEY

Space travel has been a theme in fiction since the second century AD, when the Greek satirist Lucian of Samosata described a fictional voyage to the Moon.[3] Closer to the modern era, works of extrapolative fiction provided a medium for speculation about alternative futures. Both Jules Verne and H. G. Wells wrote stories of voyages to the Moon, as well as on other subjects involving scientific and technological wonders. As science fiction grew in popularity in the United States in the 1920s and 1930s to become an established field of American writing, space travel consistently remained one of its principal subjects.

During what is sometimes called the "Golden Age" of science fiction, roughly the 1940s and 1950s, many writers emphasized space themes. Robert A. Heinlein, Isaac Asimov, and Arthur C. Clarke are perhaps the most visible of the older generation of writers in this regard, but many others have followed in the "mainstream," technology-oriented approach to science fiction, often related to space. Such writers helped prepare American culture for the idea of spaceflight. One critic went so far as to call Heinlein "perhaps more than any other person, responsible for the popularization in America of the concept of space travel and for the commitment to undertake it."[4]

Science fiction stimulated many of the pioneers and enthusiasts of spaceflight, both of the older generation and the younger. According to William S. Bainbridge, the men most responsible for the development of modern space rocketry had been inspired in adolescence by reading science fiction stories about spaceflight.[5] Several of the space activists described in this book explicitly recalled being "turned on" to space by reading science fiction, usually at an early age. Gerard O'Neill traces his interest in space to reading science fiction in his youth during World War II.[6]

People outside the science fiction field sometimes describe it as a literature of prediction. Certainly, many technological developments were foreseen in science fiction. Isaac Asimov has pointed out that space colonies had appeared in American science fiction as early as the 1920s and that he described a solar power station in a story published in 1941.[7] However, *The Science Fiction Encyclopedia* calls this "literature of prediction" thesis a false belief.[8] A recent book entitled *The Science in Science Fiction* points out that, in fact, many science fiction stories have rested on faulty scientific premises.[9] What science fiction appears to do better than predict is to entertain, to stimulate interest in science and technology, and to stretch the mind, encouraging the reader to be more receptive to unfamiliar ideas and technologies. NASA planner Jesco von Puttkamer, who has had an on and off relationship with parts of science fiction fandom for many years, describes science fiction as "a strangely addictive medium for extrapolation and consciousness enlargement." He goes on:

> ... "good" science fiction purports to popularize the technical problems of our time, to prepare the reader for the possibilities of the future, and to bring before his/her eyes the good that man really could achieve if only he would try. It stimulates his thinking and widens his horizon without boring him by being outright didactic; it warns man of the effects of his blunders without explicit exhortation, and it points him to new ideals. . . .[10]

Indirectly, science fiction can help to create the futures it describes by preparing people's minds for them. The editors of *The Science in Science Fiction* wrote that "many of the Senators and Congressmen who voted money for the conquest of the Moon must have shared, in a sense, a childhood dream: the reaching of the Moon was the central, passionate symbol of the science fiction they had grown up with."[11]

The audience for science fiction includes many serious and sober people. However, science fiction has been the subject of much derisive criticism from the "mainstream" of American culture, and science fiction readers still are sometimes regarded as "escapist" or worse. Von Puttkamer believes that many science fiction fans "drop out" of reality and get compensation from the superman idea.[12] In defense of science fiction and the related field of fantasy, the noted writer Ursula K. LeGuin argues that American culture has tended to disapprove of fantasy: if it is not "work" and if it cannot be justified as "educational" or as "self-improvement," then, in the Puritan value system, it can only be self-indulgence or escapism.[13]

There may be parallels among space enthusiasts. Many are highly pragmatic, scientifically and technically well-qualified people. However, sociologist B. J. Bluth comments that many of the people on the more extreme fringes of the space movement are dissatisfied with their own lives and seek to escape to proposed or imagined alternate worlds.[14]

Since the days of the American Interplanetary Society in the early 1930s, there have been connections between science fiction enthusiasts and American pro-space groups. In the case of the modern pro-space movement, some science fiction writers have played important roles. Robert Heinlein was an early patron of the L-5 Society, and Jerry E. Pournelle is one of its leading figures. G. Harry Stine, who writes science fiction under the name Lee Corey, was vice-president of the L-5 Society at one time. Ben Bova, known both as an editor and writer of science fiction, became the president of the National Space Institute in 1983. Charles Sheffield, who was president of the American Astronautical Society at the time the new pro-space movement began growing rapidly, also writes science fiction. Boards of directors, governors, and advisers of the larger pro-space citizens groups often include one or more science fiction writers; in 1984 those of the L-5 Society included Ben Bova, C. J. Cherryh, *Analog* Editor Stanley Schmidt, and Jerry Pournelle, and those of the National Space Institute included Isaac Asimov and Arthur C. Clarke. Some space group publications, notably the *L-5 News*, carry advertising for science fiction books; conversely, some science fiction magazines (notably *Analog*) have carried advertisements for the L-5 Society and the National Space Institute.

Peter Nicholls writes that "the most durable of science fiction 'predictions' is that of the endless Space Age – the belief that once we have emerged from the cradle of our planet, the future and the universe will be ours."[15] One finds this same line of thinking among many space advocates. How closely related are the pro-space and science fiction phenomena?

Science Fiction and Attitudes Toward Space

Sociologist William S. Bainbridge examined the influence of science fiction in his paper "The Impact of Science Fiction on Attitudes Toward Technology." His surveys showed that there is unusual and overwhelming support for the space program among science fiction fans.[16] A sampling of students revealed a very strong positive correlation between preferences for the science fiction television series "Star Trek" and "Battlestar Galactica" and support for space program appropriations.[17] "Science fiction does have the expected propagandistic influence," writes Bainbridge, who adds that the evidence suggests that there is no "hidden global variable" making some people simultaneously favor space and science fiction.[18]

One might be tempted to link this to generalized pro-technology attitudes. Bainbridge looked at this question and concluded that "all things considered, the most reasonable interpretation of the data is that science fiction strongly promotes spaceflight, while having only the weakest capacity to prevent anti-technology attitudes, and no power to produce favorable attitudes toward technology in general."[19] This may be related to findings by Robert D. McWilliams (cited in Chapter 6) that there seems to be little connection between the rise of pro-space opinion and attitudes toward technology in general. There appears to be something different about space.

Bainbridge points out that there are now three main schools of science fiction: hard science, which is the traditional mainstream; new wave science fiction, which emerged in the 1960s, reflecting greater concern with the social sciences and literary experimentation; and sword and sorcery, which grew to greater prominence during the 1970s, linked to fantasy and stories about barbarians.[20] In another paper, Bainbridge showed that those on the political right tended to favor the hard science school, while those on the political left leaned toward new wave.[21] Contrary to what one might expect, writes Bainbridge, the new wave and sword and sorcery groups do not breed disillusionment with the space program. "Contemporary science fiction," he concludes, "continues to serve the cause of interplanetary exploration."[22]

However, it tends to be the hard science, pro-technology writers of science fiction, such as Heinlein and Pournelle, who have been most closely associated with the new pro-space movement. They often also are the most conservative politically. This has led to some strains within the movement, notably within the L-5 Society. In Chapter 12, we will see that this latent conservative-liberal split has been widened by the debate over weapons in space. In 1982, at about the time when space weapons were becoming a major issue, some science fiction writers founded an organization called "Red Shift" to counter the views of writers like Heinlein and Pournelle.[23]

Analog: An Early Communications Medium

Perhaps the most characteristic publication of the "Hard Science" school of written science fiction has been the magazine *Analog*. Founded as *Astounding* in 1930 and brought to prominence in the 1940s and 1950s by John W. Campbell, the magazine virtually defined the mainstream of traditional, technology and space-oriented science fiction. Becoming *Analog Science Fiction/Science Fact,* it has enjoyed continued success under Ben Bova and his successor Stanley Schmidt. Its editorial viewpoint has been strongly pro-technology and pro-space and often politically conservative.

Analog has served as an important communications medium for people interested in the future in space, particularly those in pre-baby boom generations. In addition to its stories, often based on space scenarios, it has offered regular features on science fact and has provided an outlet for speculative ideas. "If you couldn't get a speculative paper published by the American Astronautical Society or the AIAA," says G. Harry Stine, "you put it in *Analog*."[24] For a while, *Analog* was a communications medium for writing on space industrialization.

In April 1981, *Analog* published a survey of its readers that showed similarities between *Analog* readers and the members of the major pro-space citizens groups profiled in Chapter 6. Those readers were 75.1 percent male and 24.9 percent female. There were 41.8 percent who were single and 69.8 percent who had college degrees. The most common income brackets were between $20,000 and $50,000. The largest age bracket was between 25 and 34, with the second largest between 35 and 49.[25] Overall, the results suggested that *Analog* readers tended to be slightly older and better established than those who joined the new pro-space organizations formed in the late 1970s and early 1980s.

The "Star Trek" Phenomenon

Science fiction first became popular in the literary medium and then began to be used as a theme in films. In the 1950s, a powerful new medium became available – television. Bainbridge comments "mass visual media do dilute the science fiction message. But still it gets through."[26]

The early science fiction programs on television, such as "Tom Corbett, Space Cadet," tended to be aimed at children. In 1966, at the height of the space race, Gene Roddenberry brought a new series to the public called "Star Trek." Here, space adventure was taken out of the category of children's programming and given prime-time exposure in a way that involved a large number of adults and older children. The series soon generated a considerable following.

The appeal of "Star Trek" appears to have lain not only in its interesting characters and settings but also in the optimistic and moral messages it conveyed. Commenting on a later "Star Trek" film, one reviewer noted that the characters share creator Gene Roddenberry's optimistic vision of the future, his heroic dreams seeded by the astronauts of the 1960s.[27]

Von Puttkamer comments that "Star Trek" fans were different from science fiction fans, whom he believes tend to be reclusive and oriented toward the literary medium. According to von Puttkamer, "Star Trek" fans were not alienated and were "turned on" to the future.[28]

When the National Broadcasting Company announced that it was going to cancel the "Star Trek" series at the end of the 1968 season, many fans were infuriated. In Los Angeles, "Star Trek" fans Bjo Trimble and her husband John, using a mailing list of 4,000 from a science fiction convention, mounted a letter-writing campaign in which each person asked ten others to write. The result was that NBC got hundreds of thousands of letters. "Star Trek" was given another year on the air, and the leaders of its fans discovered the potential of mail campaigns.

After the demise of the "Star Trek" series, many of its fans organized into clubs. Some are in large federations, such as Star Fleet and Star Fleet Command, and hold their own conventions.[29] "Star Trek" memorabilia, such as "Mr. Spock" plates, still were being advertised in 1984. "Star Trek" reruns have been shown on television for years, and three "Star Trek" movies had been produced as of 1985.

In 1975, a large "Star Trek" convention in Chicago asked NASA to provide a presentation on "the real space program." Von Puttkamer got the assignment and found the response to be "tremendous." From that point on, he believes, support for the real space program merged with fandom. Actress Nichelle Nichols, who played Lieutenant Uhura on "Star Trek," helped recruit women and blacks for NASA, doing commercials free of charge. "She was a symbol of the future, a symbol of hope," says von Puttkamer.[30] Nichols also was on the boards of both the L-5 Society and the National Space Institute as of 1984.

"Star Trek" fans made their influence felt politically on the naming of the first Space Shuttle orbiter to come off the assembly line. NASA proposed that it be named *Constitution*. "Star Trek" fans, reportedly including Richard Hoagland, *Space Trek* co-author Jerome Glenn, Carol Rosin (later a space arms control activist), space enthusiast Richard Preston, and "Star Trek" star George Takei ("Mr. Sulu"), came up with the idea that the first orbiter should be named *Enterprise*, after the ship in "Star Trek."[31] The Trimbles and their co-workers organized a write-in campaign that sent an estimated 500,000 letters to the White House. This action, the essence of a grass-roots operation, succeeded in its immediate objective when the Ford administration agreed to change the name to *Enterprise*. Unfortunately, the organizers did not realize at the start of the campaign that the *Enterprise* would never go into space but would be used as a passive test vehicle or display instead.[32] In the course of this campaign, the Trimbles learned about the L-5 Society and got in contact with Jerry Pournelle, who arranged for them to be invited to the first rollout of the *Enterprise* in September 1976, an event of considerable symbolic importance to pro-space people.[33]

Von Puttkamer believes that this write-in campaign was the moment when science fiction fans came closest to influencing the space program. Some space enthusiasts believe there also have been others; political scientists Michael Fulda and Nathan C. Goldman have described how "Star Trek" fans at a New York convention helped save the Jupiter Orbiter Probe/Galileo project by writing and calling their representatives in Washington,[34] and comparable claims have been made in connection with the 1984 space station decision. The *Enterprise* campaign also made a more general point, which was to send a signal to Washington that many ordinary citizens cared about the space program.

Since that time, the Trimbles and their co-workers have organized several mail campaigns on space issues. They formed an organization called Write Now! in 1979 and incorporated it in 1981 (at one time, it was called Space Write Now!). Like the L-5 Society's phone tree, Write Now! alerts the people on its mailing list (some 23,000 in early 1984) and provides guidance on how to write more effective letters. The organization is run out of the Trimble living room on a budget of less than $2,000, since it lives on small donations and volunteer labor (the L-5 Society and Spacepac donated money for paper and envelopes for the Write Now! mailing on the space station in December 1983, when Write Now! worked with other pro-space citizens groups). Like a true grass-roots operation, Write Now! provides an opportunity for participation by ordinary individuals.

Star Wars and After

The year 1977 saw the arrival of a new phenomenon, surpassing even "Star Trek" in its popularity: the George Lucas film *Star Wars*. Lucas, a talented young California film director, had not been successful commercially with the earlier film *THX 1138*, which presented a very dystopian view of the human future. He decided to focus on the hopeful side of human nature in *Star Wars*.[35]

This space adventure proved to be immensely popular with younger people, becoming what was then the biggest money-earner in film history. Unlike many earlier science fiction works, the film showed a rather ordinary young man, working on a farm and wearing ordinary clothes, participating in fantastic adventures and making a difference in a world of powerful organizations and high technology. *Star Wars* helped revive the hero, who could be master of his own destiny. Like "Star Trek," it may have been one of a new set of modern myths and fairy tales. Gary Kurtz, the producer of *Star Wars,* comments that the audience seemed primed for the film, which turned into "instant myth." In his view, enthusiasm for *Star Wars* also was a response to frustration.[36] *Star Wars* also make space adventure seem accessible and suggested that working and living in the space environment could be not just dangerous but also fun.

   Star Wars was followed by Stephen Spielberg's *Close Encounters of the Third Kind,* which showed extraterrestrial visitors to be harmless and benevolent, in sharp contrast to the interplanetary invasion scenarios of the 1950s. Black space and cold, distant stars were replaced by a welcoming environment; one reviewer wrote that "commonplace space, just beyond a star-spattered, low-hanging sky, is the legacy of Stephen Spielberg."[37] The Cold War image of outer space as a threatening, dangerous place where individual humans were helpless or irrelevant was being changed by a new generation of film-makers who themselves were products of the Space Age. (It should be noted, however, that threatening space themes were reappearing in mass media science fiction by 1984, for example, in the television series "V.") McWilliams commented in a 1979 paper:

> Beginning with the phenomenal success of the motion picture "Star Wars" in 1977, the popular media have turned increasingly toward entertainment based on outer space scenarios. ABC invested record amounts of money in the production "Battlestar Galactica," which has proven to be a highly successful series. In addition, there are no less than two planned sequels to "Star Wars." Several variously space-oriented motion pictures have been and are about to be released, including "Close Encounters of a Third Kind," a motion picture version of "Buck Rogers in the Twenty-Fifth Century," and a movie adaptation of NBC's resurgently popular "Star Trek." Also we note that the production and marketing of various types of attendant paraphernalia (toys, comic books, buttons, etc.) has become a multi-million dollar business.[38]

   The boom in space media spectaculars was by no means over in the early 1980s; the third *Star Wars* film appeared in 1983, and the third *Star Trek* film in 1984. Stanley Kubrick's 1968 film *2001,* based on a story by Arthur C. Clarke, was succeeded by *2010* in 1984. Science fiction works created for the literary medium were transformed into films. Particularly notable was Frank Herbert's sweeping, exotic epic *Dune* (itself a product of the age of disillusion). An adaptation of James Michener's novel *Space* was shown as a television series early in 1985.

SPACE ARTISTS

An even more subtle and immeasurable influence came from the growing number of "space artists," painters who depicted astronomical and space travel scenes with accuracy as well as power. The pioneer among astronomical artists appears to have been Lucien Rudaux, who was active in the 1920s and 1930s.[39] The covers of science fiction magazines provided an outlet for such artists for many years, but their influence increased during the Space Age when their works began appearing more frequently in "mainstream" media. The dean among space artists still living as of 1985 is Chesley Bonestell, who teamed with space writer and early Verein fur Raumschiffahrt member Willy Ley to produce the memorable 1949 book *The Conquest of Space*.[40] During the height of the space age, Robert McCall became a significant contributor of space paintings, doing artwork for *2001* and a dramatic mural for the National Air and Space Museum in Washington, D.C. McCall's 1971 painting of a city floating over the Arizona desert might be seen as an evolution of architect Paolo Soleri's arcologies on the one hand and an indirect precursor of Gerard K. O'Neill's space colonies on the other.[41] During the later 1970s, younger space artists found good outlets for their work in new astronomy and science fact magazines, reaching a larger audience. One of them, Don Davis, did depictions of Gerard O'Neill's space colonies, which were still being sold as posters by the L-5 Society in 1984.

THE POPULARIZATION OF SCIENCE

Mass media presentations on scientific and technological subjects, including space, were relatively infrequent in the 1950s. In March 1952, *Colliers* magazine presented the first of a series of illustrated articles on space, continuing into 1953 and 1954; many of the illustrations were by Chesley Bonestell. In March 1955, the Disneyland television program presented a show entitled "Man in Space," which was followed by programs on "Man and the Moon" and "Mars and Beyond." These educational efforts may have influenced many people. However, continuity was not maintained. Among magazines, *Popular Science* appealed to some, but did not single out space travel as a prominent theme.

The late 1970s and early 1980s saw a dramatic increase in media responsiveness to growing public interest in science and technology, marked in particular by an explosion of new popular science publications, many of which gave prominent coverage to developments in the space field. Perhaps they were competing with television; they often placed heavy emphasis on vivid art work. The forerunner of this trend may have been in the astronomy market, where there was a known and growing demand for astronomy-related products. Attempts were made to launch a popular, nontechnical magazine in competition to the long-established *Sky and Telescope*.

The first to succeed was *Astronomy*, founded in Minneapolis by the late Stephen Walther, who was then a young man. The first issue came out in August 1973 – after the Apollo program. *Astronomy* made a conscious effort to write for the average, younger, nonexpert reader, made generous use of color art work and quality photography, and gave extensive coverage to spaceflight, openly and effectively linking it with the amateur astronomy market. Its letters pages became a forum for debates over different aspects of the space program, and its editorials clearly were "pro-space." *Astronomy's* circulation rose to 179,000 in 1982, far surpassing that of *Sky and Telescope*. Another venture in the popular astronomy field, *Star & Sky*, proved to be short-lived, but it was the first magazine to publish science writer Trudy E. Bell's work on the emerging pro-space movement.

The first big success story of the new generation of popularized general science magazines was *Omni*, which, more than any other single media product, may have stimulated the popular science movement. The president of *Omni* Publications, Kathy Keeton, recalls that she and *Penthouse* publisher Bob Guccione conceived the idea in the early 1970s. "We were worried that young people were turning against science and technology," says Keeton. "In 1974 and 1975, we were hearing about interesting things on television, but there was no place to read about them."[42] They and their colleagues decided to bring out a glossy magazine mixing science fiction with popularized science fact and speculation, with a strong tilt toward the future and toward space. The first issue came out in October 1978. By early 1984, *Omni* had a circulation of about 850,000.

From the beginning, *Omni* featured space-related developments. "We felt that the space program epitomized American achievement," says Keeton. "It was as important as the first fish coming out of the water on to the land."[43] It was *Omni* that popularized the T-shirt slogan "The Meek Will Inherit the Earth – The Rest of Us Will Go to the Stars."

Omni went beyond editorial policy to support the space program and to help pro-space groups. According to Keeton, *Omni* assisted the space movement by providing free advertising space, participating in events, arranging for radio and television interviews, providing its mailing list, and running articles, notably a revised version of the Trudy E. Bell study.

Omni also has engaged in lobbying itself. The magazine sent out telegrams to mobilize space groups at the time the Galileo mission appeared to be threatened by budget cuts. It did the same when the Reagan administration took office, sponsoring a "Prospectus on Space Development" and encouraging pro-space activists to urge the adoption of a visionary space program. The magazine editorialized in favor of the space station in October 1983, before the Presidential decision to start that program had been announced. It also has alerted readers to write their congressmen on space issues, and published a congressional "black list" for the 1982 election.

Omni's relationship with the National Space Institute (described in Chapter 3) "petered out," says Keeton; *Omni* did not want to get involved with just one group, but wanted to be a clearinghouse or a catalyst. Keeton herself is a director of the L-5 Society, which she describes as "great in exciting the public and communicating with people, but split." In her view, the best job now is being done by Gerard O'Neill's Space Studies Institute, which is "the most serious" and "has the best credentials." *Omni* gave SSI its list of subscribers to help with a membership and fund-raising drive.[44]

The early 1980s saw the creation of several general popular science magazines.[45] Staffers from the respected weekly *Science* were instrumental in starting *Science 80*, which updates its name each year. In July 1981, Time, Incorporated began publishing *Discover*, a colorful and lightly written "newsmagazine of science." Even more popularized is *Science Digest*, which changed to a full-color glossy format. All three have achieved large circulations. The New York *Times* started a weekly

"Science Times" section in 1980. The pervasiveness of the trend is shown by the fact that several science publications of more modest circulations changed their appearance and sometimes even their names to appeal to wider audiences.

Another interesting venture in this field was a pair of magazines known as *Starlog* and *Future*. *Starlog*, written in a lively style and generously illustrated in color, specializes in space-oriented science fiction, particularly in the mass media of motion pictures and television, and seems to have carved out a secure niche for itself. Its companion publication *Future* concentrated more on science fact and science speculation and could be seen as a forerunner to *Omni*. Edited by Robin Snelson, *Future* gave frequent and positive coverage to space developments. For a while, L-5 Society co-founder Carolyn Meinel was a regular columnist.[46] After changing its name to *Future Life*, it was closed down by its owners in 1982.

Perhaps the most important fact about these magazines is that many have survived in the highly uncertain world of magazine publishing. Apparently, the demand was not simply a fad but reflected a longer-term trend. Many seemed to hit a peak of creation and growth between the years 1979 and 1982, about the time pro-space opinion rose to a new high and pro-space groups were at their own maximum rates of formation and growth.

Sagan and "Cosmos"

There is no better symbol of science popularization during the late 1970s than Cornell University astronomer Carl Sagan. He achieved fame first through his work on planetary exploration and his advocacy of a search for extraterrestrial intelligence. His first important popular success in the space/astronomy field was the highly readable book *The Cosmic Connection*, first published in 1973 and released in paperback in 1975. Sagan proved himself to be unusually skillful in popularizing science for mass media audiences and wrote successful books on nonastronomical subjects, such as *The Dragons of Eden* and *Broca's Brain*. It is worth noting that Sagan began his rise to prominence when attitudes toward science and technology were particularly negative and when the U.S. manned space program was at its low point.

Television, which already had brought the solar system into American living rooms through news coverage of planetary exploration missions, clearly offered a powerful medium for Sagan's messages.

Science then had a small niche in educational television, notably through the "Nova" series produced by WGBH in Boston. In 1977, Sagan began work on his own television series, called "Cosmos." Concerned about the future of the planetary exploration program, Sagan decided he had to do something, and "Cosmos" was one of the results (another was the Planetary Society, discussed in Chapter 10).

The series, which began its run by the Public Broadcasting Service in September 1980, was exceptionally popular for a science program, and Sagan's book with the same title was on the best seller list for many weeks.[47] Again, the message was an optimistic one of enjoyable wonder about the human relationship with and exploration of the Cosmos, coupled with moral and political messages about the fate of humanity. The series roughly coincided with the sharp rise in pro-space opinion prior to the first flight of the Space Shuttle in April 1981. Sagan has gone on to become deeply involved in the "nuclear winter" question and the issue of space weapons (see Chapter 11).

Related Phenomena

During the late 1970s and early 1980s, there were other signs of media awareness of heightened popular interest in space. Advertisers made growing use of futuristic themes, with a noticeable upsurge in space imagery after the first successful flight of the Space Shuttle; that vehicle became a symbol of success. "War games" based on space scenarios, first in board form and later in computer cartridges and diskettes, multiplied rapidly; while most had to do with interplanetary or interstellar conflict, some simulated such possible future activities as mining the asteroids and rescuing a space colony. Film producers placed cameras on the Space Shuttle, producing spectacular imagery that gave audiences the feeling of being in space.[48] A new Public Broadcasting Service documentary on spaceflight was shown in May 1985. It seems inevitable that such media products will stimulate young minds as comparable efforts have in the past.

FICTION, POPULARIZATION, AND THE SPACE MOVEMENT

It is tempting to conclude that "media events" created the new space movement. However, the evidence is too complex and ambiguous to allow that simple an interpretation. The earliest of the major new pro-space groups clearly were formed *before* the blockbuster science fiction movies and the explosion of popular science magazines, at a time when intellectual attitudes toward science, technology, and space were still rather negative. And they were created for other reasons: the National Space Institute grew out of NASA and aerospace industry concern, and the L-5 Society was stimulated by the ideas of Gerard O'Neill. The only one of the new successful popular science magazines relevant to space then in existence was *Astronomy*. Figure 6.1 suggests that pro-space opinion was beginning to turn upward before the release of "Star Wars" and well before the first issue of *Omni*. One could argue that the media were responding to demand as much as they were creating it. There was a ready and willing audience for the wonders of science and technology, and especially for space. Mass media science fiction and science fact provided opportunities for vicarious participation in the space adventure.

Instead of looking for a cause and effect relationship, it may be more useful to consider both science fiction/science popularization and the growth of interest in space as reflecting the same underlying factors. One may have been a widespread concern about the future of American science and technology after the cancellation of its most visible symbol, the Apollo program, and the negative attitudes of the early and middle 1970s. Interest in science, technology, and space also may have reflected the desire for a more positive future, in which Americans could use technology to open up new options. However, the most important factor appears to have been the emergence of a new generation of young people who had been stimulated by the experiences of the Space Age, who found science and space exciting and promising, and who wanted to know more about these subjects and, in many cases, to get more directly involved.

The older end of the baby boom generation was about 30 in 1977, the younger end about 13; by most accounts, it was people between those ages who were the source of much of the new demand for science fiction and science fact in the media, and particularly for materials on space. George Lucas and Stephen Spielberg were themselves representatives of that younger generation, perhaps reflecting attitudes as much as shaping them. It was that same generation that was to play a central role in the proliferation of pro-space organizations during the late 1970s and early 1980s.

8

THE SPACE GROUP BOOM

> According to Charles Chafer of the Public Affairs Council of
> Washington, D.C., the sudden proliferation of special-interest
> groups is often regarded as a leading indicator of issues that will
> be of major political importance five years ahead.
>
> Trudy E. Bell, 1980[1]

The boom in mass media science fiction and science fact occurred at
about the same time as an extraordinary proliferation of pro-space
groups, concentrated in the years from 1978 to 1982. Until the late
1970s, the older professional and industry groups – plus the National
Space Institute and the L-5 Society – had the field largely to themselves.
From 1977 through 1980, space interest groups were formed at an
increasing rate all over the United States (Figure 0.1). This surge of
group formation appears to have continued at a declining rate through
1984. These roughly simultaneous phenomena – the changing
demography of pro-space attitudes, the rising interest in space-related
science fiction and science fact, and the formation of pro-space groups –
may be interrelated.

The phenomenon of citizens forming organizations to educate the
public about the potential of space was noted by Georgetown University
Associate Dean T. Stephen Cheston in testimony presented before the
Senate Subcommittee on Science and Space in March 1977: "The
important element here is that these citizen groups did not exist until very

recently and they are growing. It is not unreasonable to expect that they will develop some political force."[2]

THE BELL SURVEYS

The principal chronicler of this space group boom is science writer Trudy E. Bell, who first called attention to it in articles published in the fall of 1980 (a revised version, reaching a larger audience, appeared in *Omni* in February 1981).[3] Using as her principal criteria nationwide activities or intentions (with the addition of a few of the most important local groups having widespread activities), Bell found that, as of May 1980, there were 39 American space interest groups (this later was raised to 42). If all local chapters and independent local groups had been included, the number would have risen to about 100. Adding on other groups with a positive attitude toward space, such as professional and amateur astronomical societies, science fiction clubs, and certain technical societies, would have brought the number of pro-space or space-sympathetic organizations to about 500. Thus Bell's list of 39 represented the bare minimum of organized space interest.

Dividing the 39 primary space interest groups into two main categories, Bell found that citizens support groups had about 40,500 members as of May 1980 (later revised to 33,800), and that trade/professional groups had about 39,500 (later revised to 47,000). This gave an initial total of 80,000 members, later revised to 80,800. The citizens support groups had an estimated total budget of over $4 million (the figure for trade and professional groups, about $10 million, is far too low because the Aerospace Industries Association declined to reveal its budget). Bell noted that the citizens groups' budgets did not include the value of volunteer labor and access to borrowed or donated equipment.

As a measure of geographic distribution, Bell used the locations of group headquarters. One third were based in Washington, D.C., and another third in California; the rest were scattered across the United States. Washington, D.C., with four of the seven trade/professional groups, had been a home for space interest groups for many years. California, by contrast, emerged as a center for pro-space citizens activist groups only after 1977.

Bell did a second survey in 1982 (it was to have been published in the second volume of the *Space Humanization Series* put out by the Institute for the Social Science Study of Space, but had not appeared as of 1985).

She found that the number of space interest groups with nationwide activities or intentions had increased to 50. Ten of the groups listed in 1980 had disappeared; 4 merged with others, 2 became defunct, and 4 could not be traced. Bell noted that most of the citizen support groups which had grown to have some influence were two to seven years old, suggesting some stability.

As of July 1982, the aggregate membership of these space interest groups had risen to just under 250,000. While trade and professional groups had grown by a respectable 15 percent, citizens groups had quadrupled in total size. The Planetary Society, which then had about 120,000 members, accounted for fully half of what Bell called "the formal space constituency." Other large groups were the American Institute of Aeronautics and Astronautics, the National Space Institute, the L-5 Society, and the Space Studies Institute. Allowing for overlapping membership, Bell estimated that between 150,000 and 200,000 people belonged to space interest groups in mid-1982. The number of people with a strong interest in space almost certainly is much larger, since formal memberships normally are only the tip of an interest iceberg. Subsequently, the numbers rose further, primarily because of the growth of the new American Space Foundation, which claimed 22,000 members in the spring of 1984.

Budgets had risen also, to an aggregate total of more than $23 million. This does not include the Aerospace Industries Association or the World Space Foundation, which declined to provide figures. Again, Bell pointed out that a number of active and influential groups, such as Spaceweek Inc. and the Chicago Society for Space Studies, had formal budgets that were deceptively low because they did not necessarily reflect the value of donated services, volunteer labor, or access to facilities such as photocopy machines. Conversely, some of the newly formed groups appeared to be projecting unrealistically high budget figures.

TYPES OF GROUPS

One of the most striking features of this surge of space interest group formation is the variety in purpose and function of the new organizations. Although the National Space Institute had been a generalized space interest organization, many of the new groups were specialized. In her 1980 study, Bell divided space interest groups into four types by primary objective:

1. Educating and informing the public
2. Conducting internal research
3. Funding external research
4. Engaging in political activities

Groups in the first category, in Bell's view, function much like typical astronomical societies with meetings, lectures, and other public events to promote awareness of space. The last three types of space interest groups, which aim to turn awareness into action, were virtually all under three years old in 1980. It is this turn toward building hardware, funding research, and lobbying that made the new pro-space "movement" different from existing space interest groups. Nathan C. Goldman and Michael Fulda have noted that the advent of more purposive groups may have limited the future for such a generalized group as the National Space Institute.[4]

In her 1982 study, Bell noted the advantages of a typology suggested by veteran space activist Stan Kent. According to Bell, Kent sees space groups divided into three "arms" of the space movement: (1) the talking arm, (2) the doing arm, and (3) the political arm.

All categorizations encounter the problem of multipurpose groups. For example, the Planetary Society is largely educational but also funds research. Bell's category "educating and informing the public" and Kent's "talking arm" include a wide variety of activities. I will use the following categories:

1. Educational and Mixed Purpose
2. Economic Interest Groups
3. Nonprofit Interests
4. Professional Organizations
5. Funding Organizations
6. Do It Yourself (Research and Technology Development)
7. Political Organizations
8. Space Defense and Space Arms Control
9. Other Special Interest Groups

In the sections below, only the most significant organizations are described. A more complete list is in Appendix A.

Educational and Mixed Purpose

The educational and mixed purpose group includes some already discussed, such as the National Space Institute and the L-5 Society, and some which will be discussed under other headings, notably the Planetary Society. Other important or interesting examples are discussed below:

Hypatia Cluster (1981)

One of the criticisms sometimes directed at the pro-space movement is that it is largely a male enterprise, generally shunned by women. After the first flight of the Space Shuttle, Amy Marsh and Marita Dorenbecher founded the Hypatia Cluster in San Francisco in May 1981 to motivate other women to participate in space science and to support space exploration, with the hope of getting more women into policy-making positions in the space field. According to Hypatia activist Mickey Farrance, a systems engineer with Lockheed Corporation, the name was inspired by Carl Sagan's treatment in "Cosmos" of the famous female scholar Hypatia, who directed the great library at Alexandria, Egypt. Sagan reportedly included this material at the urging of his wife, Anne Druyan, who is a member of the Hypatia Cluster's Board of Advisers. Other board members include science writer Richard Hoagland, Barbara Marx Hubbard, Stan Kent, space arms control activist Carol Rosin, and veteran space activist David C. Webb. More of a network than a formal organization, Hypatia conducts courses and workshops and encourages young people to prepare for space careers.[5]

Spaceweek Inc. (1980)

One of the most interesting and successful of the grass-roots space interest organizations, Spaceweek grew out of a 1979 effort in Houston to commemorate the tenth anniversary of the first Moon landing. (David C. Koch, who played a major role in starting Spaceweek, went on to found the American Society of Aerospace Pilots in 1981.) Not a membership organization, Spaceweek provides guidance and information kits to other groups organizing events in July of each year to celebrate space achievements. Spaceweek programs, organized by unpaid volunteers, spread from two cities in 1980 to 100 cities in 1983. The local committees that organize Spaceweek programs often are composed of members of the L-5 Society, AIAA, and other pro-space groups, who work together to present lectures, exhibits, and films, and to get materials into the media. Spaceweek headquarters in Houston, which operates on a shoestring budget, gives local committees virtually complete autonomy, much in keeping with the ethos of the new space movement.[6] Spaceweek President Dennis Stone, a pleasant, articulate young engineer who moved from Ford Aerospace in Houston to NASA's Johnson Space Center in 1983 to work on Space Shuttle payload integration, is the linchpin of the operation.

Economic Interest Groups

The Aerospace Industries Association and the GEOSAT Committee already have been described. One organization in this category that emerged during the space interest group boom was the Sunsat Energy Council (1978), founded by Peter Glaser and others to support the solar power satellite concept and to foster research into that idea. Sunsat began to publish *Space Solar Power Review* and *Space Solar Power Bulletin* but appears to have been inactive since funding for further satellite solar power station studies was defeated in Congress in 1980.

Nonprofit Interests

The nonprofit interests category exists essentially because the Public Service Satellite Consortium, described in Chapter 2, does not fit in any of the others. Made up of a number of nonprofit member organizations, the PSSC is not a typical economic interest group nor a typical educational organization.

Professional Organizations

We already have looked at the older professional organizations in the aerospace field: the AIAA, the American Astronautical Society, the Aerospace Education Association, The Aviation/Space Writers Association, and the IEEE's Aerospace and Electronic Systems Society. The early 1980s saw the creation of two others of interest.

American Society of Aerospace Pilots (1981)

The American Society of Aerospace Pilots (ASAP) seeks to train civilian pilots and other personnel to operate the Space Shuttle when it is commercialized and to accelerate the coming of routine commercial space operations. In the mid-1970s, United Airlines pilot David C. Koch pursued his interest in spaceflight to the Johnson Space Center, where he learned a great deal about shuttle operations and tried his hand at flying the Space Shuttle simulator. Convinced that space-based businesses would create an astronomical demand for space transportation, he persuaded his peers in the Air Line Pilots Association to form a committee to look into the possibility of flying commercial space shuttles. In September 1981, less than five months after the first flight of the Shuttle, Koch and nine other United Airlines pilots formed the ASAP, which grew to include over 200 United pilots.[7] In 1982, ASAP was reorganized to allow non-United pilots and nonpilots to join, and it later created a Pilot Division, a Spacecraft Crewmember Division, a Spaceline Operations Division, a Space Station Operations Division, and a Youth Division. By 1984, the group had over 1,000 members, 400 of them professional pilots. In that year, ASAP moved from a Chicago suburb to Grants Pass, Oregon, where it planned to create an entrepreneurial enclave including condominiums, an airstrip, and recreational and seminar facilities (an "Oregon trail to the stars," writes Koch).[8] There ASAP is setting up the first private spaceflight ground school. Twenty-one high school students participated in ASAP's first Youth Aviation and Space Camp at the Oregon Institute of Technology in Klamath Falls, Oregon, in 1984.[9] In tune with the new space movement, ASAP is dedicated to creating opportunities for its members to participate personally in routine commercial space operations and to helping to create the proper environment for participation in space operations by the broadest possible segment of society. Its literature is sprinkled with visionary rhetoric, and Koch's writings make clear that he accepts Gerard K. O'Neill's vision of unlimited growth through space colonization and the mining of extraterrestrial resources. "The challenge of conquering the virtually unlimited new lands in our solar system," he writes, "should test the capabilities of those of us who are seeking new adventures and provide all of us with new hope for the future."[10] Elsewhere he writes, "The ultimate goal of ASAP is to prepare the pioneers who will settle the frontiers of space, and to launch those pioneers on the exodus from this planet that will sow the human seed throughout this solar system, our galaxy, and beyond."[11] Picking up another strain of the new pro-space movement, Koch argues that "the American Free Enterprise System is being unleashed to bring the wealth of space home to us here in America."[12]

Society of Satellite Professionals (1983)

Reflecting the growth and professionalization of the communications satellite industry, the SSP was put together during 1982 and 1983 to "provide an international forum to increase public awareness, stimulate public discussion, distribute educational materials, and encourage professional development in the satellite industry." Emphasizing professionalism, the SSP states that "it is in no way intended to be a trade association, a lobbying group, a technical standards-setting entity, or a labor or employee representation group." The SSP, which had about 300 members in mid-1984, is based in Washington, D.C., which a *Washingtonian* article called "the satellite capital of the world."[13]

Funding Organizations

Several groups were founded in the late 1970s to provide alternatives to federal funding for space-related research. (Bell notes that funding research and development through contributions from citizens repeats the pattern of the 1840-60 American observatory-building movement.) Gerard O'Neill's Space Studies Institute is in part a funding organization for research. The other principal groups are discussed below.

California Space Institute (1979)

Growing out of discussions between University of California at San Diego space scientist James Arnold and Scripps Institute of Oceanography Director William Nierenberg, CalSpace was pushed through the California legislature in 1979 with the help of former astronaut Russell Schweickart, who had been an adivsor to Governor Edmund G. (Jerry) Brown. A statewide research unit of the University of California, CalSpace is headquartered in a white frame house idyllically located atop a seaside cliff in La Jolla, north of San Diego. The organization supports space research with an emphasis on practical applications. Its main areas of concentration are remote sensing, long-range weather forecasting, the production of energy and materials from space, and pure space science research, especially astrophysics. CalSpace, which received $900,000 from the State of California during its first full year of operation, provides "seed money" to university researchers for projects. As of 1984, the institute had three staff members who also do research themselves: (1) Director James Arnold, who is a long-time advocate of mining lunar and asteroidal resources; (2) "idea man" David Criswell, who has produced a wide variety of concepts for space industrialization; and (3) young, aggressive Executive Director Stewart Nozette. According to Nozette, CalSpace hopes to branch out into related scientific and technological fields, with an emphasis on those that will have an economic payoff. Nozette, who learned some of his lobbying techniques from Leigh Ratiner, comments that making the institute a part of the university gives it a legitimacy that some other organizations lack. CalSpace hosts conferences on space-related subjects, including one on low-cost approaches to space industrialization.[14]

Delta Vee (1980)

In 1979, members of the San Francisco section of the American Astronautical Society launched a "Viking Fund," which raised over $100,000 to help keep going the flow of data from the Viking landers on Mars ("Feed a Starving Robot," said the advertisements). The first $60,000 was turned over to a somewhat bemused NASA in a public ceremony in January 1981. The leader of this effort was English-born space activist Stan Kent, who came to the United States to get involved with the space program. Kent formed Delta Vee (a term having to do with the amount of energy needed to change the trajectory of a spacecraft), an unusual form of corporation whose "stockholders" were its contributors. He and Van R. Kane launched a Halley Fund – an appeal to support a Halley's comet mission, which the Carter administration declined to approve. After Senator William Proxmire ended NASA funding for a radio search for extraterrestrial civilizations, Kent and Kane also launched an Extraterrestrial Connection Fund to help finance public and private radio searches. When the magazine *Cosmic Search*, the only American publication devoted to the search for extraterrestrial intelligence, went out of business, Delta Vee started a successor magazine called *Astro Search*. That, too, has since folded. Delta Vee, based in northern California's "Silicon Valley" area, appeared to be inactive as of 1984.[15]

The Space Foundation (1979)

The Space Foundation came out of an amalgamation of Texas oil and business people and space science academics at Texas universities. Among its founders were Houston real estate developer David Hannah, who started the private launch vehicle company Space Services Incorporated; Samuel Dunnam, an Austin businessman; and Arthur Dula, a Houston patent lawyer well-known for aggressively advocating space development led by private enterprise. The foundation was intended to fund research in commerical enterprises, with special emphasis on projects dealing with recoverable large-scale space resources of energy and materials. It is best known for its space industrialization fellowships to students for research and prizes for achievement. Among the recipients of the fellowships have been Stewart Nozette of CalSpace and three young Harvard MBAs who founded Orbital Systems Corporation (see Chapter 12). In addition, the foundation sponsors the Space Business Roundtable, a monthly luncheon for prominent residents of Houston featuring speakers on space subjects. Foundation activists also are involved in arranging conferences on doing business in space. One of them, the dynamic Nancy Wood, runs a networking and information exchange operation called the Space Applications Network. "In Houston, we have the nucleus of a major space business center," she says. "Where else in the nation do you have this kind of venture capital and risk-taking mentality?"[16]

Do It Yourself

The do-it-yourself category, which consists of private, nonprofit organizations doing hands-on technological development or space-related research with private money, is representative of a strong ideological strain in the space movement, one that can be traced back to the experimenters of the 1920s and 1930s. These groups provide useful outlets for technically trained pro-space people, particularly younger ones, giving them a chance to participate. They also are a limited alternative to government programs, which are vulnerable to political changes and fiscal stress; they can do projects that probably would not get government funding or that had it and lost it. In 1983, most of these groups sent representatives to Cocoa Beach, Florida, for the first conference on Private Sector Space Research and Exploration and formed an Independent Space Research Projects Committee. In addition to the Space Studies Institute, described in Chapter 4, these groups include the following:

Radio Amateur Satellite Corporation (1969)

The oldest of these groups, the Radio Amateur Satellite Corporation (AMSAT) is an organization of radio amateurs ("hams") who design, build, and operate the series of satellites called OSCAR (Orbiting Satellite Carrying Amateur Radio) for experimentation and message relay by radio amateurs and amateur scientists. The first satellite, OSCAR-1, was built in California workshops for $65 and piggy-backed into space on a military satellite launch in December 1961. The Californians formed a group called Project Oscar, which still exists on the West Coast. AMSAT, formed in the Washington, D.C., area as the East Coast analog of Project Oscar, has grown to 6,000 members. Although most are in North America, AMSAT is an international organization with chapters and affiliated groups around the world. Eleven satellites had been launched as of 1984. The OSCAR program is now in phase 3 (high Earth orbit) and is aiming for phase 4 (quasi-geosynchronous orbit).[17]

Independent Space Research Group (1980)

In the latter part of 1979, three students at the Rensselaer Polytechnic Institute in Troy, New York, heard about the "Getaway Special" canisters in which private groups could send payloads into orbit on the Space Shuttle (former NASA official George Low was president of Rensselaer until his death in 1984). The students got the idea of building a six-inch diameter amateur astronomical telescope that could observe from space. After hearing about AMSAT, they expanded their horizons and founded the Independent Space Research Group (ISRG) as a membership organization in the spring of 1980, with the purpose of designing, constructing, and operating a series of increasingly advanced astronomical satellites for serious amateur astronomers, students, and professors. The group's main project is the Amateur Space Telescope, an 18-inch reflector that will orbit independently, perhaps as early as 1987. Although the ISRG wants to do real science, it hopes to design the system so that images can be picked up on home televison sets connected to amateur radio equipment, thereby widening participation. AMSAT is to help track the satellite. As of 1984, the ISRG had hundreds of members in more than 20 countries and was having work done by optical experts in Rochester, New York, as well as by students and faculty at Rensselaer. ISRG President Jesse Eichenlaub points out that such projects can be done very cheaply by using volunteer labor, donated equipment, and existing technology.[18] The ISRG's Ronald Molz is the chairman of the Independent Space Research Projects Committee.

World Space Foundation (1979)

As long ago as 1975, young aerospace engineer Robert L. Staehle was hoping to establish a practical way for people to express their enthusiasm for space. He joined the Jet Propulsion Laboratory in 1977, the year NASA suspended research into solar sailing, an aesthetically attractive form of space propulsion in which spacecraft would be driven by the impact of energy from the sun on large, thin sails stretched before them. In January 1979, Staehle and some of his colleagues formed the World Space Foundation (WSF). Two months later, the WSF began working with a group at the University of Utah doing research on solar sailing. A half-scale model of a solar sail was displayed at the Planetary Society's "Planetfest" in Pasadena in August 1981, and a full-sized prototype was completed in WSF's Pasadena workshop in 1983.[19] The group hopes to launch an engineering development model of the sail from the Space Shuttle or the European Space Agency's Ariane launcher in 1987 or 1988 and to someday use a sail to propel a mission to an asteroid. (A French group, following up on an idea in a story by Arthur C. Clarke, is advocating a solar sail race to the Moon.) The project received a boost when the Hughes Aircraft Company donated a rocket motor that will propel the sail from low Earth orbit to higher altitudes, where it will function more efficiently. The WSF, which received early help from the Charles A. Lindbergh Fund, collaborates on this project with the Jet Propulsion Laboratory, AMSAT, the University of California, the Technical University of Munich, and the Popular Astronomy Society of Toulouse. The WSF, with the help of the Planetary Society, also supports a search by astronomer Eleanor Helin and her colleagues for asteroids that approach or cross the orbit of the Earth. That group already has discovered almost half the known near-Earth asteroids, which may be the most desirable sources of extraterrestrial materials for space industrialization. As of February 1984, the WSF had 40 to 50 volunteers, supported by a much larger but undisclosed number of subscribers. The group periodically publishes

Foundation News and a series of papers intended to make up a Foundation Astronautics Notebook. Inspired by the 1984 poll that showed that 45 percent of Americans wanted to fly in the Space Shuttle, the WSF launched its "I want to GO" slogan and planned to publish a book entitled *Why America Wants to Fly the Shuttle.*[20]

Political Organizations

One of the features that distinguishes the new space movement from its predecessors is the formation of organizations explicitly intended to lobby for space or to exert related influences in the political arena (for example, through political action committees). Some groups, notably the L-5 Society and small, temporary operations like the National Action Committee for Space, had engaged in lobbying in the late 1970s. Beginning in 1980, politically oriented groups suddenly began to sprout. These included Campaign for Space (1980), Spacepac (1982), and the American Space Foundation (1982). A related phenomenon was the formation of the Congressional Staff Space Group and the Congressional Space Caucus in 1981. These political arms of the pro-space movement will be examined in Chapter 9.

Space Defense/Space Arms Control

In the late 1970s and early 1980s, groups began to be formed to advocate space-related defense concepts (for example, High Frontier and its political arm, the American Space Frontier Committee), to oppose such concepts (for example, the Progressive Space Forum, the Institute for Space and Security Studies, and the Institute for Security and Cooperation in Outer Space), or to advocate a new space-based world security system (War Control Planners, Strategic Arms Control Organization). These groups will be described in Chapter 11.

Other Special Interest Groups

This catch-all category includes some small, highly specialized interest groups with a space-related theme. Examples are Save the Apollo Launch Tower, a conservation-minded group that wants to preserve the rusting structure from which the Moon landing missions were launched, and the International Society of Space Philatelists, a stamp-collectors group.[21]

THE CALIFORNIA PHENOMENON

One of the most striking things about the post-1977 proliferation of space interest groups is that it was concentrated heavily in one state – California. Most of the California groups were small and local and could be described as grass-roots organizations; while a few proved to be durable and even significant, many turned out to be ephemeral. In order of founding, those included on Bell's 1980 and 1982 lists were as follows:

Organization for the Advancement of Space Industrialization and Settlement (OASIS) (1978)
Stanford Center for Space Development (1978)
World Space Center (1978)
California Space Institute (1979)
Citizens for Space Demilitarization (later Progressive Space Forum) (1979)
Futurian Alliance (a coalition of space interest groups in the San Francisco Bay area) (1979)
United Futurist Association (1979)
University of California Space Working Group (1979)
World Space Foundation (1979)
Delta Vee (1980)
The Planetary Society (1980)
San Francisco Space Frontier Society (1980)
Space Cadets of America (1980)
Strategic Arms Control Organization (later World Security Council) (1980)
Write Now! (1980)
Hypatia Cluster (1981)
Spacepac (1982)

This was not all. There were earlier groups that had gone out of existence by 1980. There were small private companies started by space enthusiasts (see Chapter 11). And there was the publication *Space Age*

Review, based in San Jose and founded by peace activist Steve Durst. Although that publication folded, the Space Age Review Foundation survived and started the useful publication *Space Calendar*, which was still in business in 1985.

Of the California groups on Bell's lists, nine were in the San Francisco Bay-Silicon Valley area, seven were in Los Angeles, two were in San Diego, and one was in Santa Barbara. The combination of aerospace industry activity and concentrations of educated, technology-oriented young people appears to have provided fertile ground for space group formation. Jim Heaphy, leader of the Progressive Space Forum, observes that in 1979 one could joke about "the space group of the week" in California.[22]

One key element in the California space group boom may have been the enthusiasm of California Governor Edmund "Jerry" Brown, who appears to have been "converted" at least partly through the efforts of *Coevolution Quarterly* publisher Stewart Brand and ex-astronaut Russell Schweickart, both of whom were advisers or consultants to the governor at one time. Brown was the principal speaker at an August 11, 1977 convocation in Los Angeles known as "Space Day," which was attended by aerospace executives, media people, and space enthusiasts.[23] Organized by Schweickart, the meeting was cosponsored by the state and the aerospace industry. (*New Times* magazine reported that Rockwell International Corporation, prime contractor for the Space Shuttle, paid the bill.[24]) Space Day was held the day before the first Space Shuttle "drop test," when the orbiter was released from its Boeing 747 carrier to glide to a landing in the California desert.

Brown became the first major political figure to offer a national vision of space adventure since Presidents Kennedy and Johnson. This occurred at a time when the Carter administration seemed determined to cut space spending, including the Shuttle program (Brown, of course, had been a rival to Carter for the 1976 Democratic Presidential nomination). Although it is not clear that his space advocacy did him much good politically (critics called him "Governor Moonbeam"), Brown did enunciate some of the themes of the new space movement, particularly that part of it inspired by Gerard O'Neill:

When the day of manufacturing in space occurs and extraterrestrial material is added into the economic equation, then the old economic rules no longer apply. Going into space is an investment . . . through the creation of wealth we make possible the redistribution of more wealth to those who don't have it. . . . As long as there is a safety valve of unexplored frontiers, then the creative, the aggressive, the exploitive urges of human beings can be channeled into long-term possibilities and benefits. But if those frontiers close down and people begin to turn inward upon themselves, that jeopardizes the democratic fabric. . . . As for space colonies, it's not a question of whether – only when and how.[25]

Space Day was followed up by Space Day Two, organized in the San Francisco Bay area in April 1978 by the "April Coalition," which, according to a February 1978 press release, was to bring together the following constituencies: "non-nukers," "human rightists," "radical ecologists," and "spacers."[26] Pro-space groups in the April Coalition later formed the Futurian Alliance. One of the organizations involved in creating this coalition was United For Our Expanding Space Programs (UFOESP), noted in Chapter 5. Some of the groups had a leftist flavor, and Heaphy recalls that the annual Space Days became "classic counterculture rallies." By 1980, Space Day drew 2,000 to 3,000 people in San Francisco's Golden Gate Park.[27]

The April Coalition/Futurian Alliance represented, in space activist Tim Kyger's words, one "tree" of organized space activism in the bay area. (Kyger, a space enthusiast and science fiction reader whose father was in the Air Force, had come to the bay area in 1979 from Arizona.) The other was the L-5 "tree," beginning with the L-5 Society chapter organized at the University of California at Berkeley in 1977 by David Brandt-Erichsen and others. Other L-5 Society chapters were formed, moved, merged, died, and reborn and interacted with the other organizational "tree." After 1980, says Kyger, the groups in the bay area began to drift apart,[28] but many of the individuals involved showed up at the L-5 Society's convention in San Francisco in April 1984.

Meanwhile, space enthusiast Joseph "Jay" Miller, leader of the San Francisco Space Frontier Society (a former L-5 chapter), had launched an unsuccessful campaign for a California Astronautics and Space Administration.[29] L-5 Society and related groups also sought the establishment of a space museum in Los Angeles. After the election of Ronald Reagan, several bay area space groups met in San Jose in January 1981 to draft a joint statement on space policy, but the result was a bland compromise. The groups also failed to form a coalition that could join the National Coordinating Committee for Space as one organization.[30] In 1983, space activists scored a success when the California legislature passed a joint resolution, pushed by State Senator Art Torres and his aide Jay Miller, urging the President to initiate a manned space station.[31]

FRAGMENTATION

Most of the new pro-space groups were small. Of the citizens groups, only these claimed memberships of a thousand or more in Bell's surveys:

The Planetary Society (1980)	120,000
National Space Institute (1974)	15,000
L-5 Society (1975)	7,000
Space Studies Institute (1977)	5,000
United Futurist Association (1979)	5,000
Students for the Exploration and Development of Space (1980)	4,000
AMSAT (1969)	4,000
The Space Coalition (1980)	1,200
Campaign for Space	1,000

Since Bell's 1982 research, the political action-oriented American Space Foundation, described in the next chapter, has joined these groups with a claimed membership in early 1984 of 22,000. However, most citizens groups numbered in the hundreds or less, and some were little more than mailing lists; some groups had contributors rather than members.

By contrast, the professional groups surveyed by Bell tended to be fewer and older. Here are those with memberships of a thousand or more:

American Institute of Aeronautics and Astronautics (1963)	34,000
American Society for Aerospace Education (Aerospace Education Association) (1976)	10,000
Aerospace and Electronics Systems Society of the IEEE (1951)	6,900
Aviation/Space Writers Association (1938)	1,400
American Astronautical Society (1954)	1,000

ATTEMPTS AT COALITION

During 1979, several people saw a need to try to bring pro-space groups into a coalition to allow the "movement" to speak with one voice and to make it more effective politically. The most important experiment in cooperation began that year when leading figures of the AIAA, the American Astronautical Society, the National Space Institute, the L-5 Society, Campaign for Space, and other organizations started getting together informally under the name Ad Hoc Coordinating Committee on Space. Among the individuals active in trying to bring the groups together were Jerry Grey of the AIAA, Ben Bova, then with *Omni*, and space activist and organizational entrepreneur David Webb. This Ad Hoc Committee evolved into the National Coordinating Committee for Space (NCCS).[32] Meanwhile, political scientist Michael Fulda, who had been active in getting a space plank into John Anderson's 1980 Presidential campaign platform, unsuccessfully proposed a national conference on space advocacy, with Webb's support.

Although that conference never took place, the NCCS held its first formal meeting in Washington, D.C., on June 17, 1981, two months after the first flight of the Space Shuttle. As of August 7, 1981, the NCCS Steering Committee consisted of National Space Institute Executive Director Mark R. Chartrand III, Planetary Society Executive Director Louis D. Friedman, AIAA Public Policy Administrator Jerry Grey, American Astronautical Society President Charles Sheffield, and David Webb of Campaign for Space. (Notably absent from this list was the L-5 Society.) The National Space Institute agreed to act as the secretariat for the NCCS. Membership criteria and procedures of operation were established. Meetings were to be held at quarterly intervals.

The member groups of the NCCS took their first major collective action after an important gathering of space enthusiasts in Pasadena, California, in August 1981. Called "Planetfest," the event was sponsored by the Planetary Society. By the next month, representatives of 30 space interest groups had agreed on the text of a letter to President Reagan on future directions for national space policy. The consensus that they reached included the following elements:

The President should formally recommit the United States to a vigorous and leading national space program.
A high level and broadly based space policy task force should be appointed.
NASA's mission should be clarified.
NASA must have increased, stable, and reliable funding for well-chosen initiatives.

National space policy should promote the entrance of private enterprise into all segments of civilian space activity.[33]

This statement was very general, reflecting the diverse interests represented in the NCCS. The document avoided endorsing specific hardware projects, saying only that NASA funding "should allow for a more vigorous pursuit of high orbital capabilities, astrophysics, planetary science, materials processing, biomedical research, robotics, and technology for both manned and unmanned missions." This gave something to each of the major sectors of the pro-space community, notably the space developers and the space scientists. However, Victor Reis, then handling space policy issues for the White House Office of Science and Technology Policy, believes that the statement had no effect on national space policy.[34]

According to Bell, the NCCS grew to include 33 member groups, including more than two thirds of those in her 1980 directory. However, momentum was lost over the next year; the NCCS met with decreasing frequency for the rest of 1981 and 1982 and produced few agreed statements. By the time of Bell's 1982 survey, membership had declined to about 18 groups. After November 1982, the NCCS did not meet again for a full year.

The NCCS was revived at a meeting in Washington, D.C., in November 1983, attended by representatives of 18 groups.[35] There was considerable discussion of a possible new space initiative by the Reagan administration, but no agreement was reached as to what that should be. As Chartrand was leaving his post as NSI executive director, he was replaced as NCCS chairman by David Webb, who seemed more inclined to use the NCCS as a vehicle for statements on space policy. Under Webb's leadership, NCCS participants in Washington met over the next few months to develop a stance toward the expected Reagan administration space station initiative, discussed in Chapter 13. By mid-1984, with the first year of space station funding approved by the Congress and Webb working on a study related to space commercialization, the organization once again seemed inactive.

In her 1982 paper, Bell commented that, in attempting to be a unified voice representing all the interests of its member groups, the NCCS "found itself trying to mediate between the necessity of forcing priorities for the citizen-support space community as a whole, and the individual and sometimes divergent administrative agenda (and, in some cases, egoism) of each group as a part."[36] These were among the reasons that the NCCS never was politically effective.

The Space Coalition

After their early 1980 tour seeking funds for the L-5 Society, lawyer-lobbyist Leigh Ratiner and L-5 President Gerald W. Driggers saw the need for a different mechanism. Driggers recalls that his concept was to pull the aerospace community together behind some objective. Seeing many pro-space groups with slightly different objectives, Driggers wanted to form some kind of federation.[37]

Ratiner came up with the idea of an alliance of citizens groups and aerospace companies, which he called the Space Coalition. This new organization would have the specific function of lobbying for space. Initially, it was to be associated with the L-5 Society. According to one of its own documents, the Space Coalition

> was created pursuant to the L-5 Society's Board of Directors discussion in Lincoln, Nebraska [April, 1980] for the purpose of carrying out the legislative action functions required to implement the 5 year strategy and plan of action. It is intended that the L-5 Society will be the center for conducting the public information and educational activities described in the program, while the Space Coalition will be the center for lobbying and related activities.[38]

Ratiner believed that the initial focus of the Space Coalition should be a manned space station, a position that agreed with that which Driggers had taken after he became president of the L-5 Society at the end of 1979. If the space station effort proved to be successful, the organization would come back with a larger agenda. It was hoped that the coalition would get support from other citizens groups in addition to the L-5 Society.

Driggers, seeing the coalition as a bridge as well as a lobbying arm, seemed more interested in bringing pro-space forces into alliance. Seeking to persuade NSI President Hugh Downs of the virtues of the new organization, he wrote the following:

> The Space Coalition is intended to be a broad-based alliance of citizens, industry, and associations involved in the space movement. . . . During its initial operation, the primary (but by no means the only) emphasis in its activities will be to mount a lobbying campaign for a new United States

manned space station program. . . . It is our intention that both L-5 and NSI join the new organization. . . . The chief advantages we see in the proposed collaborative effort are the opportunity to launch a new space lobbying capability, with immediate access to a large citizen membership base, and the chance eventually to minimize administrative expenses through sharing facilities, and possibly staff personnel.[39]

Clearly, this went well beyond a coordinating committee like the NCCS. The goal was a large, pro-space organization explicitly oriented toward influencing public policy. As Driggers explained in an article in the July 1980 *L-5 News*, "The L-5 Society cannot, within the legal limits of its tax-exempt status, meet this need." He concluded his plea with a certain note of desperation by writing the following:

The formation of The Space Coalition is the most important step taken in the history of the pro-space movement. This organization makes it possible to rally the resources necessary to speak with a unified voice in Washington, and, ultimately, in all the major capitals of the World. It is up to us to take the initiative. The long-term prospects for large-scale habitation of space are totally dependent on the success of the Space Coalition in achieving the results discussed above.[40]

Seeking support for their idea, Ratiner and Driggers visited a number of aerospace companies during the period April to June 1980. However, the response was skeptical. According to Ratiner, company executives were afraid that a new space program would be pulled out from under them again. With the help of Northrup Chairman (and former NASA Administrator) Thomas O. Paine, Ratiner and Driggers did manage to attract modest donations from four companies but never were able to put together a "critical mass."[41]

The Space Coalition was formally incorporated in June 1980. Trying to bring aboard other pro-space citizens groups, Ratiner and Driggers concentrated their efforts on the National Space Institute, having discussions with former NASA Administrator James Fletcher and others. Barbara Marx Hubbard tried to help through social functions at Greystone. However, says Ratiner, "we couldn't make it happen."[42]

Meanwhile, Ratiner had been approached by New York businessman Robert E. Salisbury, who wanted to play a role in the pro-space cause. Salisbury and Hubbard provided funds that allowed Driggers to move to Washington for the summer of 1980 to work full-time out of Ratiner's office, promoting the coalition. Salisbury was elected chairman of the Space Coalition in December 1980, with Driggers as president. The other two initial directors of the coalition were G. Harry Stine and ex-astronaut Philip K. Chapman, who later became president of the L-5 Society. Meanwhile, a gap was growing between the coalition and some L-5 board members who did not support the new venture or who wanted more L-5 control. Driggers became increasingly disillusioned.

Salisbury and Ratiner attended some of the early meetings of the Ad Hoc Coordinating Committee. Ratiner says he told the committee that the space movement needed one organization with 100,000 members and a lobbying arm and explained his Space Coalition proposal. However, Ratiner and his ideas were greeted coolly by the representatives of other space interest groups, some of whom suspected a takeover bid or saw Ratiner as a "hired gun."

Seeing that this was not "coming together," Ratiner said the Space Coalition would go ahead on its own and offered to have it serve as the lobbying arm for a loose coalition of space organizations. Nothing came of this.

As of April 1981, the Space Coalition was describing itself as "a not-for-profit advocate for accelerated systematic space development." Its board of directors by then consisted of Professor James R. Arnold of the University of California at San Diego; Philip K. Chapman of Arthur D. Little, Inc.; Gerald W. Driggers and Marne A. Dubs of the Kennecott Corporation; Thomas O. Paine, Robert E. Salisbury, and G. Harry Stine.[43] However, the Space Coalition never went any farther as an effective organization. Driggers and Ratiner dropped out, as did Salisbury later.

Between his work on the Moon treaty and his efforts on behalf of the Space Coalition, Ratiner had put considerable time and effort into pro-space activity, at no small cost to his earnings ("I have put more of my personal money into the space movement than anyone," he says). In the end, he concluded that there was no critical mass of serious people interested in space, and no real space movement.[44]

Other Efforts

At the end of 1984, there was yet another effort to bring the pro-space community together. Under the auspices of the Aerospace Education Association, Brian T. O'Leary invited addressees of a direct mailing campaign to join the National Space Council, which was to speak for the entire space community and to publish the *Space Newsletter*. As of this writing, it is too early to tell whether or not this effort will share the fate of the NCCS and the Space Coalition.

There have been other, more modest efforts toward merger or improved coordination. Bell noted that the World Space Federation and United for Space (both small, rather weak groups) merged into the L-5 Society and that Citizens of Space merged into OASIS. Several campus-based groups became chapters of Students for the Exploration and Development of Space. There also were plans for creating pools of resources, such as mailing lists, visual aids, and bibliographies. Two thirds of the organizations responding to Bell's 1982 queries reported that they had cooperated with other space groups in some kind of joint project.[45] One of the most positive examples was the sharing of mailing lists for specific lobbying campaigns, notably that conducted for the space station in late 1983 and early 1984. Leaders of the National Space Institute and the L-5 Society discussed the possibility of merger during 1984 and 1985.

Divisive Factors

Clearly, the diverse interests involved in pro-space activity have found it difficult to coordinate their efforts, much less to merge. Catalyzing events, such as a threat to the space program, the arrival of a new administration, or the prospect of a major new space project, brought them into temporary cooperation, but the generalized desire to do something has proved inadequate to maintain effective coordination. In striving for consensus positions, the space advocacy has risked losing the support of one or more organizations.

The most fundamental problem is divergent interests. Space is a macro-concept arching over many fields of activity and is an umbrella for very diverse user communities. Although they all favor increased activity in space, the citizens groups often have different agendas for action. One of the most basic divisions (which is as old as the space program) is between those who support manned spaceflight and space "development" and those who give first priority to unmanned applications and scientific exploration. Among the pro-space groups, this is reflected in the policy differences between the Planetary Society and most of the other groups. It is no surprise that the space "movement" often appears to be in disarray. "The fewer the purposes of the organization," observes John Kenneth Galbraith, "the greater its internal discipline will be."[46]

Another problem might be called organizational jealousy or rivalry among organizational entrepreneurs. The newness, small size, and fragility of many pro-space groups appear to make them wary of each other. In addition, space activist James Logan has written that "space enthusiasts, I have found, have a very developed sense of territory."[47] Group leaders sometimes worry about inroads on their potential memberships by other groups, as if mobilizing the pro-space community were a zero-sum game. National Space Institute Executive Director Mark Chartrand, criticizing the then new American Space Foundation, once wrote

> the new groups are having a bad effect on NSI (and presumably, on other established, active space organizations). . . . I urge you to toss out solicitations from such groups, unless they have proved themselves. Having more groups is not good for the space program, because it ensures that no group will reach a really effective number of members.[48]

Interorganizational frictions have been heightened by personality conflicts and differences in operating style. The activists of the early L-5 Society in particular sometimes grated on people in older organizations. Conflicts between technological utopians and pragmatic engineers or critical scientists remain a familiar story to observers of pro-space organizations. These factors are sometimes intensified by the fact that many pro-space organizations revolve around a small nucleus of activists, in some cases one person.

Citizens groups in general often depend on volunteers who are motivated by "psychic rewards," not only by dedication to a cause, but also by a desire for recognition. One may need a stronger than average ego drive to persevere as an unpaid leader of a citizens activist group. One analysis of earlier American protest movements may add another dimension in suggesting that personal power was a significant motivation for many of their leaders.[49]

So it is with space. "There are more egos per square inch in the space movement than anywhere else I know," says veteran space activist Charles Chafer.[50] This common phenomenon may be intensified in the pro-space movement because of the transcendental element in the pro-space cause. There is a hint of immortal reputation in being visibly associated with developments of historic significance and open-ended promise.

Leading the Coalition

Several people have aspired to lead the modern American pro-space movement at one time or another, and there have been suggestions that the right individual could have brought it together. The names usually put forward are Wernher von Braun, whose background was controversial and who died before the pro-space phenomenon really blossomed; Gerard K. O'Neill, who many believe did not have the political skills required; and Carl Sagan, whose liberal political stance alienates him from many pro-space people and whose criticisms of the manned space program have not endeared him to groups such as the L-5 Society. As for organizations, the National Space Institute, the L-5 Society, and later the Planetary Society each may have had the chance to become the nexus of a pro-space movement, but none succeeded. Pro-space citizens groups have not coordinated successfully with other parts of the space interest constituency, such as the Aerospace Industries Association. The pro-space community remains without a joint organization, a single dominant leader, or a universally agreed platform.

This is not necessarily fatal. The environmental movement, which has provided a model for the pro-space movement despite its much greater size, never coalesced into a single organization under a single leader. Its separate groups have been able to focus most of the time on separate agendas, occasionally forming tactical coalitions to deal with major questions. As of 1984, pro-space groups seemed to have a long way to go in learning coalition politics. This has reduced their impact in the political field.

9

ORGANIZING FOR POLITICS

> Whether we like it or not, the space program was born of politics, has declined because of politics, and will continue for better or worse through politics.
>
> Thomas J. Frieling, 1984[1]

> The space movement is desperately looking for a way to be effective. People are tired of paying for newsletters – and are now establishing ways of efficiently influencing public policy.
>
> Charles Chafer, 1980[2]

> I'm pro-space and I vote.
>
> Bumper Sticker, early 1980s[3]

One of the reasons for the surge of space interest group formation in the late 1970s and early 1980s, and the attempts at coalition, was the desire to influence public policy decisions; the reality of the need to lobby had intruded on the space dream. The space constituency had not been organized for the 1976 and 1978 elections, and its wishes had seemed to be largely ignored. Meanwhile, however, new technologies were promoting grass-roots lobbying and the rise of single-issue groups. In some other fields, such as environmental protection, groups already had demonstrated the potential effectiveness of citizens organizing for politics outside the traditional party and interest group structure. In addition, some space activists had acquired political experience through their work with other cause-oriented groups.

Around 1980, people in the citizens space movement "woke up to politics," as Trudy E. Bell puts it.[4] There had been growing impatience with the unwillingness of the Ford and Carter administrations to support major new projects in the space field. There also was a widespread desire to get new space goals approved by national authorities. The 1980 election seemed to offer a focal point for such an effort. With the approaching first flight of the Space Shuttle, some sections of the new space advocacy saw a need to begin campaigning for the next step in the classic agenda for manned spaceflight: the permanently manned space station. In the communications media used by the space advocacy, calls for political action were seen with growing frequency. "To have true political clout, we must first learn to apply the techniques of other already successful special interest groups," wrote one correspondent in a letter in the August 1981 issue of *Astronomy*. "Let's stop being politically naive and get our act together."[5]

These initiatives have led in the direction of more permanent, formal efforts to influence the decision-making process in Washington, although citizens pro-space groups still are relatively weak as a political force. They also have led to a growing exchange of activists between the space movement and the world of politics.

BEGINNINGS

In Chapter 2 I pointed out that the major aerospace interest groups had government relations offices at work by the mid-1970s. In the case of the pro-space citizens groups, the beginnings of their organization for politics can be traced back at least as far as 1977, when Carolyn Meinel Henson started the L-5 Society's Legislative Information Service. Ken McCormick's National Action Committee for Space, formed in 1978, might be described as the first citizens pro-space lobbying operation actually located in Washington, D.C., although it was small, underfunded, and did not last long. The L-5 Society also was the first citizens pro-space group to hire a professional lobbyist (Leigh Ratiner), in 1979. The L-5 phone tree, formally created in 1980 after three years of informal existence, also was an early attempt to form a political arm of the citizens pro-space movement, as was the Space Coalition in 1980.

Political scientists Michael Fulda and Nathan C. Goldman, themselves space activists, have written that pro-space groups emerged as a political force in the 1979 session of Congress and in the 1980 elections. They noted that the groups first made a foray into Washington lobbying, then into electoral campaigns.[6] That 1979 foray was focused primarily on the Moon treaty and the satellite power station issue, described in Chapters 4 and 5.

THE CAMPAIGN FOR SPACE

The first lasting, independent organization created specifically for the purpose of pro-space politics at the national level came from different origins. In late 1978, several individuals "actively involved in some aspect of making the citizens of this country aware that a commitment to space is absolutely vital for our future" formed an "invisible college" held together by regular correspondence and telephone calls.[7] This group included Mark R. Chartrand III, his then wife Trudy E. Bell, space writers David Dooling and James E. Oberg, photographer Karl Esch, and space enthusiasts Harrell Graham, James Logan, and Alan Fader. In early 1980, Graham told members of this group that he was thinking of forming a political action committee (PAC) for space. (Graham had been bringing lecturers on space to Antioch College in Ohio, where he was then working.)

Members of the invisible college were in contact with space activists David Webb, Charles Chafer, and Sallie Chafer. In a meeting held in March 1980, a new organization was formed called Campaign for Space, which was intended to be "a grass-roots public-interest politically active organization." The original Campaign for Space was to have several arms: member services, public information services, a space lobby, a space PAC, and an adjunct research organization.[8] In effect, it attempted to bring together into one organization the functions then distributed among different space interest groups. Graham was to be the Campaign for Space representative in Washington.

It was decided to incorporate Campaign for Space in March 1980, when the American Astronautical Society was holding an annual meeting in Washington, D.C. According to Bell, the feeling of an awakening to political awareness was particularly evident in a special AAS session on the politics of space, where Carolyn Meinel Henson gave a speech about the need to work on Congress.[9] Many space activists were very conscious of the fact that 1980 was an election year and that something would have to be done politically to get space concerns back near the top of the national agenda.

After filing papers with the Federal Election Commission, Campaign for Space put out a press release in which its Executive Director David C. Webb said, "Our national leadership is put on notice that Campaign for Space is joining the election process immediately to represent the views of a diverse constituency calling for a revitalized American space program." To achieve its goals, the new group announced that it would assist the election of political candidates sympathetic to the need for a stronger U.S. space effort and actively oppose those who had failed to recognize the importance of a stronger effort in space. Campaign for Space sought the greater involvement of private industry in all aspects of space development, the early establishment of factories in Earth orbit, and changes in regulatory and tax structures to encourage private sector space development and exploration. In addition, the group favored ongoing research into the use of lunar and asteroidal resources for space industry.[10] One of the organization's board members was Gene Roddenberry, creator of "Star Trek." Senator Adlai Stevenson III wrote a letter of support for the Campaign for Space in early 1980.

Unfortunately, Campaign for Space got off to a bad start, in part due to friction between Graham and some of the other activists. Graham left and founded the Citizens for Space PAC, which advertised in *Omni* and organized letter-writing campaigns. He then formed its educational offshoot United for Space. Both later merged into the L-5 Society.[11] Graham also invented the T-shirt design of our galaxy with the slogan, "You Are Here." David Webb became the leading figure in Campaign for Space but was less and less active as he got involved in campaigning successfully for U.S. participation in the UNISPACE conference held in Vienna in August 1982.

Campaign for Space survived these problems and donated about $2,000 to 12 candidates during the 1982 campaign, concentrating on members of the House Committee on Science and Technology and the Senate Committee on Commerce, Science, and Transportation. Contributions were doubled during the 1984 elections, when Campaign for Space contributed about $4,000 to 28 candidates.[12] This was very modest when compared with the $113 million given by PACs in the 1983-84 election cycle; the National Association of Realtors gave over $2 million.[13] On the Presidential side, Campaign for Space supported John Glenn's bid for the Democratic nomination with a $1,000 contribution.[14] After Glenn lost the nomination to Walter Mondale, Campaign for Space endorsed Ronald Reagan for his support of the civil space program but explicitly did not support the idea of a space-based anti-missile system because it opposed the "militarization" of space.[15]

As of 1984, Campaign for Space was being run by Thomas J. Frieling and his wife Barbara in Bainbridge, Georgia. According to a 1984 promotional circular, the organization supported with financial aid and in-kind contributions the election of candidates who favor a strong civil space program, researched the attitudes of Congress, and provided the legislative and executive branches with up-to-date information on space issues.[16]

Campaign for Space produces the bimonthly newsletter *Update*, and occasional notices to members on events related to space politics. A "Space Station Alert" postcard mailing in late 1983 called for supporters to telephone the White House opinion line, and an April 1984 circular called on recipients to send telegrams supporting the space station to House Appropriations Subcommittee Chairman Edward P. Boland. Campaign for Space also supports a fifth Space Shuttle orbiter.

Campaign for Space remains small, with about a thousand contributors. It also is handicapped by the fact that it has no full-time representation in Washington, D.C. A larger pro-space PAC might have been possible if pro-space groups had joined together to support it. However, other political action groups sprang up, continuing the fragmentation of the space movement into the political field as well.

Charles Chafer, who originally supported the PAC approach, had become skeptical of it by 1983, noting that space advocates got into the PAC field just as other interest groups were moving away from it and toward coalition politics.[17] On the other hand, PACs have remained a growth industry, with "nonconnected" or ideological PACS the fastest growing category. There were 3,803 PACs registered as of July 1, 1984.[18]

CAMPAIGNING FOR THE PRESIDENCY: EARLY EFFORTS

The first serious candidate for a major party Presidential nomination to speak out on space during the 1970s was Edmund G. (Jerry) Brown, who stirred a brief flicker of hope among some space enthusiasts between 1976 and 1980. However, the effort by citizen pro-space activists to support Brown was never well organized, attracted little attention outside

California, and more or less died when Brown failed to win national office.

Pro-space activists had no success in persuading either the Democratic or Republican camps to include a strong pro-space plank in their 1980 platforms. However, political scientist Michael Fulda, a supporter of third-party candidate John Anderson, played a role in persuading Anderson to adopt a strong pro-space policy statement, in what he considered a first-time attempt to mobilize the space constituency at the Presidential campaign level. The AIAA's Jerry Grey made significant contributions to the drafting process, which involved about one hundred individuals; Fulda calls it "a community effort of the space constituency."[19]

A group was organized known as Anderson Supporters for Space Science and Technology, which included two former NASA administrators on its board of directors and had chapters in 13 states. The Anderson campaign set up a special account for space-related donations. Although the dollar amount collected in the fund was small, it did account for 10 percent of the funds collected from citizens groups for Anderson, third behind womens and environmental issue donations.

During the campaign, space fan Peter Anderson, with the assistance of *Omni* magazine and Barbara Evans of the Space Studies Institute, sought to organize a debate on space among the major Presidential candidates. According to Anderson, contact was made with Reagan adviser Edwin Meese, and arrangements were made with the Smithsonian Institution to sponsor the event. However, the Reagan camp reportedly declined the invitation.[20]

AFTER THE 1980 ELECTION: A TURN TO THE RIGHT?

The pace of space-related political activity began to accelerate noticeably in early 1981, when the Reagan administration was taking office and the Space Shuttle was moving toward its first flight. This was due both to the perception of an opportunity, with the arrival of an administration that might be more sympathetic than its predecessor to the space dream, and to concerns that the Republicans might slash NASA funding as part of a budget-cutting exercise. Several groups and individuals made active efforts to influence the space policy of the new administration during the transition period and the first months of the Reagan administration.

Omni had a Space Project, whose coordinator Peter Anderson sought to put together "The Prospectus for Space Development" to influence the new administration. Anderson sent out telegrams to pro-space organizations in January and February 1981 calling for a mail campaign supporting NASA.[21] Meanwhile, the L-5 Society conducted a letter-writing and telephone campaign in late 1980 and early 1981 that it claimed was carefully coordinated with "behind the scenes negotiations in Washington."[22] The L-5 Society called on members to urge support for a permanently inhabited space station and for increased research and development on solar power satellites. The ubiquitous Barbara Marx Hubbard also was involved, writing to Vice-President Bush to urge a visionary space policy.[23]

This period also saw a notable increase in pro-space activity by political conservatives, who saw space as an arena for private enterprise, the new military high ground, a symbol of high technology leadership, and a place where the United States could compete successfully with the Soviet Union. They had forceful spokesmen in Congress, such as Senator Malcolm Wallop and Congressman Ken Kramer on space defense, and Senator Harrison Schmitt and Congressman Newt Gingrich on civil space policy issues. Some conservative groups had elevated mail solicitation into a science, and they applied these techniques to the space interest field. The subsequent emergence of conservatively oriented pro-space organizations was to change the character of the pro-space movement, bringing new strains into its already divided councils.

THE CITIZENS ADVISORY COUNCIL

After Ronald Reagan's victory in November 1980, leading figures in the L-5 Society and the American Astronautical Society decided that something should be done to influence the new administration's views on space policy, and they planned a meeting in the Los Angeles area in January 1981. Science fiction writer Jerry E. Pournelle, the local coordinator for the meeting, invited selected individuals to a "workshop" to be held on January 30, 1981, at the home of science fiction writer Larry Niven, a frequent co-author with Pournelle. The meeting was to prepare documents to be submitted to the "Space Policy Advisory Committee." When neither of the expected chairmen was able to attend the meeting, Pournelle took the chair of what was called the Citizens Advisory Council on National Space Policy, a position that he has held ever since. Well-connected in conservative political circles, Pournelle played a role in getting the council's views to decision makers in the new administration. The work of the council was supported by grants from the L-5 Society and the Vaughan Foundation and by "the generosity of individual members."[24]

The council is not a coalition of organizations but an ad hoc group of individuals invited by the chairman. Although its ranks have tended to include a disproportionate number of L-5 members, it is an intriguing cross-section of space activists and other figures with a space connection. At that first meeting, it included science fiction writers Paol Anderson, Robert A. Heinlein, Larry Niven, and Jerry Pournelle; ex-NASA Administrator Thomas O. Paine; former astronauts Gerald Carr, Philip K. Chapman, Gordon Cooper, and Walter Schirra; scientists David Criswell, Freeman Dyson, John McCarthy, Marvin Minsky, and Lowell Wood; veteran L-5 activists Randall Clamons, K. Eric Drexler, Gerald W. Driggers, Mark M. Hopkins, J. Peter Vajk, and Gordon Woodcock; von Braun rocket team member Konrad Dannenberg; Space Foundation activists Arthur Dula and astronomer Harlan Smith; space entrepreneur Gary C. Hudson; Space Coalition Chairman Robert E. Salisbury; American Astronautical Society President Charles Sheffield; aerospace engineer and writer G. Harry Stine; Bjo Trimble of Write Now!; and Barbara Marx Hubbard.

Not all these people were political conservatives by any means. However, the first report of the council had a strongly nationalistic ring. Stating that "the statesmen who lead Mankind permanently into space will be remembered when Isabella the Great and Columbus are long forgotten," it urged the President to announce a bold new space plan, with the following specific goals:

A space "industrial park" in low Earth orbit to be partially operational by the fall of 1988
Development of technology for construction of large space structures in high Earth orbit
A lunar base to exploit lunar resources, to be operational before the end of the century.[25]

According to the council, the low Earth orbit base would develop U.S. space capabilities; large space structures would lead to national wealth from communications and energy systems (notably the solar power satellite); and lunar resources would ensure U.S. raw material requirements in the next century. There was a strong implication that space science and planetary exploration were considered secondary; missions were to be subordinated to technology acquisition, and missions that increased capabilities to reach national goals were to have priority over those that did not.

Much of the language of the report was straight out of the mainstream of space activism. It stated: "We do not have to accept limits to growth."[26] However, there were new themes, which were to be increasingly characteristic of the post-1980 space movement: the importance of space for national defense and an increased emphasis on private enterprise in space. "Space has very great military potential," said the report. "Many experts believe that strategically decisive weapons can be deployed in space. . . . Space based beam weapons may develop into reliable missile defenses."[27] This last point was made more than two years before President Reagan gave implied endorsement to the concept in his speech of March 23, 1983. Titling one committee report "How to Save Civilization and Make a Little Money," the Council argued, that "the most important goal is to make space self-sustaining, which means economically profitable."[28] This was to become a central theme of the Reagan administration's space policy.

In August 1983, the council issued a report entitled "Space and Assured Survival" that called for immediate deployment of defensive systems.[29] According to Pournelle, its recommendations were the result of a compromise among different technical schools of thought within the council.[30]

Pournelle believes that the council's reports have been read at high levels in the White House and have had a real influence on policy. According to him, a copy of the first report was hand-carried to the office of Reagan adviser William Clark "within hours."[31] The L-5 Society also has claimed that Richard Allen, the first National Security Adviser in the Reagan White House, had read the first report and that a copy had been placed on the President's desk.[32] Pournelle has a letter from then Deputy National Security Adviser Robert C. MacFarlane acknowledging receipt, and letters from President Reagan and Science Adviser Keyworth complimenting the council on its work. One Keyworth letter regretted that he could not attend the meeting but said he had asked Dr. Lowell Wood to represent him. (Wood, a physicist at Lawrence Livermore Laboratories, has been active on the issue of space-based weapons.) However, Victor Reis, who was then Keyworth's assistant director for space issues, believes that the council's reports did not reach the President and had little impact on policy.[33]

Pournelle has stated that language from the council's report appeared in President Reagan's State of the Union address of January 23, 1984. A textual analysis shows some similarities.

The Citizens Advisory Council on National Space Policy, which has remained an ad hoc body of fluctuating membership, seems to have been more effective in influencing policy than either the National Coordinating Committee on Space (NCCS) or the Space Coalition (which never really began operating). It was a single organization, in which members were present as individuals and not as representatives of other groups. Its members tended to have broadly similar views about the potential uses of space. Its views were in tune with those of the new administration, and its timing was good (it took the NCCS until September 1981 to put together a much vaguer statement). The council concentrated its efforts on the White House, where major space policy decisions are made, rather than on the Congress, where most other space advocates have focused their energies. Above all, many of the individual members of the council were well connected, and some appear to have had access points at high levels in the administration, an advantage that most pro-space groups have lacked.

THE AMERICAN SPACE FOUNDATION

The American Space Foundation (ASF) grew out of a December 1980 conversation between Republican Congressman Newt Gingrich of Georgia, a young, dynamic ex-history teacher first elected in 1978, and Robert Weed, a young but experienced Republican political activist who had worked on a Gingrich campaign. Gingrich, who was to emerge as a leading spokesman for the "Conservative Opportunity Society," suggested that Weed set up a political action committee for space. During the following month, Weed brought together other young Republican political workers he knew, including Andrew Alford, William G. Norton, Melinda Farris, and Carlyle Gregory (Weed, Farris, and Gregory all were on Gingrich's staff at the time). They founded the American Space Political Action Committee to raise funds for pro-space candidates and to fill what they saw as a vacuum in U.S. space policy.[34]

When the American Space PAC failed to attract enough support, its organizers quickly abandoned it and founded the American Space Foundation, which was to be a pro-space lobbying organization. They were convinced that they were filling a gap and that no other comparable organization existed. Others were brought in, including Frank Lavin, a young Republican politico who had been with the conservatively-oriented Bruce Eberle direct-mail organization and who had joined the White House personnel office after the Reagan administration took power. It was his idea to use direct mail for a pro-space lobbying organization.

In the view of ASF's founders, the space movement needed a political arm, a way of bringing space enthusiasts together with those who knew politics. In short, it needed a formal lobby (this was somewhat reminiscent of Leigh Ratiner's views when he and his colleagues tried to launch the earlier Space Coalition). "You can't just have enthusiasts," says Lavin. "You need to involve managers and political experts."[35] ASF's intent was to be the lowest common denominator, a nonpartisan organization with a pro-space legislative agenda. Lavin, who became chairman of the American Space Foundation in the fall of 1981, argues that ASF was started to protect NASA, to focus the debate, to give legitimacy to space, to allow congressmen to be more pro-space, and to facilitate the civil and commercial uses of space.[36] In those desires, ASF was reflecting much of the pro-space constituency.

During 1981, two fund-raising mailings were sent to *Omni* subscribers in Georgia and Virginia. The letters were signed by Senator Mack Mattingly, of Georgia, another Republican. By the fall of 1981, the Bruce Eberle organization was doing fund-raising mailings for the new group.[37]

During 1982, Weed spent much of his time working on the campaign of Republican Paul Trible, who was running for (and won) a Virginia Senate seat. Frank Lavin, then an employee of the Agency for International Development, became the new chairman of ASF. The organization continued its direct mail fund-raising campaign, building up its finances and its mailing lists. According to Lavin, ASF was applying the fund-raising and direct mail sophistication of modern politics to space issues. The new organization deliberately kept its program three to six months behind its revenues,[38] in sharp contrast to the early L-5 Society, which had tended to take action first and worry about the money later. This approach generated suspicion among some other pro-space activists, who asked if ASF members were getting anything for their money.

However, L-5 Society Vice-President Mark Hopkins commented in 1984 that this is a legitimate technique.[39]

In January 1983, ASF began putting out a quarterly newsletter, *ASF News*. It sent out appeals for funds, including one signed by ex-Skylab astronaut Edward Gibson.[40] Later that year, ASF set up an office in a building on Capitol Hill owned by the conservative Heritage Foundation. Former congressional staffer Fred Whiting was hired to be ASF's first executive director. (Whiting, a former television and radio news reporter in Richmond, Virginia, had worked for Senator Trible.) In November 1983, ASF performed an important Capitol Hill ritual by hosting a reception in the Capitol building, thereby announcing the existence of the new organization and giving it some political credibility. By then its advisory board included former astronauts Edward Gibson and Brian T. O'Leary.

During 1983, ASF took on its first lobbying campaign, in support of the Space Commerce Act introduced by Democratic Congressman Daniel Akaka of Hawaii. Lavin believes that ASF was the first pro-space organization to use a postcard campaign as a technique, stimulating about 1,000 cards to Congress.[41] Akaka aide Diana Hoyt was quoted as saying that the ASF postcard campaign provoked a considerable response on the part of many congressional offices.[42] The foundation also made a point of publicly giving awards to members of Congress whom it considered pro-space.

ASF became active again in the fall of 1983 when it joined other pro-space groups in campaigning for the space station. ASF also conducted a survey of opinion on space-based defense systems and a congressional poll on the space station and released an "issue brief" on why the station should be manned.[43] For the L-5 Society and the World Science Fiction Convention, the organization prepared a membership guide for student space groups.[44]

In January 1984, ASF's board of directors set out the objectives of its lobbying for the next year:[45]

1. The establishment of a permanently manned space station
2. Significant increases in NASA funding, to a level of 1 percent of the federal budget
3. The fostering of private sector development of space and its opportunities
4. Serious consideration of the issue of a space-based defense system to prevent nuclear war.

Except for the second item, these positions supported those already taken by the Reagan administration. In other initiatives, ASF sought support for the Young Astronaut Program backed by the Reagan administration, circulated a petition for a stamp honoring astronauts Grissom, White, and Chafee, who died in the Apollo fire of January 1967, and worked on the Tax Status of Space Act.

ASF renamed its PAC the ASF Candidates Committee in 1983 to raise money that it could give to pro-space candidates in the 1984 elections.[46] Like the L-5 Society and its offshoots, ASF seemed to be attempting to perform at least three major functions: (1) education, (2) lobbying, and (3) the PAC type of political action.

As of 1984, ASF seemed to have been quite successful in attracting members through direct-mail techniques. With only 1,100 members at the end of 1982, ASF grew to 22,000 by the spring of 1984. In its 1983 *Annual Report*, ASF published a graph of its membership growth (Figure 9.1).

On the other hand, ASF's first annual conference, held in Washington, D.C., in September 1984, attracted only about 75 attendees. By the end of that year, Fred Whiting had left the organization to join the National Space Council and Robert Weed became executive director and later president.

On the face of it, ASF looks like a useful step toward a permanent, professionally staffed, pro-space lobbying organization. However, several space activists interviewed during 1983 and 1984 expressed reservations about ASF because of its close ties with conservative Republican activism.

ORGANIZING CONGRESS

Until 1981, the only institutional advocates of the space program in Congress were the authorizing committees responsible for NASA. Individual members such as Congressman George E. Brown and Senator Barry Goldwater also spoke out in support of increased space activity outside the context of those committees.

New phenomena appeared in 1981. One was the emergence of new conservative Republican spokesmen who took high profile positions in favor of increased American activity in space. The most visible were Newt Gingrich of Georgia, and Robert Walker, of Pennsylvania, both also spokesmen for the Conservative Opportunity Society. The other

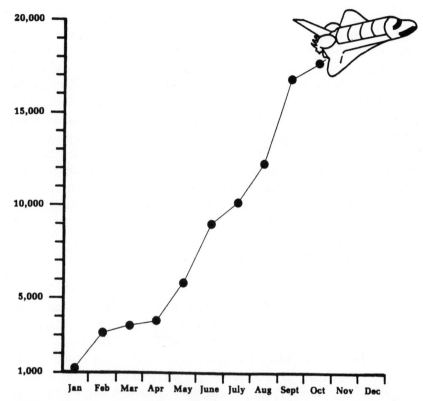

Figure 9.1 – American Space Foundation Membership
(*Source: American Space Foundation 1983 Annual Report,* p. 5. Reprinted with permission.)

phenomenon was the formation of organizations of members and staffers within the Congress around the pro-space theme.

In addition to spawning the American Space Foundation, Gingrich and Weed also discussed ways to increase interest in space within the Congress. One idea, also discussed with Senator Paul Trible, was a Congressional Space Caucus, a coalition of members "pushing" space.[47] But other events occurred first.

In February 1981, Gingrich and Trible sent out a pro-space "Dear Colleague" letter to other members of Congress. The letter supported a space operations center (a form of manned space station), a space fighter, and a satellite solar power station. The idea of Gingrich and Weed, the letter asked other members to write if they were interested.[48]

Inspired by the first flight of the Space Shuttle in April 1981, Gingrich sent out another "Dear Colleague" letter to collect signers for a letter to President Reagan urging a visionary space policy. Seventeen members, including some liberal Democrats, signed the letter.[49] Gingrich also began talking to others about a bill on space policy. On July 28, 1981, Gingrich and Congressman David Emery (later Deputy Director of the U.S. Arms Control and Disarmament Agency) introduced a bill entitled the National Space and Aeronautics Act. According to James Muncy, who did much of the drafting, this was a rewrite of Senator Harrison Schmitt's space policy act. If anything, it was even more visionary, containing language about using space to enhance freedom and foreseeing the addition of extraterrestrial colonies to the Union. Although it drew 17 co-sponsors, the bill went nowhere.[50]

THE STAFF GROUP

Meanwhile, in May 1981, several Republican congressional staff aides began meeting to discuss ways to work for space within Congress. In July, they formed the Congressional Staff Space Group (CSSG). The key organizers were Robert Weed, the peripatetic young space activist James Muncy (then an aide to Gingrich), William Norton (aide to Congressman James Jeffries), Karl Pflock (aide to Congressman Ken Kramer), and Fred Whiting (press secretary to Congressman Ray McGrath). Norton was elected chairman, John Bosma, vice-chairman, Melinda Farris (then an aide to Gingrich) secretary-treasurer, and James Muncy, executive director. Diana Hoyt, the dynamic young aide to Congressman David Akaka, also became an early activist in the CSSG and later chaired the organization. According to Muncy, Barbara Marx Hubbard supported the effort from the sidelines.[51]

Karl Pflock, who had been a member of the L-5 Society and the National Space Institute as well as being a science fiction writer, recalls that the motivations for the CSSG were (1) a coincidence of interests among people who already knew each other and (2) the catalyzing effect of the first Space Shuttle mission in April 1981 (he notes, however, that the idea had been discussed before then). According to Pflock, he, Norton, and Weed thought the defense side was important; in their view, the United States was falling behind the Soviet Union and could lose its access to space. All strong supporters of free enterprise, they also wanted to see space commercialization happen.[52]

The CSSG was essentially an informal mechanism for exchanging views and information among congressional staffers interested in space, sometimes with the intent of influencing members. Muncy says it was intended to be the infrastructure for a space caucus.[53] It also became something of a public forum on Capitol Hill, cosponsoring lunches with the American Astronautical Society. Although useful as a network, it did not have the potential political clout of a caucus, which was the next step.

THE CAUCUS

The CSSG people, particularly Muncy, approached Gingrich and liberal Democratic Congressman Timothy E. Wirth, who agreed to be co-chairmen of a congressional space caucus. When Wirth lost interest, the CSSG activists approached Akaka, who had become interested in space matters after NASA refused to fund a support facility for an observatory in Hawaii. Akaka agreed to become co-chairman.[54]

On November 20, 1981, Gingrich, Akaka, and six other members of the House sent out a "Dear Colleague" letter to all members of the House of Representatives. This letter, drafted by Diana Hoyt and James Muncy, said the Congressional Space Caucus would provide legislative support on space issues, serve as a clearinghouse on space information, and assist its members in promoting their common goal of revitalizing America's space program. Proposed activities of the caucus included special briefings and presentations for members and staff and the publication of a newsletter and issue briefs. "By joining the Congressional Space Caucus," it was stated, "you can demonstrate your concern and interest in this growing political issue."[55]

Hoyt became executive director of the caucus in 1982. Largely through her efforts, the caucus remained active and continued to build its membership. It sponsored briefings and debates on a wide variety of space-related subjects. Carl Sagan spoke on planetary exploration in 1982,[56] and other speakers have addressed space commercialization, the Space Shuttle, the Centaur launch vehicle, and Ken Kramer's space defense-related "People Protection Act" (see Chapter 11). The caucus sponsored a debate on the issue of anti-satellite weapons.[57] Materials on the High Frontier space defense proposal, public opinion about space, and other subjects have been circulated to members of Congress. The Congressional Space Caucus joined with the Congressional Black Caucus and the New Jersey Congressional Delegation to honor black high school students from Camden, New Jersey, whose experiment flew on a Space Shuttle in 1983.[58] The increase in caucus membership was impressive. The original eight signers grew to 35 by the time the caucus was registered as a legislative service organization in January 1982. With Hoyt in charge of recruiting, membership took off: 80 in February 1983, 130 in July 1983, and 166 in February 1984. This made it one of the largest caucuses on Capitol Hill. As of mid-1984, its leaders hoped to expand to the Senate. The caucus remained bipartisan, about half Rebublicans and half Democrats. As of 1984, the caucus was a loose, unstructured organization, without dedicated office space, permanent staff, or its own budget.

The caucus has avoided taking positions on specific issues. Like the pro-space movement, it would have difficulty in achieving consensus on such divisive issues as weapons in space or manned versus unmanned space exploration. The caucus also has to be careful about stepping on political toes on Capitol Hill. The more it sought to take positions, the more it would risk offending the House Committee on Science and Technology, which regards civil space issues as falling within its jurisdiction.

It has to be a boost for the pro-space cause that over a third of the members of the House of Representatives publicly committed themselves to saying that space is important to the nation. Perhaps even more interesting is the fact that so many professional politicians see no political risk in doing so. The caucus helps members of Congress feel comfortable about being pro-space.

A caucus can be a useful political instrument within the Congress, a kind of internal lobby that can support or oppose specific positions or legislation. However, the Congressional Space Caucus had not reached that stage as of 1985 and appeared to be inactive. William Norton comments, "The Caucus never fulfilled its potential."[59]

SPACEPAC

Throughout the late 1970s, the L-5 Society had been the most politically active of the citizens pro-space organizations. In lobbying so actively, it ran a risk of bringing into question its tax-free status as an educational, nonprofit organization under Section 501 (c) 3 of the Internal Revenue Code. Under the law, such organizations may not spend more than 20 percent of their funds on lobbying. Responding to the opportunity of the 1980 election, L-5 wanted to divert more of its funds to political action. The emerging leadership of L-5, headed by Mark Hopkins and Jerry Pournelle, saw the need for a separate organization.

Early in 1982, an L-5 Society mailing announced the formation of a new organization called the L-5 Spacepac, or simply Spacepac. Mark Hopkins, signing as L-5 Legislative Action Coordinator, explained that Spacepac had the same general goals as the L-5 Society. It also had much of the same board of directors; Spacepac's board included Hopkins, L-5 President Philip Chapman, Jerry Pournelle, and L-5 phone tree Coordinator David Brandt-Erichsen. In the best L-5 tradition, Spacepac was decentralized, starting with five local chapters. At the national level, Spacepac was to engage in fund-raising, donating money to candidates, and lobbying. "Donating money to political campaigns is the key reason for the effectiveness of political action committees," wrote Hopkins. "Let us show the politicians that the pro-space constituency understands this basic fact of political life."[60]

Using direct-mail techniques, Spacepac first raised money from L-5 members, then from outside the society. Arrangements with some other space interest groups allowed Spacepac to use their mailing lists for its appeals. Spacepac received cooperation from the National Space Institute, Spaceweek, High Frontier, *Omni*, and others.[61] Spacepac circulated letters of support from former NASA Administrator Thomas O. Paine, author James A. Michener, and science fiction writer and editor Ben Bova, then vice-president of the National Space Institute.

Spacepac, like Campaign for Space, did a Congressional Report Card, which rates all members of Congress on their positions on space issues. In 1984 it issued *The Space Activist's Handbook*,[62] inspired by Robert A. Freitas Jr.'s 1978 book *Lobbying for Space*. The handbook included advice, voting records, and statistical "ammunition"; the American Space Foundation provided a section on "membership technologies." Spacepac also has access to the L-5 phone tree of 17,000 to 18,000 names, which it claims can place at least 2,000 telephone calls on anyone's desk within 48 hours.[63]

During the 1982 election campaign, Spacepac contributed to 11 pro-space candidates. Volunteers, through the network of local Spacepacs, were active in several campaigns. Congressman Newt Gingrich, of Georgia, who received a $1,000 contribution, is said to have publicly thanked Spacepac when he addressed the spring 1983 L-5 Society Space Development Conference in Houston.[64]

In its summer 1983 mailing, Spacepac said it was organized by the leaders of the L-5 Society to "make politicians listen to the hopes and dreams of the millions of space enthusiasts in America." The primary focus was to be the space station. Noting that Walter Mondale had been a leader in the fight against the Space Shuttle, Spacepac Chairman Mark Hopkins wrote "Mondale's past record on space issues marks him as a disaster if he is elected."[65] Spacepac was left with only one option for the Democratic nomination, which was to support Senator (and ex-astronaut) John Glenn. The organization also endorsed Ronald Reagan for the Republican nomination, stating in a January 1984 letter, "There is a consensus emerging in Washington that Ronald Reagan is strongly pro-space."[66]

Late in 1983, Spacepac and L-5 drew closer to the Glenn campaign, contributing $5,000 and volunteers. Glenn in turn joined the board of governors of the L-5 Society. To make it easier for supporters to contribute to the party of their choice, Spacepac set up a Republican fund and a Democratic fund.[67] Spacepac also was involved in what Mark Hopkins calls the "Space Station War" (see Chapter 13).

Politics was a central theme at the L-5 Society Space Development Conference in San Francisco in April 1984. Since John Glenn was by then out of the Democratic candidate race and Walter Mondale was considered unacceptable, L-5 and Spacepac leaders decided to throw their support to Ronald Reagan. At the conference, Hopkins announced the formation of a group called "Space Advocates for Reagan."

In an August 1984 article in the *L-5 News*, Spacepac Director of Public Affairs Scott Pace described the group's policy platform, which showed how far the L-5 Society had moved toward the mainstream of space advocacy, NASA, and the Classic Agenda for manned spaceflight, as well as toward the private enterprise approach to space development favored by the Reagan administration:[68]

1. Complete full development of the National Space Transportation system
2. Make the establishment of a permanently manned space station an immediate national priority
3. Establish at least one new planetary exploration mission start each year
4. Develop, publicly or privately, orbital transfer vehicles for a wide range of missions
5. Increase support for both pure and applied space science programs

6. Increase efforts to facilitate the transfer of government developed technology to the private sector
7. Rekindle the enthusiasm of America's students and educators for science and technology education
8. Create a favorable tax environment for new commerical ventures and products in space
9. Streamline the government regulation of private sector launches and payloads
10. Ensure reliable and sufficient funding to carry out the points of this policy outline

Spacepac appears to have a broader base of financial support than Campaign for Space, and therefore a bigger fund from which to make contributions. According to its Washington representative Gary Paiste, the organization hoped to raise $100,000 for the 1984 campaign.[69] However, this remains modest compared with the funds raised by some of the larger PACs.

As of 1984, Spacepac appeared to be more of a lobbying organization than a typical PAC. Of the roughly $30,000 Spacepac spent in 1984, most went into mail and telephone campaigns, primarily in support of full funding for the space station.[70]

THE 1984 CAMPAIGN

The January/February 1984 issue of Campaign for Space's *Update* commented that the 1984 Presidential campaign had the potential of being a favorable one for the U.S. space program. For the first time in years, space policy was being debated on the national level. One major candidate (Glenn) had announced a detailed space policy (in 1983, including a call for a space station),[71] and the incumbent President might be in the process of launching the United States on its first new manned space program in more than a decade. This situation, said *Update*, "is a veritable embarrassment of riches compared to the previous Presidential election, when neither Reagan nor Carter mentioned the civil space program, and only very late into the campaign did third-party candidate John Anderson release a space policy statement."[72]

The 1984 platforms of the two parties revealed a sharp difference of opinion on space-related issues. The Democrats had nothing to say in support of the dreams of most pro-space activists and attacked proposals to put weapons in space.[73] By contrast, the Republicans supported the space station as a stepping stone to a multi-billion private economy in space. Their platform specifically stated: "The permanent presence of man in space is crucial both to developing a visionary program of space commercialization and to creating an opportunity society on Earth of benefit to all Mankind." The Republicans also supported the President's Strategic Defense Initiative, which might include space-based weapons.[74] Reportedly, Newt Gingrich played a major role in drafting the platform.[75]

The 1984 election posed the clearest choice in years for the advocates of manned spaceflight. This polarization suggested that the Republicans had seized the initiative on pro-space issues, leaving the mainstream of space advocacy little choice but to incline toward the Republican approach in the civil field (this was complicated by the opposition of many space advocates to space-based weapons). Many Democrats seemed to have abandoned the space field. Diana Hoyt put it succinctly in 1983: "Where are the Democrats on this issue?"[76]

THE TRANSMISSION BELT

One of the features of the space advocacy's increasing involvement in politics in the early 1980s was the interchange of individuals between pro-space groups and congressional staffs. Those staffs are a kind of transmission belt that brings in untrained young people, gives them a chance to experience political life and to play a role in influencing legislation and policy, and then often sends them on to other positions in their areas of interest.

By 1984, this clearly was happening in the pro-space world. Fred Whiting, for example, moved from a congressional staff position to the American Space Foundation, and then to the National Space Council. John Bosma went to High Frontier and then to the editorship of the weekly newsletter *Military/Space*. Diana Hoyt moved from Congressman Akaka's office to the staff of the House Committee on Science and Technology.

One particularly fascinating example is James Muncy. In December 1980, he was a student at the University of Virginia, finishing a degree in mathematics. Meeting other people interested in the future of American science and technology, he founded the Action Committee on Technology (ACT), whose purposes somewhat resembled those of the recently defunct FASST. Noting that this occurred as the citizens space groups were becoming known, Muncy comments that ACT was to be "more political," a lobbying organization. Becoming a registered lobbyist, Muncy was at work lobbying in May 1981.

Meeting Robert Weed in February 1981, Muncy became involved in the discussions that led up to the formation of the Congressional Staff Space Group and the Congressional Space Caucus. In April 1981, ACT, which consisted mostly of Muncy himself, endorsed this effort. Muncy was the executive director of the CSSG until the end of 1981.

Muncy also worked on "Project Skyport," an attempt in late 1981 and early 1982 to get President Reagan to include a manned space station in his January 1982 State of the Union address. Muncy drafted a letter for Congressman Gingrich, which appeared in *Omni* in December 1981, calling for people to write to Vice-President Bush rather than to President Reagan (the theory was that Bush's office would be more likely to notice the letters).[77] This effort failed to produce the desired commitment to a space station that year.

In January 1982, Gingrich created a new position on his staff for Muncy, whom he describes as "a phenomenon."[78] Gingrich and Muncy came up with the idea of an organization called Using Space for America (USA), visualizing it as a lobby, a political action committee, and a foundation. USA was formed in May 1983, describing itself as a coalition of labor, education, defense, science, business, and other interests, all of whom benefit from space and who see space as part of their future, united for political activities (this, of course, is reminiscent of earlier efforts). As Muncy puts it, USA was formed to "market space."[79] Like ACT, USA is primarily Muncy himself.

In 1983, Muncy joined the staff of the Office of Science and Technology Policy in the Executive Office of the President. There he played a role in writing speeches on space issues, notably President Reagan's address in October 1983 on NASA's 25th anniversary, when the President called on the space agency to be more visionary. Muncy says he wanted to get the President to articulate a vision to capture the hearts and minds of the American people, not just a bureaucratic compromise. Meanwhile, Muncy has written for publications such as *USA Today* and the Washington *Times*.[80] He also has spoken on space politics at the L-5 Society's Space Development Conferences.

Muncy is the first of the new space advocates to serve in a citizens group, as a congressional staffer, and in the White House, giving him a unique combination of experiences. If more pro-space people gain this kind of experience, they could enhance their impact on national policy.

PART II
ISSUES

10
SCIENTISTS, CITIZENS,
AND SPACE   *

Scientists are the legislators of our possibilities. They alter our
sense of where the boundaries are.

Horace Freeland Judson, 1984[1]

The Space Age introduced science and technology to the political
arena, but it did not transform politics or usher the scientists to
power.

Walter A. McDougall, 1982[2]

The research community may well be the last important sector of
our economy to organize itself to participate in the political
process.

Congressman George E. Brown, Jr., 1983[3]

THE UNVEILING

Until the Space Age, scientists observed the universe beyond the
Earth through a filtering veil of atmosphere. The rocket made it possible
to place instruments outside that obsuring gas; as early as October 1946,
an ultraviolet spectrum of the Sun was taken from above the ozone layer
by the Naval Research Laboratory.[4] By extending the tools of traditional
scientific disciplines such as physics and astronomy into space, the rocket
helped create the fields of endeavor known collectively as space science.
Workers in these new areas became space scientists, a new interest group
created by the Space Age.

Space scientists, and the space engineers with whom they have worked on many projects, have done nothing less than change our view of the universe through sensitive instrumentation, sophisticated data analysis, and near miracles of interplanetary navigation. Space astronomers, observing in wavelengths not observable through, or heavily filtered by, the Earth's atmosphere, have shown the Cosmos to be more complex, dramatic, and violent than we had thought. Discovery has followed discovery, making space-based astronomy one of today's most exciting sciences. "Astronomers will look back on this period as a golden age," University of Wisconsin astronomer Robert C. Bless told the New York *Times.* "And there is no way it could have happened except by getting above the atmosphere."[5] In what may be a related phenomenon, public interest in astronomy appears to be at an all-time high, as indicated by the increased circulation of popular astronomy magazines and the success of suppliers of amateur astronomy equipment.

Planetary exploration through unmanned spacecraft has revealed other worlds to us in unprecedented detail, shortening their psychological distance from us and allowing comparisons with the Earth, one of which contributed to the "nuclear winter" theory. Jet Propulsion Laboratory Director Bruce Murray pointed out in 1980 that we had received new images of another celestial body every year since 1964.[6] Astronomer Carl Sagan often has compared this age of exploration to the Western European voyages of discovery in the fifteenth and sixteenth centuries, which revealed much of the Earth's surface to the West.

Hundreds of space scientists and engineers have made heavy career commitments to their fields (about 500 attend the Lunar and Planetary Science Conference, held annually near Houston). Some have achieved previously undreamed of public recognition. However, these fields are vulnerable to political events because space science has depended heavily on funding from the federal government, primarily from NASA. When space funding was rising, space science prospered; when space funding declined, space science funding declined with it. After the successes of early missions, space scientists sought approval for new projects, which often were more complex, larger in scale, and more expensive, sometimes falling into the category of "big science." As NASA budgets were cut back in real terms, getting approval for new starts in the space science field became increasingly difficult. Even for continuing space science activities such as the analysis of data, the level of funding became uncertain as NASA responded to budgetary restraints by protecting its largest programs, particularly the Space Shuttle. Among other things, this revived a debate as old as the space program, between advocates of manned spaceflight and those who believe that priority should be given to exploration by unmanned spacecraft.

Space scientists generally were slow to organize themselves for systematic lobbying or to sustain a lobbying presence over long periods of time, although former senior House committee staffer James E. Wilson notes that individual scientists always have lobbied for their projects.[7] At first, space scientists relied primarily on formal advisory mechanisms, individual contacts, and the presumed respect for prominent leaders in their disciplines. Professional societies generally abstained from political activity. Former Senate space committee Chairman Frank Moss recalls that scientists were not well organized during his term in that office – 1973 to 1976.[8]

The mid-1970s saw more lobbying campaigns by ad hoc coalitions for specific projects. By the end of the Carter administration, there was a growing feeling within the space science community that something more was needed (this roughly coincided with the emergence of political activity among the new pro-space citizens groups). Not until the first year of the Reagan administration, however, did the American Astronomical Society's Division of Planetary Sciences become consistently active in Washington, and only in 1982 did the first organization to be a permanent lobby for space science interests – the Space Science Working Group – establish an office in the capital. The year 1980 saw the public appearance of the first citizens group formed to support one space science field: The Planetary Society.

Several knowledgeable people have offered explanations for this reluctance to lobby. The most frequent is that many scientists believe that engaging in "politics" is improper or distasteful, know little about the lobbying process, and resent having to invest their time in such efforts. Congressional staffers comment that most scientists do not understand that an interest group cannot just make its case once but must go on making it year after year, reminding the administration and the Congress of its interests. "They think that once they have spoken, they can expect continuing support," says Senate space subcommittee staffer Stephen H. Flajser. "They don't understand the reality of keeping your presence known."[9] "Scientists have always gone to Capitol Hill and presented the scientific facts in their testimony, thinking, 'because we're right, we'll prevail,'" says Marc Rosenberg of the National Coalition for Science and Technology. "They usually didn't realize they had to present political arguments as well as the facts."[10]

Scientists often have presented a divided front to Congress, with each discipline or subdiscipline advocating the priority of its projects. However, members of Congress are not qualified to determine which project is more valuable as science or to establish priorities. Marc Rosenberg observes that this invites Congress to cut back all of them.[11] Several congressional and administration staffers also have noted deficiencies in the lobbying style of some scientists, pointing to what they perceived as condescending attitudes and failures to recognize efforts that had been made on their behalf.

In the end, scientists involved in space-related research had to get their political act together. They were not getting the number of new starts they wanted, the continuity of their work was being interrupted, and their employment opportunities were threatened. They responded the way one might expect people with shared interests to respond. By the early 1980s they had begun to display a wide variety of modern American interest group behaviors, including more effective use of existing advisory committees and the formation of new ones, increased political activism by professional societies, the formation of a permanent lobbying operation for academic space scientists, appeals for public support, and the creation of a citizens support group.

INSTITUTIONALIZED LOBBYING

Advisory committees have become typical of the organized relationship between U.S. government agencies and private citizens expert in related fields. They came early in the Space Age. The National Academy of Sciences formed a Space Science Board in 1957 by combining the functions of the IGY Technical Board on Rocketry with the IGY Technical Panel on the Earth Satellite Program. That board provides comments and advice to NASA on its space science program, often in response to requests from the agency. A Space Applications Board was established in 1972, in response to a heightened emphasis on the practical benefits of space technology.

NASA itself established advisory committees soon after its formation, continuing a tradition established by its predecessor agency, the National Advisory Committee for Aeronautics. Advisory committees were established for the newer, and soon larger, programs of space science, space applications, and manned spaceflight. In 1971 the committee structure was revised into two separate committee groupings, (1) the Space Program Advisory Council and (2) the Research and Technology

Advisory Council, each with its subsidiary committees. These two structures were consolidated in 1977 into a NASA Advisory Council and six standing advisory committees, which went into operation in 1978. The standing committees, which are recommended by and which work with particular offices in NASA, are listed below:

Aeronautics Advisory Committee
History Advisory Committee
Life Sciences Advisory Committee
Space Applications Advisory Committee
Space and Earth Science Advisory Committee
Space Systems and Technology Committee

NASA considers these to be "internal" advisory bodies, since they are chartered by NASA and provide their advice directly to NASA management. NASA considers as external the Space Science Board, the Space Applications Board, and the Space Engineering Board, which are administered by the National Research Council for the National Academies of Science and Engineering, since they provide advice to the entire U.S. government and are, to a much greater degree, independent.[12]

The advisory committee system is an example of the two-way communications process that is central to the success of most interest group relations with government agencies. A particular advantage of this system is that it allows scientists to work out their priorities so that they can present a united front within a particular field, or as American Astronomical Society Executive Officer Peter Boyce puts it, "do their bloodletting in private."[13] These advisory councils provide a legitimized channel for lobbying by their members, although this is done within constraints of propriety and avoidance of clear conflict of interest. The Space Science Board, for example, has issued several reports recommending new NASA programs that would be in the interest of space scientists. Advisory boards also provide a communications channel from NASA to the scientists, so that they can learn about NASA plans and problems.

UNIVERSITIES SPACE RESEARCH ASSOCIATION

In the years before the first Moon landing, NASA saw a need for a more formalized interface with academic research scientists, and particularly for a means to handle research on the Moon rocks that were soon to be returned to Earth. With the aid of the National Academy of Sciences, NASA set up a Lunar Science Institute near Houston. In 1969, a group of universities formed the Universities Space Research Association (USRA), which became the manager of the institute, now called the Lunar and Planetary Institute. USRA, which had 55 member institutions as of August 1984, now also manages three other types of institutes under contract for NASA, dealing with computers, space biomedicine, and space processing. Although this is essentially a funding and administrative mechanism and not a lobby for space scientists, Executive Director David Cummings states that the USRA works to ensure the continued health of the space science program.[14]

A SHRINKING PIE

The existence of these legitimized lobbying mechanisms did not prevent a decline in funding for space science between the late 1960s and late 1970s. That decline was roughly proportional to the decline in space program funding in general. Since 1969, there has been a recurrent pattern of new administrations wanting to cut spending on space because it is "discretionary," and space science has suffered along with other space programs.

The focus of the problem in the 1970s was the Space Shuttle, which was to become the primary national launch vehicle for science missions as well as others. In 1977 NASA began to phase out the expendable launch vehicles that had carried scientific satellites and planetary probes into space. However, the Shuttle encountered technical problems and delays that held up some scientific missions. The Shuttle also needed additional funding; when some of this was not forthcoming from Congress, NASA reprogrammed its own resources to protect its highest priority program, taking funding away from other areas, including space science. These events inflamed resentment among many space scientists toward the Space Shuttle and the manned space program in general. James A. Van Allen, a critic of the Shuttle from the beginning, wrote in 1981, "it is time to recognize that the dominant element of our predicament is the massive national commitment of the past decade to development of the space shuttle and the continuation of manned flight."[15] Because the Shuttle has military as well as civilian uses, criticisms by some scientists also were directed at the apparent connection with defense interests. Carl Sagan added his view that current and near-future manned programs are not exciting to the public because they are not exploratory. "Guys in a tin can in low Earth orbit," he says, "are where the excitement isn't."[16]

In fact, space science suffered no worse proportionately than NASA's overall budget. The graph in Figure 10.1 suggests that space science actually did slightly better than a constant percentage would have suggested.[17]

However, there were significant distinctions within the space science budget, with planetary science funding showing more of a roller-coaster pattern than other sectors of space science (Figure 10.2).

While astrophysics and solar-terrestrial science enjoyed a long-term increase after 1974, funding for the planetary program entered a steep decline after fiscal year 1973 (Figure 10.3). The planetary scientists were hit hardest; by the late 1970s and early 1980s, they had become the most

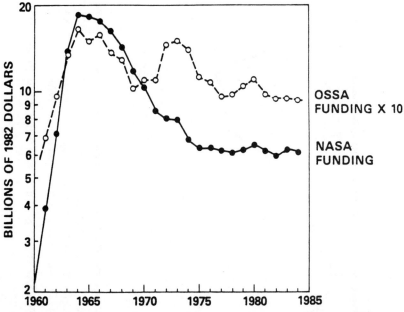

Figure 10.1 – NASA Funding Compared with Office of Space Science and Applications (OSSA) Funding (*Source:* The Universities and NASA Space Sciences, "Initial Report of the NASA/University Relations Study Group, July 1983," Appendix 2, p. 1.)

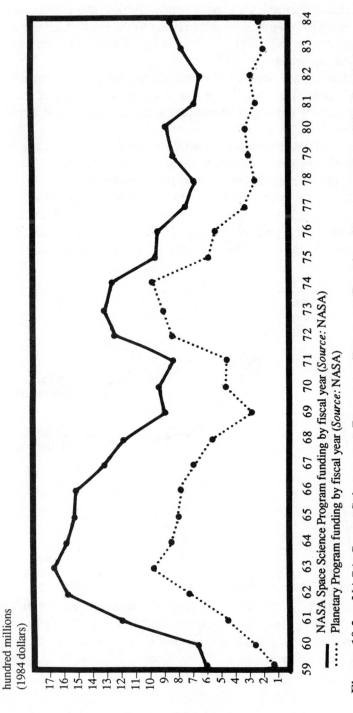

In dollars
hundred millions
(1984 dollars)

——— NASA Space Science Program funding by fiscal year (*Source:* NASA)
····· Planetary Program funding by fiscal year (*Source:* NASA)

Figure 10.2 – NASA Space Science and Planetary Program Funding (*Source:* Adapted by the author from figures provided by NASA.)

194

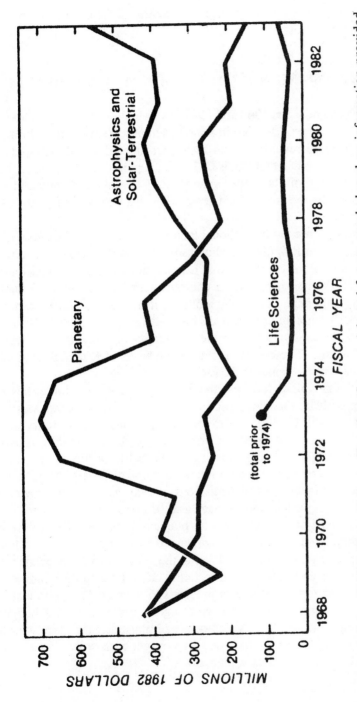

Figure 10.3 – NASA Space Science Funding (*Source*: Adapted from a graph, based on information provided by NASA, which appeared in *Sky and Telescope*, March 1982, p. 245. Reprinted with permission.)

vocal sector of the space science community and the first to form a citizens support group.

THE EXPLORATION OF THE PLANETS, INTERRUPTED

The period between 1962 and 1981 has been described as the Golden Age of planetary exploration, in which Mercury, Venus, Mars, Jupiter, and Saturn were revealed to us in unprecedented detail. (Planetary Society Executive Director Louis D. Friedman points out that it so far has been the *only* age of planetary exploration.) Carl Sagan has noted many times that we are the only human generations that will be the first to have this experience.

The most important institutional locus for this effort in the 1960s and 1970s was the Jet Propulsion Laboratory (JPL) in Pasadena, California. Unlike other NASA centers, the JPL is administered by the California Institute of Technology under a contract with NASA, giving it more autonomy – and more freedom to lobby for its interests.[18] Another, more recent concentration of professionals interested in planetary exploration is in Tucson, Arizona, notably at the University of Arizona's Planetary Science Institute. Tucson also is the headquarters of Kitt Peak National Observatory and the home of the L-5 Society.

After some initial failures, the JPL enjoyed a stunning series of successes with its lunar exploration programs and with the Mariner spacecraft, which flew by or orbited Mars, Venus, and Mercury. The laboratory proposed major new programs for the 1970s that, mostly because of budget cutbacks, never became reality or did so only after delays or in altered form. In addition to cost, one of the arguments made against such projects was that missions already had been approved to survey the target planet. Why go back a second or third time?

The first major setback was the planned Voyager Mars mission, which would have used massive unmanned spacecraft to orbit and land on the Red Planet (this was unrelated to the two Voyager spacecraft that flew by Jupiter and Saturn more recently). Its estimated $2 billion cost ran afoul of the funding strains of the late 1960s, and the program met its demise in the summer of 1967 when Congress pared the NASA budget. John Naugle, who was NASA's associate administrator for space science from 1964 to 1974, recalls that "we lost Voyager because of a division in the science community over what to do."[19]

Acting on the recommendations of groups of scientists, NASA created a Lunar and Planetary Missions Board and an Astronomy Missions Board in late 1966 and early 1967. These boards were to address themselves to the priorities of missions, recommending what NASA should do next. According to Noel Hinners, NASA's associate administrator for space science from 1974 to 1979, the Lunar and Planetary Missions Board produced an "advocacy package" that included the late Mariner missions, Pioneers 10 and 11 and Pioneer Venus, and the "Grand Tour" of the outer planets. Of the major missions, all eventually became reality except the Grand Tour, which met its demise in 1972.[20] The Grand Tour mission was partially replaced by the less expensive Voyagers 1 and 2.

NASA found an alternative to Voyager Mars in the less expensive Viking, which it sold to Congress in 1968. However, opposition was rising in the scientific community, where some preferred parceling the money out among several smaller missions (this was one of the factors leading to the cancellation of the Grand Tour). Under pressure from the Office of Management and Budget, NASA stretched out the Viking launch date from 1973 to 1975.

Orbiting Mars and placing landers on its surface in 1976, Viking was a highly successful mission. Carl Sagan recalls that President Gerald Ford called the scientists at the JPL to express his congratulations and to ask what they thought should be done next. (At the time, there was a fairly well developed plan for a Mars roving vehicle that could be done relatively cheaply by using Viking technology, although there was no consensus that it was the best next step.) The senior NASA official present told the President that the matter needed further study. In Sagan's view, the political moment was lost.[21] Only two months later, Naugle was testifying before a House subcommittee that Viking could be followed in 1986 by a roving vehicle.[22] However, the planetary scientists have not been back to Mars since. Not until January 1984 did the administration request funding for a new Mars mission, which will not fly until 1990.

THE SPACE TELESCOPE

The success of astronomers in getting a very large space project approved at about the time of Viking provides an interesting contrast. For over two decades, the desires of astronomers for new projects funded by the federal government have been given a stamp of legitimacy, and some degree of priority ordering, through reports of committees of astronomers and space scientists. As early as 1962, a group of scientists organized by the National Academy of Sciences proposed a space telescope as a natural long range goal; this recommendation was repeated in similar studies in the following years.[23] Under the aegis of the National Research Council of the National Academy of Sciences, an Astronomy Survey Committee headed by Jesse L. Greenstein of the California Institute of Technology was appointed in 1969 to consider the broad range of astronomy, including lunar and planetary exploration. In its 1972 report *Astronomy and Astrophysics for the 1970s*, the committee listed as its ninth recommendation "An expanded program of optical space astronomy, including high-resolution imagery and ultraviolet spectroscopy, leading to the launch of a large space telescope at the beginning of the next decade."[24]

The Large Space Telescope (LST) was a symbol of big ambitions in the space astronomy field. Some believed it would revolutionize astronomy because of its ability to resolve much finer detail and see more distant objects than the largest ground-based telescopes. However, the LST was vulnerable to budget cutting and delay because it was such a large and highly visible project, the most expensive orbiting observatory proposed up to that time.

The LST ran into trouble in 1974 when the House Appropriations Committee questioned its priority and, after nearly eliminating the proposed program, recommended that its funding be cut and that NASA consider a less expensive alternative.[25] Scientists John Bahcall and Lyman Spitzer led an effort to explain the importance of the program to congressional members and staffers. A House-Senate compromise restored sufficient funds to allow studies to continue. Meanwhile, the Space Science Board issued two reports strongly endorsing the LST as a new start.[26] The most serious threat to the program came in 1976, when the LST was removed entirely from the Ford administration's budget submission to Congress. Bahcall and Spitzer organized an extensive campaign by scientists to restore the project to the budget, at one time including as active lobbyists 30 to 40 scientists and industrial people (the principal contractors were Lockheed Corporation and Perkin-Elmer Corporation). Bahcall testified several times before congressional committees, drafted a senator's speech 24 hours before its delivery, and recalls being interviewed by a congresswoman in the lounge of the women's room near the floor of the House. In one span of "a couple of days," Bahcall had to organize a unanimous response from the members of the Greenstein committee to the charge that the committee had not ranked the LST as one of its highest priorities.[27]

This might be described as the first highly visible and vocal lobby for a scientific space project organized outside of NASA and the aerospace industry. Former House staffer James Wilson, who was helpful to this effort, recalls that it was done on "a high ethical plane,"[28] and Bahcall believes that "this unprecedented scientific lobbying was the result of the merits of the program and could not have been generated by commercial lobbying for a project that did not have ST's appeal."[29]

The LST was going forward at the same time as the Pioneer Venus mission and later the Jupiter Orbiter Probe. In 1974 and 1975 this led to some friction between planetary scientists and deep space astronomers, signs of competition for pieces of the budget pie. Several prominent astronomers tried to avert this by supporting restoration of funding for Pioneer Venus after it was cut back in 1975.[30]

The lobbying for the LST eventually paid off in 1977 when the project made a new start for fiscal year 1978. Its size had been reduced (the primary mirror was cut back from a planned 120 inches to 94 inches), and it was called the Space Telescope instead of the Large Space Telescope. But it survived its first severe budget test in 1978. By this time, notes Bahcall, the program had many good friends and supporters on Capitol Hill. "I believe the experiences we had in this activity," he wrote later, "have been useful to scientists interested in having other major programs approved."[31] Subsequently, informal lobbying networks were set up for the Advanced X-Ray Astronomy Facility and the Shuttle Infrared Telescope Facility, drawing on supporters willing to make telephone calls.[32] This, of course, is reminiscent of the L-5 Society's phone tree.

As things have turned out, the Space Telescope has encountered technical and management problems and has gone far over budget, suggesting a space science parallel to the Space Shuttle program.[33] Until the Space Shuttle accident on January 28, 1986, however, it appeared that the Space Telescope would fly in 1986.

CARTER AND PLANETARY EXPLORATION

In October 1975, the Committee on Planetary and Lunar Exploration (COMPLEX) submitted to the Space Science Board a report proposing a long-term strategy for exploring the outer solar system for the period

1975-85.[34] A subsequent report, published in 1978, did the same thing for the inner solar system.[35] These reports appear to have had little impact in terms of getting new starts approved by the Carter administration. Van R. Kane wrote in 1982, "since 1975, the White House – and sometimes Congress – has rejected each year's proposed mission."[36] One astronomer attending the 1979 St. Louis meeting of the American Astronomical Society's Division of Planetary Sciences was quoted as saying, "There's no perceived political advantage to making new starts, and there are no penalties for not making new starts."[37] Reporting on cuts in science funding in the fiscal 1981 budget, *Congressional Quarterly* observed that federal contractors and the academic community had conducted little lobbying to head them off.[38]

By the time Bruce Murray became director of the JPL in April 1976, the planetary program had entered the most ominous funding decline in its history, with appropriations dropping to less than a fourth of their fiscal year 1973 level by fiscal year 1978. For a time, this was to challenge the very survival of the JPL and was a significant factor in the beginnings of a more active lobby for planetary exploration.

Early in the Murray administration, the JPL responded to cuts in planetary exploration by conducting what was known as the "Purple Pigeon" exercise, in which a group of JPL experts sought to propose politically attractive interplanetary missions. Press reports from 1976 described its recommendations as a Mars Rover, a Solar Sail, a Venus Radar Mapper, a Saturn-Titan orbiter/lander, a mission to Jupiter and its Galilean satellites, an asteroid tour, and an automated lunar base.[39]

Much of the JPL planning during the Murray era focused on three planetary exploration projects: (1) the Jupiter Orbiter Probe (JOP), later named Galileo; (2) a mission to Halley's comet; and (3) the U.S. spacecraft for the International Solar Polar Mission. The laboratory came close to losing all three.

The Jupiter Orbiter Probe had been ranked by NASA in 1975 as its top priority new planetary mission. A proposed new start was contained in the fiscal year 1978 budget request. However, the project nearly died in the summer of 1977, when the House Appropriations Committee recommended against it. A major factor was that another expensive project, the Space Telescope, had gone forward in the same year.

Some planetary scientists had lobbied before, when there was an attempt to cancel the Pioneer Venus project. However, the fight over the JOP was a turning point for the planetary exploration lobby. With 1,200 jobs at stake at the JPL, planetary scientists launched a frenetic lobbying effort, mostly by telephone and mail. Although the scientists did most of the work themselves, they for the first time got active help from the emerging citizens pro-space movement, including the L-5 Society, *Omni* magazine, and "Star Trek" fans, who were contacted at a convention in New York.[40] This added hundreds, perhaps thousands, of letters and telephone calls.

The JOP also got help from pro-space members and staffers of the House Science and Technology Committee. In an unusual move, they challenged the Appropriations Committee in a floor fight on July 19, 1977 and succeeded in getting the money restored. In 1979, Senator William Proxmire tried to kill the JOP (renamed Galileo) but was voted down.[41] Galileo went through another crisis in late 1981 when the Reagan administration's Office of Management and Budget took it out of the NASA budget request, but it was reinstated after a "reclama" (appeal) to the White House by NASA.[42] Unfortunately, this was at the expense of another mission, the Venus Orbiting Imaging Radar.[43] Galileo was to go on to have other troubles because of delays in the Shuttle and problems with upper stages, and its planned launch date slipped to 1986.

In January 1980, the Office of Management and Budget effectively killed the plan to fly by Halley's comet and rendezvous with another comet by deleting funding for Solar Electric Propulsion from the fiscal year 1981 budget.[44] Funding for a Halley intercept was deleted from the fiscal year 1982 budget, reportedly to allow funding for the Venus Orbiting Imaging Radar, described by NASA as a better mission.[45] Despite cries of anguish from space scientists and a letter-writing campaign by the Planetary Society, the Halley project never was restored.

The participation of a U.S. spacecraft in the International Solar Polar Mission was first delayed by cuts in the administration's fiscal 1981 budget request, then threatened by the House Appropriations Committee, and finally cancelled in the administration's budget request for fiscal year 1982.[46] Despite protests from the European Space Agency and its member countries, funding never was restored. Bruce Murray reportedly said later that the planetary exploration constituency had been divided between ISPM and the Halley mission.[47]

The Venus Orbiting Imaging Radar (VOIR) mission also had a tortuous history. This project first was deferred to a later launch date, then was stricken from the Reagan administration's fiscal years 1982 and 1983 budget requests.[48] JPL scientists and engineers did a thoughtful redesign for a scaled-down and much cheaper mission called the Venus Radar Mapper, which became a new start in the fiscal 1984 budget.

This series of emergencies provoked a variety of responses from those interested in planetary exploration. Bruce Murray and others with a direct interest in these programs actively lobbied in Washington. They got support from professional and citizens groups. In the first issue of the *Planetary Report*, Planetary Society Executive Director Louis Friedman reported that during 1979 congressmen received telegrams and letters from persons associated with the L-5 Society, *Omni*, American Institute of Aeronautics and Astronautics, the American Astronomical Society's Division of Planetary Sciences, universities, and the aerospace industry. However, he noted that this was an unorganized effort, with no central leadership.[49]

More organized approaches were needed. In an article published in January 1981, Richard F. Hirsch of the Virginia Polytechnic Institute argued "space investigators have become one of many interest groups, vying for a slice of the federal budget. In the absence of a jolt such as experienced in 1957 and 1961, space astronomers must engage in the politics of open debate like other interest groups."[50]

Addressing the American Astronomical Society's Division of Planetary Sciences at about the same time, Bruce Murray identified the lack of a serious space lobby group as a major cause of the "crisis" in planetary exploration.[51]

THE SOLAR SYSTEM EXPLORATION COMMITTEE

By the end of the 1970s, NASA and the planetary scientists realized that the earlier recommendations of the Space Science Board were not politically realistic and that they must plan for a more constrained program. There was serious concern about the long hiatus between the Voyager program and any expected new starts in the planetary exploration field. John Naugle, then NASA's chief scientist, proposed something like the old Lunar and Planetary Missions Board to design a program that the scientists could sell. In November 1980, as Voyager 1 was sweeping past Saturn, NASA formed an ad hoc committee to formulate a long-range program of planetary missions and called the group the Solar System Exploration Committee (SSEC).[52]

The first two chairmen of the SSEC, John Naugle and Noel Hinners, had been associate administrators of NASA for space science, and one of the members of the committee, Thomas Donahue, was chairman of the Space Science Board. It was as authoritative and well connected a group of planetary scientists as one could find.

The SSEC completed the first phase of its work in October 1982, presenting its initial, "stopgap" recommendations to NASA Administrator James Beggs the following month.[53] In its formal report, the committee proposed a program of lower cost missions, using standardized parts and existing technology wherever possible. It suggested a roughly constant budget level (about $300 million) rather than the ups and downs of previous years. The Core Program recommended by the SSEC gave temporal priority to these four missions: (1) Venus Radar Mapper, (2) Mars Geoscience/Climatology Orbiter, (3) Comet Rendezvous/Asteroid Flyby, and (4) Titan Probe/Radar Mapper.[54]

This carefully coordinated report was well received by both NASA and the Congress. A well-thought-out, systematically justified long-range plan making more efficient use of resources had considerable credibility. "The Committee is inclined to follow it," said Harold Volkmer, then chairman of the House Subcommittee on Space Science and Applications.[55]

The SSEC also was well received in the White House. Victor Reis, a former assistant director of the Office of Science and Technology Policy, comments that the report met a need: "It did our job for us."[56] To the surprise of some, the Reagan administration included the Mars Geoscience/Climatology Orbiter in its budget request for fiscal year 1985, presented to Congress on February 1, 1984. The Venus Radar Mapper was already underway by the time the SSEC study began; the committee urged that it be carried out on its current schedule, leading to a launch in 1988.

The SSEC plan was off to a good start. By 1983 some commentators were saying that the fortunes of the planetary program had been turned around, to a significant degree, by the SSEC.[57] But this success had required lowered expectations as well as sophisticated use of the advisory committee mechanism.

THE TRAUMA OF 1981

Things had seemed to turn up in the Carter administration's last, "lame duck" budget, for fiscal year 1982, which included new starts on the Venus Orbiting Imaging Radar and on the Solar Electrical Propulsion System. However, the arrival of the Reagan administration in January 1981 appeared to pose a new threat to the planetary program. Budget Director David Stockman, who regarded the space program as discretionary, sought major cuts in the Carter administration's NASA budget request. The revised budget request slashed $468 million, including $196.5 million from planetary science. No new starts were proposed; the Venus Radar Mapper was deferred, and U.S. participation in the International Solar Polar Mission was not funded.[58] Bruce Murray testified on March 21, 1981 that the U.S. deep space program was in jeopardy and might even face extinction.[59] "If we hadn't had the launch vehicle problem, we could survive the budget problem," he said later, "but with both we are now in a desperate situation."[60]

During late 1981 the Reagan administration prepared its fiscal year 1983 budget request, the first that would be fully its own. Reportedly, NASA Deputy Administrator Hans Mark signed a memorandum in October 1981 that suggested that deep space missions might best be "de-emphasized until we have a space station that can serve as a base for the launching of a new generation of planetary exploration spacecraft."[61] An instruction from the Office of Management and Budget to NASA of November 1981 reportedly ordered NASA to virtually cease its planetary activities and called for the cancellation of Galileo and VOIR.[62] House Space Science and Applications Subcommittee Staff Director Darrell Branscome recalls, "It was widely known that the planetary program was almost eliminated."[63]

This crisis triggered a series of press stories with alarming titles, such as "Planetary Science in Extremis" in *Science*. In an editorial entitled "Bean-Counting the Solar System," *Aviation Week and Space Technology* wrote that "Planetary exploration is foundering under the Reagan Administration."[64]

According to *Time's* story "Clouds Over the Cosmos," Bruce Murray told worried JPL employees that he would seek new contracts outside the space area to keep the laboratory going. In response to the JPL's plight, the California congressional delegation organized to send a letter to the White House.[65] The crisis also sparked a new surge of organized activity by space scientists.

THE DIVISION OF PLANETARY SCIENCES AND THE NEW ACTIVISM

One response to this challenge was increasing activism by the American Astronomical Society's Division of Planetary Sciences (DPS). Founded in 1969 by about 30 scientists in this new field (one of the early members was O'Neill associate Brian T. O'Leary), the DPS initially followed the tradition of the much older American Astronomical Society of being very conservative about getting involved in politics. When the Pioneer Venus mission was in trouble in the mid-1970s, the DPS did very little because American Astronomical Society rules restricted lobbying. There was an informal activist group within the DPS at the time of the 1977 JOP crisis, and people in both Pasadena (JPL) and Tucson (Planetary Science Institute) participated in a telephone network. However, the DPS was not given autonomy in representing its interests to the government until later.[66]

The turning point was in the fall of 1981, notably at a meeting of the DPS in Pittsburgh in October. It was unusually well covered by the media, partly due to efforts by Richard Hoagland, then a reporter for the Cable News Network. There were calls at the meeting for a lobbying campaign and a growing awareness of the need for a long-range program of generating political support. A motion was passed authorizing the development of an action program. The DPS formed an alliance with the Planetary Society, a citizens group supporting planetary exploration. At a press conference addressed by DPS Chairman David Morrison and Planetary Society President Carl Sagan, it was announced that the two organizations were sending a joint letter to Presidential Counselor Edwin Meese to express deep concern about the future of planetary exploration.[67] According to planetary scientist Clark Chapman, contacts with the White House continued through December 1981.[68] The planned budget cuts reportedly were reduced by one third.

The DPS then shifted its attention to Congress, cranking up a multipronged effort including a telephone network and press releases. Continuing to stay in contact with the Planetary Society, the DPS subsequently worked in concert with other groups, including the SSEC. This effort succeeded in getting some funds restored. Since then, the DPS has focused more on individual line items, concentrating on funding for research programs rather than on new starts.[69]

The DPS has been assisted in its recent efforts by the American Astronomical Society itself, whose Executive Officer Peter Boyce has given the organization's headquarters office a more activist character since it moved to Washington in 1979. Formerly a very conservative professional organization, the AAS had become increasingly concerned about the job market for astronomers. Boyce stays in contact with administration officials and congressional members and staff and puts out a newsletter to AAS members describing events in Washington. Like other groups, the AAS encourages members visiting Washington to call on their congressmen, focusing on those on the relevant committees.[70]

THE SPACE SCIENCE WORKING GROUP

The fiscal 1983 budget for space science, in the opinion of *Astronomy* magazine, was a "disaster" for planetary exploration.[71] As *Science* reported, the ensuing uproar produced "a lot of bad vibes and a lot of good dialogue," and was "a catharsis that forced people to take a hard look at what they were doing and how much it cost."[72] The cuts also provoked the formation of the first permanent space science lobbying operation in Washington.

Seeing the Reagan administration's budget of January 1982 as a disaster for research scientists, Professor John Simpson of the University of Chicago called together a number of his colleagues to discuss the problem. Together they formed the Space Science Working Group, a unique combination of university faculty and governmental relations people. The Association of American Universities, a group of 52 research universities concerned mostly with research policy and graduate education, hired former Harvard University governmental relations consultant Geraldine Shannon to represent the SSWG and gave her office space in the AAU's Washington headquarters.

Shannon, whose cousin James Shannon was then a congressman from Massachusetts, quickly established an effective, low-key lobbying operation. Concentrating mostly on congressional staffers, she provides information on space science programs and escorts visiting experts to meetings on Capitol Hill. "Most members are predisposed to be sympathetic," says Shannon, "but are uninformed."[73]

In cooperation with the American Astronomical Society and its Division of Planetary Sciences, the SSWG conducts analyses of the administration's space science budgets and prepares recommendations after consulting with its members. The SSWG also testifies on space science budgets, and congressional committee requests for help in choosing witnesses on space science matters often come through Shannon. She and the American Astronomical Society's Boyce are invited to meetings of the Space Science Board.

Of the congressional staffers interviewed on this subject, everyone who was asked about the SSWG praised it as a very effective operation. Senate space committee staffer Stephen Flajser described the SSWG as one of the most impressive lobbying efforts in the whole science community, let alone the space community. According to him, the SSWG not only evaluates budgets and transmits its views to Congress but also comes back later to thank the staffers – one of the marks of a professional lobbying operation.[74] Two staffers emphasized that the SSWG was useful because it provides independent budget analyses on which the committees can draw; it is a helpful participant in the two-way process. The downward trend in space science funding has been reversed, and the space scientists now get a respectful hearing.

The SSWG also stays in close touch with NASA, sometimes getting half a loaf there and the other half from Congress. Yet SSWG's permanent Washington presence consists of one full-time professional supported by a secretary, with experts on call when needed. This efficient use of limited resources could be an example for other groups.

THE PLANETARY SOCIETY

Cutbacks in planetary exploration also led to the formation of the first citizens support group in the space science field. A significant reason for its success was the burgeoning reputation of Cornell University astronomer Carl Sagan as a science popularizer.

Sagan recalls that, in December 1977, he was invited to brief President Carter and Vice-President Mondale at the Vice-President's house in Washington, D.C. With Presidential Science Adviser Frank Press operating the slide projector, Sagan gave a presentation on planetary exploration that the fascinated Carter would not let him leave for other subjects. When Sagan pointed out that Carter himself could take action on behalf of planetary exploration, the President indicated that while he and Sagan understood the importance of the planetary program, the public did not. Sagan describes this as a "crystallizing insight," showing that political leaders need to be convinced that the public cares.[75]

Sagan was at the JPL in late 1978 and early 1979 to work as a principal investigator on the Voyager mission that flew past Jupiter. After conversations in which they contrasted the media attention and public interest Voyager was generating to the absence of new starts in the planetary program, Sagan and Bruce Murray decided to form a new organization, whose basic purpose would be to prove that planetary exploration was popular.

Murray contacted Louis D. Friedman, a JPL engineer then on leave to work as a staffer in the Senate, and asked him to do research on forming such an organization. Remaining in Washington through the summer of 1979, Friedman discussed this with a wide variety of people, including pro-space Senators Harrison Schmitt and Adlai Stevenson III, and John Gardner, founder of Common Cause. The reaction was positive. As Friedman puts it, he found that "the whole world was interested in space, not just scientists and engineers."[76]

Friedman returned to Pasadena in the fall of 1979, went on half time from the JPL, and set up an office in his home. In the meantime, Sagan had used the royalties from his book *Murmurs of Earth* to set up a foundation to support research in the field of Communication with Extraterrestrial Intelligence (CETI). Friedman asked the leading figures in the CETI Foundation (Frank Drake and Bernard Oliver) for a no-interest loan to get his new operation going and then began approaching other people for money. Early contributors included science fiction writers Isaac Asimov and Larry Niven, Joseph W. Drown, Michel Halbouty, and Charles Brush for the Explorers Club.[77] Meanwhile, Friedman was learning how to use direct mail techniques to build membership. The new organization, called The Planetary Society, was incorporated in December 1979.

A few months later, Sagan pointed out that he would be on the "Tonight Show" in April 1980. A decision was made to go public with the Planetary Society, and a press release was issued in May 1980. Meanwhile, letters were going out to potential donors. Scientific colleagues were invited to join in July 1980 and again in September, when the new society began a major direct mail operation.

During the summer of 1980, a decision was made to produce a quality magazine rather than the usual newsletter. The founders of the Planetary Society concluded that they were selling the excitement of exploration and that pictures would be an important medium. Charlene Anderson was hired as the editor. At the time this book was written, the *Planetary Report* remained the most professionally produced publication put out by a pro-space citizens group.

The new society, like some of its counterparts in the emerging citizens pro-space movement, had multifunctional aspirations. In the first issue of the *Planetary Report*, dated December 1980/January 1981, Sagan wrote, "If we are as successful as at least some experts think we are likely to be, we may be able to accomplish not only our initial goal of demonstrating a base of popular support for planetary exploration, but also to provide some carefully targeted funds for the stimulation of critical research."[78] In those early days, the range of subjects of interest to the society was sweeping:

Among the Society's most important objectives will be to dramatize and advocate the strongest possible case for planetary exploration and to focus broad popular support for the systematic development of programs such as solar sailing; planetary sample returns; a probe into the Sun; robot rovers on the terrestrial planets and on the Galilean moons of Jupiter; radar mapping of Venus; utilization of extraterrestrial resources, especially from the asteroids; probes beyond our solar system; the search for other solar systems, for extraterrestrial life and for galactic civilizations . . .[79]

The Planetary Society, starting from zero in July 1980, grew with spectacular rapidity. As of June 1984, the society had about 130,000 members and renewals were running at a rate unusually high for a voluntary membership organization. For a time, the Planetary Society advertised itself as the fastest growing membership organization in the United States in the past decade. In 1984, it was roughly equal in membership to all other pro-space groups put together.

Friedman agrees that much of this growth was due to the drawing power of Carl Sagan, whose television series "Cosmos" began appearing on educational television in September 1980. Exceptionally popular for a science program, "Cosmos" may have attracted as many as 15 million viewers. However, Friedman comments, "This is not just the Carl Sagan fan club."[80] The Voyager discoveries at Jupiter and Saturn probably were an additional stimulus. "Star Trek" creator Gene Roddenberry may have helped by encouraging "Star Trek" fans to join the new organization.[81] The Planetary Society also may have reached a new audience, possibly the science-oriented public described by Jon Miller.

The Planetary Society has continued to broaden the range of its activities. Its three-day Planetary Festival ("Planetfest") in Pasadena in August 1981, at the time of the Voyager 2 encounter with Saturn, drew an estimated 15,000 people and was one of the turning points in the growing self-awareness of the pro-space community. Consciously internationalist (reflecting the liberal political biases of its leaders), the society has opened offices in other countries and established an International Space Cooperation Fund. The society supports the search for extraterrestrial intelligence, specifically the observing program being conducted by Dr. Paul Horowitz of Harvard University with an 85-foot radio telescope in Massachusetts. The society has given

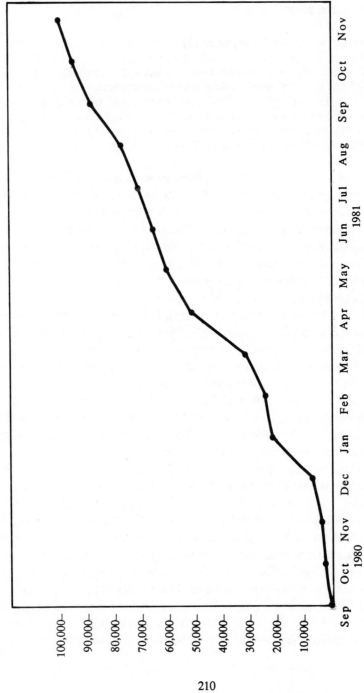

Figure 10.4 – Planetary Society Membership Growth, 1980-81 (*Source:* Adapted from "The Planetary Society Membership Count," in *The Planetary Report*, January/February 1982, p. 14. Reprinted with permission.)

money to the World Space Foundation, which is supporting research on asteroids.

In a brochure that appeared at the end of 1984, the society presented an impressive list of its projects:

Project Sentinel, the world's most comprehensive radio Search for Extraterrestrial Intelligence (SETI), using an advanced computer built with society funds and a Harvard University telescope refurbished by the Society

Search for planets around nearby stars

Continuing discovery of near-Earth asteroids that could become objectives for future manned or unmanned missions

Mars Institute: a multiuniversity study emphasizing identification and definition of requirements for a Mars base or future colony

Construction of an 8.4 million channel receiver and computer detection system (the most sophisticated yet to be built) for SETI in the Multichannel Extraterrestrial Assay (META)

Explorers maps to new worlds visited thus far only by spacecraft, with illustrations and descriptions of each world's new features

Curricula based on strong planetary exploration interest to improve the quality of science education for junior and senior high students

International conferences for scientists to discuss cooperation efforts among all nations – east and west, north and south

Student Getaway Special experiment using artificial "rings" to study planetary rings during a shuttle flight.

Production of Halley's Comet Amateur Observer's Bulletin to help coordinate worldwide activities prior to the comet's appearance in 1985-86.[82]

SUCCESSES AND FAILURES

The Planetary Society has been criticized by some other space activists for focusing its interests too narrowly on planetary exploration; it has not even done much to support space astronomy projects such as the Space Telescope. Yet its early promotional literature had some of the ring of other pro-space rhetoric, with one circular saying, "Among the struggles for minor political advantage and the ebb and flow of impoverishment and oppression here on Earth, this quest may provide a major aperture to a hopeful future."[83]

Some see the Planetary Society as another missed opportunity to build a single, general pro-space organization – an implicit parallel to the National Space Institute and the L-5 Society. "The Sagan thing could have become the nexus of the space movement if he had chosen to move in that direction," says Leigh Ratiner, "but he never politicized it." Ratiner comments that the Planetary Society, like some other pro-space organizations, was not willing to put its individual mission aside.[84]

The Planetary Society does not lobby as an organization and does not even maintain an office in Washington; Sagan notes that he personally has reservations about institutionalized lobbying.[85] However, its leading figures do make representations to administration officials and members of Congress (Sagan and Murray both did so before the society existed). Sagan, for example, met with Senator William Proxmire to encourage the reinstatement of NASA's SETI (Search for Extraterrestrial Intelligence) program. Sagan and Friedman testified before the Senate Foreign Relations Committee in 1984 on space cooperation with the Soviet Union. The society also encourages its members to contact members of Congress and other government officials about space issues.

The society's political efforts suggest the limitations of grass-roots activism without a professional lobbying arm. In June 1981, the society launched an appeal to members for funds to encourage a U.S. mission to Halley's comet and to finance a SETI project. This brought in over $70,000. Sagan signed a circular letter to the then 70,000 society members asking them to write or cable President Reagan about the Halley's comet mission. According to Sagan, this effort generated about 10,000 letters, which were weighed, boxed, and delivered to the NASA basement unread. He describes this as "a major failure."[86]

The society later stated in one of its circular letters that it had played a role in encouraging Congress to restore funds to the sharply reduced planetary research and spacecraft tracking budgets. Some of this new activism may have come close to being counterproductive. American Astronomical Society Executive Officer Peter Boyce recalls that letters about the budget, stimulated by the Planetary Society in the fall of 1981, caused an angry reaction in some quarters within the administration because negotiations on the budget were still underway and were regarded as confidential.[87]

Generally, however, the society's numbers give its leading figures credibility when they speak of popular support for planetary exploration. "The Planetary Society was created," says Spaceweek President Dennis Stone, "so that when Carl Sagan goes to Washington he can point to a hundred thousand members."[88] "Its political clout," says Sagan, "is merely by being in existence."[89]

By 1984, the Planetary Society was firmly established as the principal liberal voice in an increasingly politicized pro-space movement, perhaps the first major American pro-space organization to take a strongly liberal stance on space-related issues. In this sense, it counterbalanced other significant parts of the space movement that had taken on an increasingly conservative cast in the early 1980s. These other groups were divided from the Planetary Society by several old space policy issues. The most powerful divisive force was whether or not weapons should be placed in space.

11

EAGLES AND DOVES

Will space merely become an extension of our air, land, and sea arenas for fighting wars, or does the vacuum of space offer us a new breath of life – a place where all Humankind could breathe a little easier?

Leonard W. David, 1976[1]

The militarization of space is inevitable as the Sun coming up in the East.

Ben Bova, 1983[2]

Many space advocates have seen space as a "clean slate," an area from which earth-bound political and military rivalries could be excluded. "The crossing of space – even though only a handful of men take part in it," wrote Arthur C. Clarke in 1968, "may do much to reduce the tensions of our age by turning men's minds outward and away from their tribal conflicts."[3]

Despite this idealistic view, military interests have played a major role in space policy and space activity in both the United States and the Soviet Union since the beginning of the Space Age. The launch vehicles that put the first men and machines into space were designed as or adapted from military missiles. Space has been "militarized" for a generation, in the sense that some satellites have been used for military purposes. For the first two decades of the Space Age, however, the military uses of space by the United States were confined essentially to unmanned satellites that observed the Earth, gathered information, and relayed communications.

In the late 1970s and the early 1980s, military activity in space became an increasingly visible public policy issue in the United States because of a confluence of events, including a new space capability (the Space Shuttle), the maturing of anti-satellite and anti-missile weapons development programs started earlier, rising concern about Soviet military capabilities, and the arrival in 1981 of an administration more ready to speak out on the military uses of space technologies.

Although these were essentially defense and arms control issues, space advocates played interesting roles in advancing or opposing space-related defense concepts. By 1983, the question of weapons based in space or directed against objects in space had become the single most divisive issue for the American pro-space movement. It also held the potential of altering public attitudes toward the entire U.S. space enterprise.

THE MILITARY DETOUR

In *The Spaceflight Revolution*, William S. Bainbridge described how some early advocates of spaceflight, particularly in Germany, took a "military detour" to speed up the coming of the Space Age. "If Von Braun had not succeeded in building the V-2 rocket for Hitler," wrote Bainbridge, "the Spaceflight Revolution would probably have failed."4

Since the 1950s, some American advocates of spaceflight have supported increased military activity in space. In some cases, this was done out of genuine conviction that defense functions could be performed more efficiently with space systems. In others, it was because defense arguments gave added leverage to those seeking to expand human activity in space. The military detour has remained tempting, particularly in times of reduced political and public support for spaceflight.

The creation of NASA in 1958, and the Moon landing program, may have forestalled a military detour by many American advocates of spaceflight. By the late 1970s, however, frustration with the unwillingness of the Ford and Carter administrations to enunciate new space goals or to approve major new civil space programs led some American space advocates to consider a new military detour. This coincided with growing interest in the potential of defensive space systems and with a campaign within the U.S. Air Force by space advocates who wanted a separate Space Command.

The question of military activity in space was given a higher public profile in the early 1980s for a variety of reasons. By far the most important was a paradigm shift as bold as that of Gerard K. O'Neill's space colonies: the Reagan administration's endorsement of the concept of using weapons systems in space as a possible means to defend against ballistic missiles. This was done primarily for defense reasons, but space advocates were very much involved, with some seeing this as a means of speeding space development and space humanization.

THE NEW HIGH GROUND

Early in the Space Age, some argued that space offered a new military high ground. Wernher von Braun, transplanted to the United States, was arguing as early as 1952 that an orbiting station equipped with nuclear missiles would allow the United States to dominate the Earth.[5] In the 1960s, American space visionary Dandridge M. Cole described his "Panama theory," in which control over certain positions in space (such as geosynchronous orbit) would give a nation a decisive military advantage. Cole also suggested that a captured planetoid in orbit around the Earth could be the ultimate deterrent.[6] In his 1970 book *War and Space*, aerospace engineer Robert Salkeld proposed that the United States should deploy strategic weapons, and their command centers, in space to solve the then foreseeable problem of land-based missiles becoming vulnerable to preemptive attack.[7] (During the 1960s, the Soviet Union actually tested a nuclear Fractional Orbital Bombardment System that, in wartime, would complete most of an orbit around the Earth before descending on its target.)

The U.S. Air Force has long contained a school of thought that argues that space is a new operational medium with unique characteristics, requiring a separate doctrine and a separate command structure, and possibly the creation of a U.S. Space Force. The resistance these arguments have met remind many of the early resistance to the idea of a separate Air Force with its own doctrine.[8]

Early in the Space Age, the Air Force had plans for an aerospace plane called the X-20 Dyna-Soar, to be launched into orbit atop a large rocket to perform military missions. There also were plans for a military space station called the Manned Orbiting Laboratory (MOL). One of the military astronauts chosen for the mission was James Abrahamson, who later headed the Space Shuttle program and the office responsible for the Strategic Defense Initiative. However, both projects were cancelled and were succeeded indirectly by NASA systems: the temporary civilian space station Skylab and the joint civil-military Space Shuttle.[9] As of 1984, the Defense Department was disclaiming any interest in a space station, but the aerospace press was carrying stories about military interest in a possible "Trans-Atmospheric Vehicle."[10]

In April 1981, as the Space Shuttle was about to fly its first mission, the United States Air Force Academy hosted a symposium on military space doctrine. This event was a response to a challenge hurled by then Secretary of the Air Force Hans Mark when he visited the academy in January 1980. Thomas Karas, who reported on the symposium in his book *The New High Ground,* saw it as an important symbolic moment for Air Force space advocates led by General Bernard Schriever (USAF, retired).[11] That same year, Colorado Congressman Ken Kramer introduced a bill calling for a U.S. Aerospace Force. In September 1982, under the more receptive Reagan administration, advocates won the creation of a USAF Space Command, which is located at Colorado Springs.

A Consolidated Space Operations Center, which also will be located at Colorado Springs, is scheduled to become operational in 1986. In late 1984, President Reagan decided to form a combined, multiservice Space Command (also located in Colorado Springs), which was to be operational by October 1, 1985.[12]

In the early 1980s there was growing media concern about the alleged "militarization" of NASA.[13] Because the Shuttle is used both for civil and military missions, its introduction has to some extent blurred the long-standing separation between civil and military space systems. "The Shuttle is bringing high visibility to the military role in space," said a senior NASA official in 1982.[14] The first all-military Space Shuttle mission took place in January 1985, amid controversy about press stories concerning the satellite it was to place in orbit.[15] Meanwhile, a new Space Shuttle facility has been under construction at Vandenberg Air Force Base on the California coast to allow launches into polar orbits, with the first Shuttle launch planned for 1986.[16] By the late 1980s, the United States should have well-developed organizational structures and physical facilities for expanded military operations in space.

The Shifting Balance

When most Americans thought of space in the 1960s, they thought of the civilian space program, which then was not only more visible but much larger in budgetary terms; military space spending never approached a comparable peak. Trends began to change during the Carter administration in the late 1970s, when defense spending turned upward while NASA funding stayed approximately level in real terms. The trend lines crossed in fiscal year 1982; for the first time since 1961, the United States officially spent more on its defense space programs than on NASA (Figure 11.1). All reports in the open literature suggest that the gap has continued to widen since, a factor in stimulating the debate about military activity in space.

The Anti-Satellite Question: A Little-Noticed Issue

For most of the groups that made up the post-1972 American space interest movement, military activity in space was not a front-burner issue until after the arrival of the Reagan administration in January 1981. Yet Paul Stares reports in his book *The Militarization of Space* that the United States tested an anti-satellite (ASAT) weapon as early as 1958, and President Johnson announced in 1964 that the United States had the means to destroy satellites.[17] The Soviet Union reportedly began testing an ASAT system in 1967.[18]

Generally, the subject received little attention in the media until late 1975, when press reports claimed that the Soviet Union might have illuminated U.S. satellites with a ground-based laser.[19] (According to some reports, the United States also was doing research on the possible use of lasers as anti-satellite weapons.)[20] In 1976, the Soviet Union resumed testing of an ASAT system of the "co-orbital" variety, designed to place an explosive warhead in an orbit that would bring it near the target satellite. This caused some alarm in Washington.[21]

In October 1977, Secretary of Defense Harold Brown called the Soviet ASAT system "operational."[22] Following up on one of the last decisions of the Ford administration, the Carter administration decided to move forward on two tracks: the United States would propose discussions with the Soviet Union on possible ASAT arms control measures but would proceed with the development of its own ASAT system.[23] The American ASAT was an aircraft-launched, direct ascent

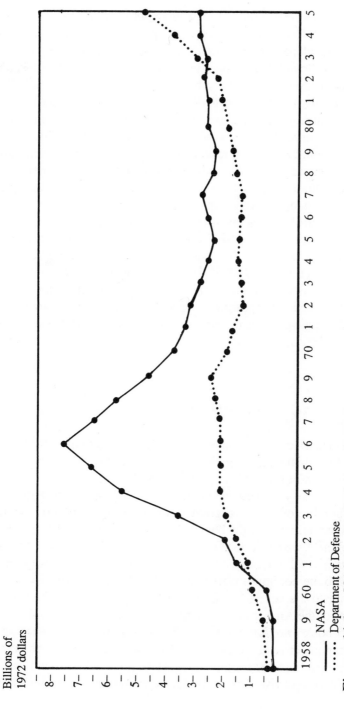

Billions of
1972 dollars

Figure 11.1 – NASA and Defense Outlays on Space Activities (*Source*: Adapted from data in *Aerospace Facts and Figures 1985/86* [New York: Aviation Week and Space Technology, 1985]. Used with permission.

NASA
Department of Defense

system, whose small nonexplosive warhead would seek out and crash into the target. Reportedly, this Homing Intercept Technology (HIT) was originally developed for an anti-ballistic missile system.

U.S. and Soviet delegations first met to discuss ASAT arms control in Helsinki in June 1978, with subsequent meetings in Bern and Vienna in early 1979. According to Robert Buchheim, who headed the U.S. delegation during most of these sessions, "both sides were much closer to an ASAT agreement than was publicly known."[24] Whether final agreement could have been reached on an ASAT arms limitation treaty became a moot point when these talks, along with others, were suspended due to deteriorating U.S.-Soviet relations, particularly after the Soviet invasion of Afghanistan in December 1979.

Arms control groups not involved in space advocacy did a limited amount of work on the ASAT question, and the Stanley Foundation included it as a subject for discussions at its "Strategy for Peace" conferences in the late 1970s.[25] Pro-space groups seemed to take almost no notice of this issue in their publications during the 1970s, with only a few scattered articles appearing.[26] Space interest groups became actively involved in the space weapons issue not because of ASAT systems but because of the paradigm-breaking idea of basing anti-missile systems in space.

DEFENSE AGAINST MISSILES

The idea of defending against missiles was not new. According to Benson D. Adams, thinking about the problem began as early as 1944, when German missiles hit targets in suburban Paris.[27] But most early thinkers saw no way to escape vulnerability to these new weapons. "The flying bomb portends as great a revolution in warfare as those successively effected by the bow and arrow, the musket, the cannon, and the airplane," wrote Major General J. E. C. Fuller in 1944. "Possibly greater, for as all those inventions aggravated Man's propensity for war, it seems to me not improbable that this winged projectile may at length bring him to his senses."[28] G. Edward Pendray, one of the founding members of the American Interplanetary Society, wrote the following in his 1945 book *The Coming Age of Rocket Power*: "No defense against such weapons has yet appeared. Probably no defense will be possible, except preparation for prompt and instant retaliation in kind and with greater power, provision of shelter underground for civilian populations, principal manufacturing plants and centers of command and communication."[29]

A serious U.S. ballistic missile defense program was underway by 1955, and the Soviets probably were pursuing a parallel course by that time.[30] The United States had no deployed ballistic missile defense system until 1972, although it did have a Nike-Zeus spinoff with some ASAT capability.[31] However, using nuclear warheads had serious disadvantages, including damage to friendly space systems and to electronic equipment because of the electromagnetic pulse created by nuclear explosions. There also were doubts about the ability of anti-ballistic missile (ABM) systems then feasible to defend against a major missile attack.

A long debate on ABM systems took place in the United States in the late 1960s and early 1970s. In 1972, the United States and the Soviet Union signed a treaty limiting the ABM systems of each side. In effect, this codified a situation that American defense planners called mutual assured destruction (MAD). Each side remained vulnerable to an attack by the other. Some defense intellectuals argued that deterrence would be maintained because an attacker would be hit with a devastating second strike in retaliation.

MAD was disliked by many Americans for various reasons. Faced with the possibility of a nuclear holocaust, some called for dramatic cuts in the number of nuclear weapons, or even their abolition. Others argued that it was unnatural for a nation to surrender the right to defend itself against attack and pressed for the deployment of an ABM system. Among the latter were several space advocates.

Space Advocates, Beam Weapons, and the Space Based ABM

"Death rays" had been a feature of science fiction for a long time before the laser was invented in 1959. (They may have first appeared in H. G. Wells' *The War of the Worlds*.) The laser, which generated a coherent beam of light, held the potential of being an entirely new kind of weapon, which could damage a distant target almost instantaneously. There also was speculation about weapons using beams of subatomic particles, like those used in accelerators for scientific research. It did not take long for thinkers to make a connection between these new technologies and defense against ballistic missiles.

In his 1958 book *Spaceflight – A Technical Way of Overcoming War*, space visionary Eugen Sanger proposed that the problem of destroying ICBM warheads might be solved through the use of energy rays.[32] In the November 1965 issue of *Astronautics and Aeronautics*, physicist Arthur Kantrowitz (later chairman of the board of the L-5 Society and a member of the Citizens Advisory Council on National Space Policy) wrote the following:

> I believe that before we can balance the enormous power which the offense now has in nuclear weapons, we will have to resort to something as revolutionary as radiation weapons which are capable of striking at great distance with the speed of light or thereabouts. It seems to me entirely possible that radiation weapons will be of more importance in the space environment than on Earth.[33]

In 1969, Stefan T. Possony and Jerry E. Pournelle noted the possibility of a space-based laser ABM system in their book *The Strategy of Technology*.[34] Pournelle, a space advocate as well as a science fiction writer, later became a leading figure in the L-5 Society and the chairman of the Citizens Advisory Council on National Space Policy. However, neither Kantrowitz nor Pournelle pursued the idea as far as a Lockheed Corporation engineer named Maxwell W. Hunter II.

In 1966, Hunter had published a book called *Thrust into Space*, in which he speculated that expanding the basic operations of the human race throughout the solar system could well create a new Renaissance. "Space is the one place," he added, "where we can obtain natural resources without damaging either the earth's ecological balance or its natural beauty."[35] This was a decade in advance of the O'Neill/L-5 phenomenon.

Hunter, who had helped design the Nike-Zeus ABM system, has always disliked MAD. In 1984, he described it as "illegal, immoral – and fattening."[36] According to Hunter, he started thinking about space-based ballistic missile defense in 1967 and began serious formulation in 1970. After doing initial work at Lockheed, he and his colleagues briefed officials of the Defense Advanced Research Projects Agency, the U.S. Air Force, and the U.S. Army, and got a contract in 1971 to do further work. In Hunter's phrase, the idea "percolated through the system" from 1971 to 1977. Hunter comments, "You cannot hasten the process of getting people to accept an idea"[37] – an interesting contrast to O'Neill and his space colonies.

In 1977, Hunter learned that an introduction he had written for Lockheed proposals on this subject had been declassified. Realizing that it could make a provocative unclassified paper, he rewrote it and in early 1978 began privately circulating a paper entitled "Strategic Dynamics and Space-Laser Weaponry."[38] Challenging MAD, Hunter pointed out that high energy lasers were proliferating and that space transportation was about to become sufficiently economical that, if it were used to place such lasers in space, an effective defense against even massive ballistic missile exchanges would be possible. "This," wrote Hunter, "is the only new strategic concept to present itself in a number of decades, and the only one which merits the words . . . potentially decisive." Hunter expected his ideas to draw fire both from arms controllers and from advocates of offensive systems.

Hunter's paper was photocopied and passed around and was mentioned in *Newsweek* in February 1978.[39] In the summer of 1979 it got into political channels when Hunter met Angelo Codevilla, staff aide to Senator Malcolm Wallop of the Senate Select Committee on Intelligence.[40] Codevilla arranged for Hunter to meet Wallop, who became a strong and highly visible advocate of the space-based ABM. Drawing on Hunter's work, Wallop wrote an article on the subject in the fall 1979 *Strategic Review*.[41]

Invited to give briefings to other senators, Hunter put together a team of industry experts that became known as "the Gang of Four": Joseph Miller of TRW Incorporated, Norbert Schnog of Perkin-Elmer Corporation, Gerald Ouelette of Draper Laboratories, and Hunter himself. This group briefed a number of senators during late 1979. The heart of their presentation was the argument that chemical laser weapons using known technology could be deployed in space in the 1980s, offering a near-term way to have an impact on the strategic situation.

The issue broke into the open in the spring of 1980, when Wallop and Congressman Ken Kramer of Colorado advocated increased funding for research into directed energy weapons that could be used in space-based ABM systems. Some defense officials took sharp exception to the views of Hunter and his colleagues, and reportedly asked their employers to make the Gang of Four ease off. *Aviation Week* later singled out the Gang of Four for praise for risking their professional reputations and resisting Pentagon pressure in telling Congress that U.S. industry had the capability to do the job.[42]

Meanwhile, another space advocate named John D. Rather was pursuing a parallel track toward a similar goal. Rather worked for 15 years as an experimental physicist at Kitt Peak National Observatory near Tucson, Arizona, at Oak Ridge Laboratories in Tennessee, and at Lawrence Livermore Laboratories in California (the last being a major center of directed energy research). In 1974, having changed to a career of "think-tanking," he worked on a major Department of Defense study in response to Secretary of Defense James Schlesinger's call for a long-range research and development program to find a way to "defuse nuclear war." After writing the unifying final report on new technologies, Rather put together a "white paper" on the military uses of lasers. In 1975, he was asked by the Air Force to participate in the definition of future laser weapons.[43]

Rather comments that he always had been intrigued by space and was looking for a way to get humans into space on a wholesale basis – a philosophy much in agreement with the ethos of the new pro-space movement. When satellite solar power stations began to attract attention in the early 1970s, Rather concluded that energy could be transmitted to Earth more efficiently by laser than by microwave.

In December 1976, when he was with a small firm in the Washington, D.C. suburb of Arlington, Virginia, Rather distributed an unclassified paper entitled "Space Transportation, Solar Power from Space, and Space Industrialization: A Better Way."[44] Clearly stimulated by O'Neill's ideas about space colonies, in the paper he attempted to show that the development of space as a natural resource could be accomplished on a time scale much shorter than most people would expect to be economically possible. Ground-based lasers used to energize spacecraft propulsion systems could dramatically reduce the cost of getting into orbit (this idea had been suggested by Arthur Kantrowitz in 1972). High energy lasers in space could provide the best means for the distribution of electric power derived from solar energy and could be used to propel vehicles around the inner solar system. Although this paper did not advocate a space-based ABM, Rather noted that the devices could defend themselves against attack by focusing laser energy at long range.

During the latter 1970s, Rather became increasingly active in advocating laser ABM systems using technologies more advanced than the chemical laser. Through political contacts, notably the Alabama congressional delegation led by Senator Howell Heflin, Rather proposed hearings on lasers, which were held by the Senate Subcommittee on Science, Technology, and Space in December 1979 and January 1980.

There Senator Wallop testified on behalf of the Hunter school of thought and Rather in favor of newer technologies.[45] The subsequent political battle, in which Wallop was supported in the House by Kramer, resulted in more funding for work on laser weapons. Although the emphasis on chemical lasers continued, there also was growing support for more innovative approaches, particularly after 1982 when House Armed Services Committee staffer Anthony Battista took up the cause.

Other space advocates were in the picture by the late 1970s, although they did not play as direct a role. H. Keith Henson had initiated a discussion of a space-based defense system with his article "Military Aspects of SSPS Power" in the May 1976 *L-5 News*, before either Rather or Hunter had circulated his paper. Writing that ground-based lasers for ballistic missile defense had been under study for several years, Henson pointed out the advantages of basing such weapons in space and noted the possibility of converting a satellite solar power station into a defensive system.[46] Two months later, a letter to the *L-5 News* raised the other side of the issue: "Extensive discussion of the military uses of space colonization in the News Letter may create the impression that the Society is advocating this possibility. I am sure that most members will not be favorable to this use."[47]

The debate has continued intermittently in the pages of the *L-5 News* ever since. Meanwhile, the National Space Institute, which always has supported the use of space technology for national defense, took a much lower profile, carrying only three articles on military activity in space in its newsletter between 1975 and 1980.

Space Forts

Having linked space weapons to the proposal for satellite solar power stations, some advocates of O'Neill's vision also linked them to the mining of the Moon and the asteroids. Responding in the June 1979 *L-5 News* to arguments that space-based lasers would be vulnerable to attack, Keith Henson proposed space forts made of extraterrestrial materials.[48]

Carolyn Meinel, initially opposed to weapons in space, changed her views after conversations in 1978 with Maxwell W. Hunter and John Rather. She, too, became an advocate of space-based defense using directed energy weapons and pursued the subject of space forts made of nonterrestrial materials in her own writings.[49]

A conference on "Defense Applications of Near-Earth Resources" was hosted by the California Space Institute in August 1983. This was in response to a request from former NASA Administrator James Fletcher, then heading a group studying the prospect for President Reagan's Strategic Defense Initiative. Among those who participated were Maxwell Hunter and L-5 Society activist J. Peter Vajk.[50]

The Reagan Opportunity

By the time the Reagan administration took office in January 1981, the subject of directed-energy ABM systems had been discussed in press stories for years. *Newsweek* reported, in 1976, that "scientists are looking into the practicality of incorporating lasers into U.S. anti-ballistic missile defenses."[51] Retired U.S. Air Force General George Keegan made news in 1977 with allegations that the Soviet Union was ahead of the United States in developing a particle beam weapon.[52] A year later *Aviation Week and Space Technology* Defense Editor Clarence A. Robinson, Jr., initiated a series of major articles on beamed energy weapons.[53] By July 1980, that magazine was reporting a Defense Department assessment that chemical lasers based in space could destroy ballistic missiles during launch, although the decision to use these as part of a mixed ballistic missile defense force was not expected until 1985-87.[54] Senator Harrison Schmitt reportedly said in May 1981 that President Reagan fully recognized that this technological revolution was providing new strategic policy options that "will in the not-too-distant future make weapons of mass destruction obsolete."[55]

Advocates of space-based defense systems, including some pro-space activists, seized on the opportunity presented by the pro-defense Reagan administration. The first report of the Citizens Advisory Council on National Space Policy devoted a section to space-based beam weapons, arguing that "strategic-scale war in the closing sixth of this century is likely to conclude with the total and quite bloodless triumph by the nation owning the space laser system."[56] However, the first group founded primarily to advocate space-based weapons came from different origins.

HIGH FRONTIER

The ABM Treaty and the Strategic Arms Limitation Talks (SALT) agreements limiting strategic arms have been thorns in many conservative sides, particularly as Soviet offensive capabilities increased. During the Carter administration, some conservative activists were seeking to develop alternative defense strategies that could be put into effect when the Republicans regained the White House. The core of this effort was the Strategic Alternatives Team.[57]

Many conservatives believed that the development of U.S. technology had been throttled during the 1960s, and most fundamentally disagreed with MAD, believing that the United States should deploy defenses against the strategic weapons of the Soviet Union. Instead of simply refining existing systems, some saw a need for a "technological end run." By 1981 there was a natural conjunction between a new emphasis on ABM systems, where little was being done in their opinion, and on space, which critics of the Carter administration's defense policies saw as an area of U.S. advantage.[58]

One of those who became intrigued by the idea of space-based ABM systems was Lieutenant General Daniel O. Graham, who had retired from the U.S. Army in 1976 after serving as Director of the Defense Intelligence Agency. Becoming Director of Special Projects for the conservative American Security Council, Graham also served as a national security adviser to Presidential candidate Ronald Reagan.[59]

After Reagan was elected in November 1980, Graham and retired Air Force General Robert Richardson put together a study team under the auspices of the conservative American Security Council Foundation, with funding from corporations and wealthy individuals. Industry experts gave help and advice. After a five-month study, the group concluded that a space-based ABM system was feasible and that early deployment would be possible if kinetic energy (impact) technology were used.

Graham revealed some of his thinking in the spring 1981 issue of *Strategic Review*, where he argued for a shift by the United States from the race with the Soviet Union in offensive capabilities to a thrust toward defensive capabilities, specifically a space-borne defense against ballistic missiles. Graham noted a potential symbiosis between defense deployments in space and solar power satellites, saying that his proposal would encourage and support the exploitation of space for the solution of another key strategic problem – energy. Graham thereby linked strategic defense advocacy with space development advocacy. In what proved to be a prescient comment, he wrote:

> The chances are that any public debate over this option would find most Americans in favor of the space option. Some public support would be based on a well-founded displeasure with a business-as-usual approach to

defense. . . . But much would arise from an excitement of the public imagination regarding the potential of space, as has been engendered already by highly successful futuristic books and films.[60]

Graham's study group evolved into a project called High Frontier, perhaps unwittingly borrowing the name of Gerard O'Neill's conception. Sponsorship was transferred to the conservative Heritage Foundation. The "steering group" for High Frontier included four members of President Reagan's "kitchen cabinet," Dr. Edward Teller, and Presidential Science Adviser George A. Keyworth as a "White House observer."[61] Lowell Wood later described Teller as a "major participant" in the steering group.[62] Meanwhile, the Reagan administration had released a policy statement including increased research and development on anti-ballistic missile systems as part of its plan to improve the strategic posture of the United States.[63]

Frank Barnett of the National Strategy Information Center provided High Frontier with an opportunity to present its ideas to an invited audience. One of those present, former Secretary of the Army Karl Bendetsen, spoke to Graham about his enthusiasm for the concept, and "contributed substantially to the ongoing momentum of the project and to definition and consensus."[64]

In March 1982, the group went public and published a report titled *High Frontier: A New National Strategy*. Described as a project of the Heritage Foundation, this book argued strongly for a layered strategic defense system that would be deployed in phases. The initial space-based system of 432 satellites would use kinetic energy weapons similar to the U.S. ASAT system then under development (a later High Frontier study by John Bosma showed that a similar system had been under consideration as early as 1959).[65] Other layers, added later, might include directed energy weapons and terminal defense around missile silos. The study's other proposals included a high performance space plane, improved space transportation, and a space station. The defense concepts were linked loosely to space industrialization, primarily solar power satellites. Jack Manno later wrote in his book *Arming the Heavens* that this offered something to almost every part of the aerospace constituency.[66]

Several well-known space advocates were involved in the High Frontier study, including Jerry Pournelle (who says he wrote the preface), solar power satellite inventor Peter Glaser, and Peter Vajk, who contributed to the economic section. Another space activist who took part was John Bosma, who co-founded the Congressional Staff Space Group in 1981, married Carolyn Meinel, and became editor of the Washington-based newsletter *Military Space.*

High Frontier got the support of the Conservative Caucus and money from conservative donors to, as General Robert Richardson puts it, "take the idea on the road" to determine its political feasibility and to build a constituency for it.[67] The organization received a largely positive reaction and found that the ABM treaty was a "non-issue." Graham, who personally has campaigned tirelessly for the concept, is fond of pointing out that most Americans were surprised to learn that they were not already protected by some kind of ABM system.

This was occurring at a time of revived public concern about the danger of nuclear war, expressed most visibly by the nuclear freeze movement, Jonathan Schell's book *The Fate of the Earth,*[68] and the television drama "The Day After." In addition, there was serious expert concern that the vulnerability of U.S. and Soviet land-based missiles could encourage movement toward "launch on warning" postures or even encourage a preemptive first strike in a crisis. The desire for a solution to the dilemma posed by strategic nuclear weapons was strong. In 1984 former National Security Adviser Brent Scowcroft wrote, "What seems to be emerging in the United States is a reaction at both ends of the political spectrum against deterrence and the despair which in the current situation it tends to promote."[69]

The "Star Wars" Speech

There have been several published accounts of how President Reagan came to endorse accelerated research into strategic defense; most suggest that it was the culmination of a process that had gone on for some time.[70] Yet despite the numerous press reports that had appeared over the years about beamed energy weapons, despite revived public advocacy of ABM systems and the appearance of High Frontier, most observers were startled on March 23, 1983, when President Reagan said at the end of a defense policy speech:

> Let me share with you a vision of the future which offers hope. It is that we embark on a program to counter the awesome Soviet missile threat with measures that are defensive . . . what if people could live secure in the knowledge that their security did not rest upon the threat of instant U.S.

retaliation to deter Soviet attack; that we could intercept and destroy strategic ballistic missiles before they reached our own soil or that of our allies?[71]

Reagan called on U.S. scientists to "turn their great talents now to the cause of mankind and world peace to give us the means of rendering these nuclear weapons impotent and obsolete." Although the President did not mention space-based weapons, press briefings given by officials later made clear that such systems, particularly those using directed energy, were a major option under consideration. This led critics in the media to label the concept "Star Wars."

Two days later, the President directed an intensive effort to define a long-term research and development program aimed at the ultimate goal of eliminating the threat posed by nuclear ballistic missiles. Committees of experts established to do preliminary studies issued reports in the fall of 1983 that were essentially positive, although they foresaw that the effort would take many years.[72] The President and his national security advisers reportedly decided on November 30, 1983, to proceed with an expanded program of research and development.[73] The Strategic Defense Initiative program was set in motion formally on January 6, 1984.[74] Lieutenant General James Abrahamson, formerly in charge of NASA's Space Shuttle program, was made Director of the Strategic Defense Initiative Organization in March 1984.[75] Secretary of Defense Caspar Weinberger wrote in a report issued the following month that "recent advances in technology offer us, for the first time in history, the opportunity to develop an effective defense against ballistic missiles."[76]

Stimulating the Advocacy

All this gave the High Frontier organization a huge boost, even though the technological approach it preferred did not top the list. High Frontier became a membership organization, using direct mail techniques to attract large numbers of average citizens; it grew from 13,000 or 14,000 in November 1983 to about 40,000 subscribers in early 1984 and published a newsletter.[77] Polls sponsored by conservative organizations showed that most of the respondents favored strategic defense.[78]

Devoting much of its energy to supporting and defending the President's decision, High Frontier frequently called on its subscribers for contributions or for letters and postcards on strategic defense issues. In December 1984, Graham urged subscribers to support Secretary of

Defense Weinberger's position and to oppose proposals that would make the Strategic Defense Initiative a "bargaining chip" in U.S.-Soviet arms control talks.[79] One mailing, announcing the formation of "Americans for the High Frontier," provided preprinted postcards addressed to specific senators by the state of the contributor.

However, High Frontier encountered the kinds of limits on political action by nonprofit organizations that had caused the L-5 Society to establish Spacepac. Graham decided to create a political action committee (PAC) called the American Space Frontier Committee. He asked former Congressman Robert Dornan (sometimes known as "B-1 Bob" because of his advocacy of the bomber produced in his district) to head the organization, and Dornan took the job in July 1983. Dornan notes that he was in touch with Jerry Pournelle, who had helped him in his 1970 campaign.[80] The first executive director of the new PAC, during July and August of 1983, was former L-5 Society leader Carolyn Meinel, who sought to make it a broad-based, nonpartisan organization. In her opinion, however, the group was taken over by the "New Right," and she departed.[81] People on High Frontier's mailing list received direct mail solicitations from conservative organizations such as the National Conservative Political Action Committee.

In September 1983, a breakfast reception was held in Washington, D.C., to announce the formation of the new PAC. Those invited were almost entirely well-known conservatives, including Clare Booth Luce, Jerry Falwell, Jesse Helms, Malcolm Wallop, Phyllis Schlafly, Jack Kemp, and "kitchen cabinet" member Justin Dart. Invitees with a strong space advocacy connection included Robert A. Heinlein and Newt Gingrich.[82] According to Robert Dornan, the PAC's first priority as of the spring of 1984 was to support the President's decision, regardless of which technology is chosen to implement it. (Dornan admitted that the economic dimension of High Frontier has not progressed.[83]) The American Space Frontier Committee sent out a mailing before the 1984 elections listing those members of Congress who had supported the High Frontier idea.[84]

The administration got strong support for its initiative from some conservative members of Congress, notably Senators Wallop; Paul Laxalt, of Nevada; Jake Garn, of Utah; and William Armstrong, of Colorado. In May 1983, Congressman Ken Kramer of Colorado introduced a "People Protection Act," which called for the creation of a directed energy systems agency, a military Space Shuttle fleet, a unified Space Command, a manned space station program, and consideration of new arms control regimes using strategic defenses. Armstrong introduced a parallel measure in the Senate. Newt Gingrich also was supportive, including the idea of space-based ABM in his 1984 book *Window of Opportunity*.[85]

The Strategic Defense Initiative also drew strong support from the Fusion Energy Foundation and from the publication *Executive Intelligence Review*, both of which are associated with the Presidential candidacy of Lyndon LaRouche. These organizations, which had begun campaigning for a space-based missile defense system in May 1982 (possibly in response to the High Frontier report), emphasized the economic as well as the military benefits of the beamed-energy research that would be conducted. One of their slogans was "Beam the Bomb."

The Citizens Advisory Council on National Space Policy met in August 1983 to draft a reaction to the President's initiative and to reconcile differences among advocates of different technical approaches to strategic defense such as Maxwell Hunter, Daniel Graham, and Lowell Wood of Livermore Laboratories. The report, entitled *Space and Assured Survival*, concluded that "The President's proposal to change the defensive posture of the United States from Mutual Assured Destruction to Assured Survival is morally correct, technologically feasible, and economically desirable."[86] John Rather, speaking at a seminar at the Heritage Foundation in April 1984, said, "we'll stake our reputations on the fact that it's possible to do it soon."[87]

The Strategic Defense Initiative also stimulated more expressions of opinion from citizens space advocates in their own publications. The *L-5 News* carried an article by General Graham in December 1983 and a rebuttal by David Webb the following month. Jerry Pournelle added his own views a month later.[88]

Meanwhile, technological developments based on long-standing research programs achieved increased public visibility. The United States conducted the first test of its new ASAT system in January 1984.[89] In June of that year, the U.S. Army conducted its first successful interception of a dummy missile warhead under its Homing Overlay Experiment, reportedly capping a six-year, $300 million program. A spokesman reportedly said, "We hit a bullet with a bullet."[90]

By the end of 1984, strategic defense advocates still held the initiative in the ongoing debate. By most accounts, the Strategic Defense Initiative had been effective in defusing the nuclear freeze movement and was an important factor in bringing the Soviet Union back to the arms control negotiations in Geneva.[91]

THE ARMS CONTROL RESPONSE

Judging by the open literature, established arms control groups showed relatively little interest in space weapons during the 1970s. Occasional articles and papers appeared, notably during the brief surge of media interest in the ASAT question and directed energy weapons during 1977 and 1978.[92] However, most liberal intellectuals appeared not to take space-related issues seriously.

As in the past, criticism of military activity in space often was led by individual scientists, some of whom also were critics of the manned space program. Pro-arms control groups with a scientific base, such as the Federation of American Scientists and the Union of Concerned Scientists, began to get more interested after the Reagan administration took office in January 1981. Several scientists and weapons experts issued a statement opposing laser weapons in space. *Scientific American* became an important medium for critiques of space-related weapons, particularly by Massachusetts Institute of Technology physicist Kosta Tsipis.[93] However, criticisms tended to reflect distaste for weapons buildups rather than interest in space exploration or development.

The Progressive Space Forum

The first pro-space group to campaign actively against space weapons appeared before groups actively supporting space defense, growing out of a confluence of leftist politics and the California space group boom described in Chapter 8. Jim Heaphy, who had been active in the anti-war movement in the late 1960s and early 1970s, discovered the space movement in the San Francisco Bay area in January 1978 through *Space Age Review*, a publication put out by pro-space and pro-peace activist Steve Durst. An article about Space Day 2, to be held in Sacramento in April 1978, got Heaphy involved in the rally, where he spoke on the dangers of an arms race in space. Heaphy, by then a member of the L-5 Society, saw disarmament as an unfilled niche in the pro-space movement.[94]

Meetings held in the summer of 1979 in connection with preparations for Space Day 3 led to the formation of a group called Citizens for Space Demilitarization (CFSD). In January 1980, CFSD published the first issue of *Space for All People*, a quarterly newsletter that remained in publication as of 1985. An editorial by Heaphy in the June 1980 issue described CFSD as a bridge between two movements – that favoring space development and that opposed to the arms race. Noting that the L-5 Society and other space groups were trying hard to establish closer ties with the aerospace industry and American business in general, Heaphy wrote that CFSD would like to see the space movement build bridges to trade unions as well. Mentioning that CFSD included several active environmentalists, Heaphy commented that "we also want to promote the concept of democracy in the economy."[95] Heaphy later said that CFSD tried to reach out to other pro-space groups but often encountered a wary response.

Placing advertisements in leftist papers, CFSD got a response from young Georgia space enthusiast John Pike, then involved with Barry Commoner's Citizens Party and with the anti-draft organization CARD. Pike formed a CFSD chapter in Atlanta and later went on to become a key player in the space arms control effort.

In 1981, the group changed its name to Progressive Space Forum (PSF), partly so that the public would more readily realize that the group had things to say about the social benefits of the peaceful uses of space technology.[96] Heaphy circulated a draft PSF program that favored, in addition to space arms control measures and a nuclear freeze, an international civilian space station, a passenger module for the Space Shuttle, space science programs, and active recruitment and training of civilians of all nations to serve as Space Shuttle pilots and crew members.[97]

After the election of Ronald Reagan, the PSF shifted toward a stronger emphasis on disarmament and tried to improve liaison with other arms control groups. However, Heaphy found that other groups did not take space arms seriously and that they tended to be condescending until the President's speech of March 23, 1983.

With just over 200 members in 1984, the PSF has very limited resources for influencing policy. Heaphy sent out postcards in June 1984 urging people to write the President and members of Congress on space arms control issues and later sent a personal letter supporting pro-arms control and pro-space California Congressman George E. Brown against conservative efforts to unseat him.[98] (Brown won.)

Although there is little evidence of influence on Congress or the administration, PSF leaders are proud of having been the pioneers in pushing the space weapons issue into public debate. "The perspective we have had for five years," said Heaphy in April 1984, "was adopted by the big organizations overnight." Heaphy believes that the PSF's greatest achievement was the development of John Pike into a leading lobbyist for space arms control.[99]

The Origins of a Lobby in Washington

Based in California, the PSF had no lobbying presence in the national capital. As of 1980, it could be said that no space arms control lobby existed in Washington, D.C. That year, Congressman George Brown (who describes himself as "a bonafide space nut") introduced a National Space Policy Act emphasizing the peaceful uses of space, but the bill never got out of committee. In 1981, Senator Larry Pressler introduced a resolution calling on the President to resume anti-satellite arms control negotiations with the Soviet Union, but his bill suffered the same fate as Brown's. The Soviet government submitted a draft space arms control treaty to the United Nations in August 1981. However, there was no organized American interest group response to space weapons issues.

The space arms control lobby that developed in Washington during 1982 and 1983 was the result of informal contacts between like-minded people. One of the first of those contacts was made in the summer of 1980 at the American Institute of Aeronautics and Astronautics' "Global 2000" conference in Baltimore, where Robert M. Bowman met Carol S. Rosin.[100]

Rosin, as a schoolteacher in a suburb of Washington, D.C., had attracted attention when her use of space themes as a teaching device was reported in the local press in 1972. Noticed by Fairchild Industries Incorporated, she was hired by the company as corporate manager of community relations and became a personal assistant to Wernher von Braun until his death. Unwilling to work on weapons projects, she left in 1977 to work for other firms.

Bowman, an Air Force colonel, had been manager of advanced space programs for the U.S. Air Force Space and Missile Systems Organization before his retirement in 1978. He was manager of the General Dynamics advanced space programs division at the time of the Baltimore conference, where he was chairman of a session called "Orbital Systems 2000." Rosin questioned panelists about the fact that none had raised the danger of weapons in space. Bowman later sought out Rosin and suggested that the two of them try to come up with an alternative to an arms race in space. At the Rome conference of the International Astronautical Federation in September 1981, they gave a joint presentation entitled "The Socioeconomic Benefits of an International Communications and Observation Platform Program," including a proposal to use satellites for international peace-keeping (the PSF endorsed the same idea).[101]

Meanwhile, preparations were under way for a United Nations conference on space, called UNISPACE, to be held in Vienna in August 1982. Finding that the administration was disinclined to participate as of late 1981, space activist David Webb put together an informal lobbying effort under the name U.S. Space 82. Congressman George Brown, Senator Alan Cranston, Senator Adlai Stevenson III, and others intervened with the administration. The United States hastily put together a delegation during the spring of 1982.[102]

By then, interest in the military uses of space was rising because of a variety of events. The High Frontier report had been issued in March 1982. In June 1982, the press reported that Secretary of Defense Weinberger had directed the U.S. Air Force to deploy ASAT weapons within five years.[103] The July 4, 1982, statement of U.S. national space policy made it clear that the United States intended to continue developing its anti-satellite capability.[104]

Bowman left General Dynamics at the end of 1981 to take a job as vice-president of the Space Communciations company, but it lasted only six months. One of the events that precipitated his departure was the company's refusal to grant him leave to attend UNISPACE. Bowman went to Vienna representing the Friends World Committee. Rosin was there representing the International Association of Educators for World Peace. Both wanted the conference to address the issue of weapons in space.

In fact, the "militarization" of space was already an issue. In drafting a conference document, the U.N. Committee on the Peaceful Uses of Outer Space had included four paragraphs on the subject. However, the U.S. delegation opposed the inclusion of this language, or more extreme language suggested by some other delegations. After much negotiating, a compromise was worked out in which the conference did express concern about the issue.[105]

Space weapons also were a major subject at the parallel meeting of nongovernment organizations, part of which was chaired by David Webb. Bowman chaired a session on the militarization of space. Bowman and Rosin discussed the formation of a space arms control group but had a falling out and went on to form two small groups in the Washington, D.C., area in late 1982 and early 1983: the Institute for Space and Security Studies (Bowman), and the Institute for Security and Cooperation in Outer Space (Rosin). Bowman has a broad war prevention agenda, extending beyond space weapons. Rosin not only opposes space weapons but also supports commercial and cooperative ventures in space.

Crystallizing the Lobby

Seeing a television report about ASAT systems during the summer of 1982, Congressman Joseph Moakley, of Massachusetts, asked staff aide James McGovern to research the subject.[106] In September, Moakley introduced a resolution that called on the President to resume bilateral talks with the Soviet Union for the purpose of negotiating a comprehensive and verifiable treaty banning space weapons (the goals to be pursued in that treaty were drawn from the Stanley Foundation's Strategy for Peace Conference reports).[107] Among those endorsing the bill were the three leaders of the Planetary Society.[108] In February 1983, Senators Paul Tsongas, Claiborne Pell, and Gary Hart introduced a companion bill in the Senate.

Although Moakley's bill was never passed by the House, it provided a rallying point for space arms control activists, particularly liberal Democratic members and staffers on Capitol Hill. Moakley and McGovern began working closely with George Brown and his aide Sybil Francis.

The resolution also brought people "out of the woodwork," as McGovern puts it.[109] One was Daniel Deudney, a Worldwatch Institute researcher who in August 1982 had published a critical paper on U.S. activities in space. (Deudney had no sympathy for the attitudes of space developers like those leading the L-5 Society, denouncing "the lawlessness and escapism of the frontier mentality."[110]) By December 1982, Deudney was meeting informally with other interested people in Washington, including John Pike of the Progressive Space Forum (who worked briefly with Carol Rosin), British scholar Paul Stares (then completing a book on U.S. ASAT and space arms control policy at the Brookings Institution), and former Center for Defense Information staffer Thomas Karas (then working on his book *The New High Ground*). A critical mass of activists favoring space arms control began to form.

Allying with the Scientists

This small group of activists gained considerable leverage by allying with science-based arms control groups. In January 1983, the Federation of American Scientists (which had shown interest in the subject as early as 1981) hired John Pike as its Staff Assistant for Space Policy to be a lobbyist and to coordinate space arms control efforts. The Union of

Concerned Scientists (UCS) took on Peter Didisheim to help UCS lobbyist Charles Monfort.

President Reagan's speech of March 23, 1983, had a galvanizing effect on this nascent lobby. A group of scientists first assembled by Cornell University professor Kurt Gottfried in February 1983 drafted a space arms control treaty and presented it in a UCS report published in June, and some testified in April and May.[111] In June, Carl Sagan and Richard Garwin, a scientist with International Business Machines, brought a petition calling for negotiations on space arms control to Capitol Hill; this led to a letter to the President signed by over 100 members of Congress, urging a moratorium on testing ASAT systems in space. A similar letter was sent to the President by 40 scientists, arms control experts, and former defense officials.[112]

Meanwhile, Pike began convening a mix of congressional staffers and lobbyists, allying people inside and outside the congressional institution who shared similar views. By June 1983 they were meeting weekly, initially in the office of Congressman Moakley. The list of organizations represented grew to include Physicians for Social Responsibility, Council for a Livable World, and the institutes led by Bowman and Rosin. This loose coalition, whose meetings were chaired by John Pike, came to be known informally as the Space Working Group. Lobbying efforts were divided up among those from organizations outside the Congress.

Deudney and Rosin made a special effort with Senator Claiborne Pell, who was to lead a delegation of U.S. senators to Moscow in August 1983. Deudney proposed language for a draft space arms control treaty, similar to the UCS draft. After the Pell delegation met with Soviet leader Yuri Andropov, the latter called for a complete ban on space weapons and offered to have the Soviet Union observe a moratorium on ASAT testing if the United States would do the same.[113] The Soviets submitted a draft space arms control treaty to the United Nations later that month.

John Pike achieved a major increase in the space arms control constituency in December 1983 when he persuaded the Nuclear Freeze movement to make space arms control one of its priorities for 1984. Pike noted later that most of the people active against space weapons had come from the freeze movement and knew nothing about space.[114]

Legislative Moves

The new space arms control lobby changed tactics during 1983. Instead of trying to legislate general statements of policy, those opposed to space weapons drew on a technique used by other arms control groups and went after funding for U.S. weapons systems. They began with the U.S. ASAT system because of speculation that its development offered a convenient cover for anti-ballistic missile research.[115]

In June 1983 a complex series of legislative moves was begun to delete procurement funding for the ASAT system and to block or limit tests. Members of Congress active in this effort included George Brown, of California; Joseph Moakley, of Massachusetts; Jim Leach, of Iowa; John F. Seiberling, of Ohio; Matthew F. McHugh, of New York; Lawrence Coughlin, of Pennsylvania; and Senators Paul Tsongas, of Massachusetts, and Larry Pressler, of South Dakota. The result was twofold: (1) an amendment to the defense authorization bill barring a flight test of an ASAT system against a target in space until the President certified that he was endeavoring in good faith to negotiate an ASAT arms control agreement and (2) an amendment to the defense appropriations bill that directed that the funds for ASAT procurement could not be obligated or expended until after the administration had submitted a report to Congress on U.S. policy on arms control plans and objectives in the field of ASAT systems.[116] The administration's report of April 2, 1984, expressed serious doubts about the verifiability of an ASAT arms control agreement and did not see the pursuit of such an agreement as being in the national interest.[117]

Just before the administration's report was received, Brown, Moakley, Seiberling, and their allies inside and outside Congress announced the formation of a Coalition for the Peaceful Uses of Outer Space. (Pike describes this as the formal superstructure of the Space Working Group.[118]) At a Capitol Hill press conference, Brown announced that the coalition's goals were to seek (1) a mutual ASAT test moratorium; (2) a reduction in funds for "Star Wars" research; and (3) a reaffirmation of the U.S. commitment to the ABM treaty.[119]

Meanwhile, individual scientists and groups continued to produce critiques of the Strategic Defense Initiative. The Union of Concerned Scientists put out a report in March 1984 entitled *Space-Based Missile Defense* (a parallel report by a group of Soviet scientists appeared not long after).[120] Carl Sagan and IBM scientist Richard Garwin sent letters on space arms control to world leaders in April, drawing a favorable response from Soviet leader Konstantin Chernenko the following month.[121] However, the United States conducted another test of its ASAT system in November 1984.[122]

As of the end of 1984, the space arms control lobby remained only a limited success. Legislation for fiscal year 1985 held to three the number of tests of the U.S. ASAT system that could be conducted during the fiscal year.[123] Funding for the Strategic Defense Initiative was cut below the administration request, although not as much as opponents wished.[124] Meanwhile, the United States and the Soviet Union engaged in an exchange of messages and statements about space arms control negotiations, finally agreeing in January 1985 to resume negotiations on offensive nuclear arms and to include space and defense arms in the talks.[125]

In the view of some observers, it was President Reagan's Strategic Defense Initiative that brought the Soviets back to the negotiating table.[126] Whatever the reasons for reviving the negotiations on space weapons, some space advocates remained skeptical about the long-term outcome of space arms control talks. Said former American Astronautical Society President Charles Sheffield, "The arms controllers should have started about 1975."[127]

Reviving Cooperation

The U.S.-Soviet agreement on space cooperation was not renewed when it expired in 1982. In February 1984, Senator Spark Matsunaga, of Hawaii, with the support of Senators Charles Mathias and Claiborne Pell, introduced a joint resolution "relating to cooperative East-West ventures in space as an alternative to a space arms race." Congressman Mel Levine of California sponsored a similar bill in the House. During hearings on the bill in the fall of 1984, Matsunaga got support from Carl Sagan and Louis D. Friedman, leading figures in the Planetary Society. Matsunaga suggested several future cooperative missions, including a manned mission to Mars, as an alternative to "Star Wars."[128] Matsunaga aide Harvey Meyerson believes that such cooperation opens up a new constituency for space – those interested in relations with the Soviet Union.[129]

On October 30, 1984, President Reagan signed a joint resolution on East-West space cooperation, noting that the United States had proposed a simulated space rescue mission.[130] A few months later, AIAA

President John McLucas gave his support in print, writing, "For only space offers an arena, a theme, and an organizing principle grand enough to permit us to transcend our differences and set humanity on a new and more hopeful course."[131]

The Planetary Society had been particularly active in supporting revived U.S.-Soviet space cooperation. Louis Friedman wrote in the August 1984 *Aerospace America* that extending human civilization to space could bring a new era of global security, and he suggested a lunar base, a manned expedition to Mars, or a prospecting journey to some asteroids undertaken by an international team.[132] Meanwhile, the film "2010," which was released late in 1984, proved to be a thinly disguised plea for U.S.-Soviet space cooperation.

International Peacekeeping from Space

Long before this debate, some of those interested in space technology saw it as an opportunity to create a global security system, beginning with internationally controlled observation satellites. Former airline executive Howard Kurtz and his late wife Harriett began looking into this over two decades ago, forming a small organization called War Control Planners. In a May 1969 article in *Military Review,* they spelled out a step-by-step approach to global security, beginning with surveillance systems for experimental installation in orbiting laboratories; all nations would have access to the data. Eventually, the Kurtz team wanted to see global command and control of the world's military power.[133]

During 1973 and 1974, an executive of the ITEK Corporation introduced the idea of an internationally managed "verification satellite," stimulating a study group at the Massachusetts Institute of Technology that issued a report in 1978 on a "Crisis Management Satellite."[134] At the first Special Session of the U.N. General Assembly devoted to disarmament, held in New York in May-June 1978, the French government introduced a proposal for an International Satellite Monitoring Agency, and a U.N. study group produced a generally favorable report on the subject in 1983.[135] Testifying before a congressional committee in 1981, Arthur C. Clarke endorsed the concept, which he calls "Peacesat." Daniel Deudney pursued the theme of a space-based global security system in two reports written for the Worldwatch Institute.[136] During 1984, this theme was taken up by new AIAA President John McLucas, who foresaw remote sensing satellites evolving into a peacekeeping system.[137] Carol Rosin's ISCOS has picked up on some of these ideas, supporting a global security satellite monitoring system and an international military space command post.[138]

Others have internationalized the idea of space-based strategic weapons, turning it into a proposed global defense system. In his 1946 article "The Rocket and the Future of Warfare," Arthur C. Clarke proposed that a World Security Council should be given long-range atomic rockets as an ultimate deterrent to war.[139] In the *Spaceflight Revolution*, William S. Bainbridge suggested an international "Space Patrol" designed to shoot down any object that rose above the atmosphere without special permission. "Thus," he wrote, "the Space Patrol would revolutionize war by making defense superior to offense." The Space Patrol was the only military project Bainbridge could think of that was both humanely desirable and would further space technology.[140]

In San Francisco, Kenneth Largman founded the Strategic Arms Control Organization (later the World Security Council) in 1980 to encourage the coordination of U.S. and Soviet laser and particle beam space defenses, eliminating the dangers associated with one-sided deployment; this would involve interlocking communications systems, inspection teams, and controlled deployment. Board members included Peter Vajk and Stan Kent.[141] After President Reagan's "Star Wars" speech, Princeton Professor Richard H. Ullman suggested in the New York *Times* that the United Nations be given authority to operate space based anti-ballistic missile defenses.[142] Ben Bova pursued the idea in his 1984 book *Assured Survival,* suggesting an international peacekeeping force. "The technical means for warfare suppression," he wrote, "are at our fingertips."[143]

THE NEW MILITARY DETOUR

Space-based ballistic missile defense presents the temptation of a new military detour for the pro-space movement. Some space advocates already have seized what they see as an opportunity to speed up space industrialization and settlement, which share the military's need for certain enabling technologies such as heavy lift launch vehicles, compact energy sources, and the assembly of large structures in space. To these space advocates, weapons in space are not only a solution to the nuclear dilemma but also a means to break through the cost and political barriers to large-scale space development – in effect, a successor to the solar power satellite. Science fiction writer Frederick Pohl has been quoted as saying that what most upsets him "is that most of the authors supporting 'Star Wars' are not cold warriors but people who want to trick the military into spending money on space."[144]

No one has been more active in this regard than L-5 Society co-founder H. Keith Henson. "It's a long way from military space bunkers to what we are really interested in, the human habitation of space," he wrote in the October 1979 *L-5 News*, "but other than the SSPS project, the space bunkers are the only other extraterrestrial resources project that might be economically justifiable."[145] "Hitching a ride into space on the back of the military may not be very dignified, but it beats walking," he wrote four years later. "Setting up the pipeline for ETM [extraterrestrial materials] would get us 90% of the way to space colonies."[146] Arguing that we should seek to substitute the goal of international military stability in place of national military advantage, Henson's article "Weapons for Peace" in the July 1984 *L-5 News* said that the issue "may be far more important than the Moon Treaty. It may make the difference between slavery and freedom. It may make the difference between peace and nuclear war. It might be our key to the solar system."[147]

One of the appeals of "Star Wars" is that it joins the space dynamic with the strategic defense dynamic. It couples the appeal of futuristic space technology with a promised solution to the threat of nuclear war.

On the other hand, many space enthusiasts oppose weapons in space. To a large degree, this opposition seems to stem from the fact that space weapons would conflict with their optimistic, even idealistic vision of space humanization. Many want to go into space to put the conflicts and tensions of Earth behind them, not to carry them outward into the solar system. In the March 1983 issue of *Astronomy*, space activist Jon Alexandr of the Progressive Space Forum explicitly opposed a new military detour by the pro-space movement, finding this attitude "abhorrent, unethical, and dangerous."[148]

Dividing the Pro-Space Movement

This issue of weapons in space threatens to divide the new pro-space movement along left-right lines for the first time, cracking the bipartisan consensus that has underlain most pro-space activism. By 1984, some pro-space organizations were beginning to line up on one side of the issue or the other. The American Space Foundation, for example, appeared sympathetic to the Strategic Defense Initiative, while the Planetary Society's leaders clearly opposed it. This debate seemed to be closely related to the manned versus unmanned, exploration versus development issue, since many of those opposed to space weapons also are critical of the manned space program, while some of the most ardent advocates of putting more humans into space also support the space-based ABM. In the January/February 1982 issue of *The Planetary Report*, planetary scientist Clark Chapman wrote, "Will a space station become a military fortress in the sky after millions of space enthusiasts have supported its development for more lofty goals?. . . Let's hope the Shuttle becomes more than a truck to be filled with military cargo."[149]

Nowhere is this divisiveness better seen than in the L-5 Society, where "pro-space" feelings may run highest. Several observers have predicted an exodus of the more liberal members of the society if its leaders endorse space-based weapons. Some feared an open split at the society's April 1984 convention in San Francisco, where Keith Henson, Jerry Pournelle, and some others sought society support for the space-based ABM. However, the issue was papered over when the society's board decided to not take a position. A debate also has appeared in the conservative pages of the American Astronautical Society newsletter.[150]

The issue also has divided another part of the pro-space constituency: science fiction writers. "There used to be a feeling that we were all pushing for space," said one editor. "Now there is a feeling of conflict over the imminent dangers of nuclear war."[151]

Undermining the Support Base

In the long run, the greater danger is that the controversy will rub off on the perception that Americans have of the whole space enterprise, dividing the potential pro-space constituency along left-right lines. Trudy E. Bell wrote in her 1982 paper that virtually all the pro-space groups she contacted reported a dramatic jump (since 1980) in public awareness about the militarization of space. Those groups that reported some opposition toward space activities indicated that the antagonism was specifically directed toward military activities.[152] "The idea that NASA expenditures are a quasi-military form of defense spending has been a major cause of the erosion of liberal support for the U.S. space program," wrote one correspondent in *Astronomy* in 1981.[153] Says former *Saturday Review* editor Norman Cousins, a liberal who has supported space exploration: "If space will be used for military purposes, some of us will be forced to oppose the space program."[154]

The space weapons issue already has become a new battleground between conservatives and liberals in American politics. Republicans have seized the initiative on space weapons as they have on space commercialization and the space station, and a Republican President is leading the way in all three cases. In their platforms for the 1984 elections, the Republicans and Democrats took diametrically opposite positions on space issues. The Republicans supported the Strategic Defense Initiative, the space station, and space commercialization. The Democrats blasted the idea of weapons in space and said nothing about the space station or space commercialization. Walter Mondale, who had been a leading opponent of the Space Shuttle, told an audience in April 1984 that "if you help me to get nominated, I can make the 1984 election a choice between Star Wars and a space freeze."[155]

Riding piggyback on space weapons clearly is tempting to some space developers, despite some fears that it could be a Faustian bargain. Politically, they could run the risk of a major interruption if a liberal Democratic administration took office in the near future. To some space advocates, a better and more permanent approach is to unleash the energies of private enterprise through space commercialization.

12

SPACE COMMERCIALIZATION
AND THE
NEW ENTREPRENEURS

> If history is any example . . . the entrepreneur will eventually arise to enliven the future of space exploitation.
>
> E. P. Wheaton and M. W. Hunter, 1966[1]

> I'm going to be a billionaire. A lot of us are.
>
> H. Keith Henson, 1977[2]

> It would be easier to start a private space program in the Soviet Union than in the United States.
>
> Klaus Heiss, 1983[3]

For the first two decades of the Space Age, the U.S. space enterprise was dominated overwhelmingly by government agencies and their contractors. Only the government could afford the investment in some of these new technologies. Only the government had launch vehicles.

Things started to change in the late 1970s, at roughly the same time that a citizens pro-space movement was beginning to flower. Many of the technologies of space had matured, and knowledge of them had diffused widely. Some systems that NASA had developed had been operational for years, and there were questions as to whether they still fell within NASA's mandate of research and development. The Space Shuttle, originally scheduled to fly in the late 1970s, promised reusable transportation able to revisit facilities in orbit. The concepts of space manufacturing and space industrialization were no longer alien; the idea of their being accomplished by private investment became a topic of public discussion, particularly after the landmark October 1977 meeting of the American Astronautical Society in San Francisco. There was a growing realization that space need not be a government monopoly and that the private sector could be an independent actor, not just a contractor. The failure to get government funding for major space projects such as satellite solar power stations and space colonies also may have been a factor.

Meanwhile, a policy evolution had begun tentatively late in the Carter administration, with moves toward "privatizing" some government-developed space systems, and with the development of new mechanisms for joint NASA/commercial activity in space. This trend accelerated as the pro-private enterprise Reagan White House, committed to curbing government competition with business and industry, endorsed privatization and changed the policy, regulatory, and economic climate for private space ventures.

The turning point occurred between 1981 and 1983. The first flight of the Space Shuttle made reusable space transportation a reality. After a spectacular failure, launch vehicle entrepreneurs scored a success. Congress, stimulated in part by a small but growing lobby for space commercialization, began acting in parallel with the administration, and sometimes ahead of it.

Some existing aerospace companies and other established firms saw opportunities in these changes. However, the idea of space commercialization also attracted bright young people of a new generation, impatient with the old, politically dependent ways of getting into space. There was a sudden flowering of new, entrepreneurial commerical space ventures between 1980 and 1983, at roughly the same time as the peak of the space group boom. In part, this reflected a new wave of entrepreneurship that was emerging throughout the American economy, led by a younger generation. The emergence of new centers of venture capital in Houston and in northern California's "Silicon Valley" was a factor. However, many of these new entrepreneurs also had connections to pro-space activism and were extensively intertwined with the new pro-space movement. Space commercialization was another way of making the space dream real.

THE FIRST COMMERCIALIZATION

The great success story of commercializing space technology is the communications satellite. "Never in history has there been so rapid a commercialization of a new technology," says former NASA Administrator Thomas O. Paine.[4]

First proposed by Arthur C. Clarke in 1945, extraterrestrial relays in geosynchronous orbit became an operating reality with SYNCOM II in 1962. There was a vigorous debate in the early 1960s over whether such a vital utility should be a commercial operation. Compromises led to the Communications Satellite Act of 1962, which enabled the creation of a privately funded corporation regulated by the federal government called the Communication Satellite Corporation (COMSAT). The new company was to be the U.S. participant in an international satellite telecommunications system (INTELSAT), which was created in 1964.

With its monopoly on overseas commercial communication by satellite, COMSAT has been highly successful. As of August 1982, it was allowed to enter fields other than satellite communications.

Meanwhile, the satellite communications industry had diversified. In the deregulatory environment of the early 1970s, private firms were allowed to establish domestic satellite systems in the United States, despite the opposition of American Telephone and Telegraph. Beginning with Satellite Business Systems in 1975, domestic communications satellite companies have proliferated remarkably. As of late 1984, it appeared that other companies had been at least partially successful in challenging COMSAT's monopoly on overseas traffic.[5]

Satellite communications now is a mature industry, with a global satellite market worth about $4 billion in 1985.[6] In addition to the numerous private firms that are suppliers, operators, and contractors in the field, it has a specialized nonprofit organization, the Public Service Satellite Consortium. It has its own conferences (one is called the annual "Satellite Summit") and its own specialized publications, such as *Satellite News*, the *International Journal of Satellite Communications,*, and the annual *Satellite Directory*. It also has its own professional organization, the Society of Satellite Professionals (SSP). Initially formed in August of 1982 and formally launched in August 1983, the SSP had 325 individual members and three corporate sponsors as of January 1984.[7]

The success of satellite communications occurred because the circumstances were right: the communications industry already was highly developed, the technology was ready, satellites offered more economic and efficient ways of performing services, and there was a large and growing market. The situation is less clear for other areas of space commercialization.

DIRECT BROADCAST SATELLITES

Another application of space technology that was proposed early in the Space Age was the direct broadcasting of television signals from satellites to home receivers. NASA tested direct-broadcast satellite (DBS) concepts with its Applications Technology Satellites in the 1970s, but the commercialization of this technology was delayed in the United States for a variety of reasons, including commercial and institutional resistance and the need for further technology development. Meanwhile, some other nations were planning actively for DBS systems.

Because of rising interest in the United States, the Federal Communications Commission (FCC) began an inquiry into regulatory policy in this area in 1980. Despite opposition from the National Association of Broadcasters, the FCC decided in April 1981 to endorse direct broadcasting from satellites and accepted for expedited consideration a plan by Satellite Television Corporation (a subsidiary of COMSAT) to begin the service as early as 1985.[8] Other companies entered the field, with some choosing to lease capacity on the satellites of other firms rather than waiting for dedicated satellites of their own.[9]

The companies clearly believe that the technology is mature enough and that a market exists. There is no equivalent to COMSAT or INTELSAT in this field, which is open and competitive. DBS already has its own trade association, the Direct Broadcast Satellite Association, which was formed in the summer of 1983. However, COMSAT dropped its plan for a DBS network in late 1984, citing "unacceptable risks."[10] As of 1985, the future for commercial DBS in the United States remained unclear.

PRIVATIZING LANDSAT

In its space policy statement of June 1978, the Carter administration made a gesture toward space commercialization when it established this as one of its principles: "The United States will encourage domestic commercial exploitation of space capabilities and systems for economic benefit and to promote the technological position of the United States."[11] However, the only specific action mentioned in the Private Sector section of the October 1978 statement on civil space policy was that NASA and the Department of Commerce were to prepare a plan of action on how to encourage private investment and direct participation in civil remote sensing systems.[12]

What this meant was that the administration, particularly the Office of Management and Budget, wanted to turn the government's remote sensing satellites over to the private sector and get them out of the federal budget. This is perhaps the best illustration of one form of commercialization: privatization, in which a government-developed asset is turned over to private industry.

Observing the Earth was one of the earliest successful applications of space technology, made most familiar through weather satellites. NASA began an Earth Resources Survey Program in 1965 and launched its first Earth Resources Technology Satellite in 1972. Renamed LANDSAT, this series continued, with the fifth satellite being put into orbit in March 1984.

These spacecraft have returned large quantities of useful and sometimes beautiful imagery of the Earth.[13] Ground stations to receive the data were built in the United States and other countries, and a data center was established in Sioux Falls, South Dakota. The United States encouraged other nations (particularly in the developing world) to purchase their own Earth stations to take advantage of LANDSAT imagery for such purposes as land use planning. LANDSAT data were used in a cooperative project between NASA and the U.S. Department of Agriculture to see if crop estimates could be improved. A "value-added" industry grew up around LANDSAT, composed of small companies that processed the data to meet the needs of users. Some university scientists made use of the data in their research. The result was a constituency that wanted continuity of data and that regarded LANDSAT as a kind of public service.

NASA managers and engineers, who saw LANDSAT as a technology development program of the kind NASA is supposed to conduct, wanted to move on to newer technologies, and they placed a relatively low priority on data continuity. The Office of Management and Budget complained about the cost of each LANDSAT (about $300 million a copy in the early 1980s) and saw no reason for the government to continue putting up more of them. If the system had become operational in effect and was performing a valuable service, one argument ran, let the private sector do it on a commercial basis.

Under a November 1979 directive, the Department of Commerce was given temporary stewardship of the LANDSAT system beginning in 1981. That department's National Oceanic and Atmospheric

Administration was to operate LANDSAT along with its existing weather satellites (which had been transferred to NOAA when they became operational), while charging enough to recover its costs. Once NASA finished proving out the second-generation satellites, they, too, would be turned over. Meanwhile, NOAA was to work out a plan for the timely transfer of remote sensing technology to the private sector.

The incoming Reagan administration, in its budget cuts of March 1981, eliminated funding for two more LANDSAT spacecraft and accelerated the planned pace of transfer to the private sector. Once LANDSAT-D died in 1985 and LANDSAT-D prime in 1987, that would be the end. Thereafter, the remote sensing program would be in the hands of the private sector.[14]

Numerous objections were raised by the user community and by members of Congress, with some arguing that remote sensing data are a "public good." Others said that commercialization of the whole system was premature; revenues from the existing LANDSAT system were only about $10 million a year. Some people, like then American Astronautical Society President Charles Sheffield, argued for phased commercialization, with private industry doing the analysis and distribution of data until the market grows large enough to support private sector remote sensing satellites.[15]

The situation was complicated further when COMSAT offered to take over both LANDSAT and NOAA's weather satellites, if the government guaranteed the company enough business. President Reagan directed the Cabinet Council on Commerce and Trade to study the feasibility of transferring both LANDSAT and the weather satellites to the private sector. Studies reportedly showed that the demand for LANDSAT-type services would be insufficient to support a commercially viable industry. After an initial decision in April 1982 that the weather satellites should remain with the government, the Cabinet Council was asked to reconsider, and subsequently it recommended commercialization of land, weather, and ocean satellites. Of the private companies that responded to a Department of Commerce request for information, only COMSAT recommended commercialization of both the land and weather satellites. In March 1983, the President announced his decision to transfer the government's civil operational remote sensing satellites to the private sector, including LANDSAT and the weather satellites, as well as the responsibility for any future ocean observation systems. This was to be carried out by a competitive process in which U.S. private firms would enter bids. However, opposition to the sale of the weather satellites was strong in Congress. Both the House and the Senate passed resolutions sending a clear message to the administration that they would not accept the sale of those satellites.

In November 1983, the administration dropped the idea of selling the weather satellites, but in January 1984 it issued a request for proposals for the takeover of the LANDSAT system and continued program operation. The original seven bids were narrowed to one by late 1984 and the EOSAT Corporation took over the system in 1985. However, the outcome was still in some doubt because of disagreements about the size of the federal subsidy to the private operator. Meanwhile, bills providing plans for the commercialization of the LANDSAT system became law in late 1984.[16]

The situation has been complicated further by the emergence of foreign competition in the form of the French SPOT remote sensing satellite and a transnational joint venture called SPARX, which planned to use a remote sensing system taken into space on the Space Shuttle. A leading figure in the SPARX enterprise was the ubiquitous space entrepreneur Klaus Heiss, who had developed important economic arguments for the Space Shuttle a decade earlier. However, the future of SPARX appeared to be in some doubt as of 1985.[17]

The LANDSAT story has raised questions about the commercial feasibility of privatizing government space systems not developed for commercial purposes. Some space entrepreneurs prefer that new enterprises be private from the beginning, notably in the field of expendable launch vehicles.

COMMERCIAL LAUNCH VEHICLES

The idea of private groups or individuals building their own vehicles to go into space has a long history in science fiction, dating back at least as far as Jules Verne's *From the Earth to the Moon*. The earliest rocket experimenters, private citizens without government funding, dreamed of building vehicles that could carry humans beyond the Earth. Since the arrival of a real manned space program, it has been a matter of concern to many space advocates, particularly those of a libertarian bent, that access to space is controlled by government agencies.

As of the late 1970s, it was national policy that the government's expendable launch vehicles would be phased out after the Space Shuttle was declared operational, an event that occurred after the fourth mission in July 1982. However, the Shuttle, a developmental vehicle, proved to be an expensive way of launching some payloads. It was not ideal for all launching tasks, and niches appeared for potential commercial operators. NASA's expendable launch vehicles were to become redundant, presenting an opportunity for privatization of these vehicles. Meanwhile, the European Space Agency was developing a competitor to the Shuttle in its expendable Ariane launch vehicle, which lifted its first commercial payload into orbit in 1983. This situation raised complex policy questions, including whether private operators should be allowed to compete with the government launch system, and whether the cost of putting a commercial payload into space on the Shuttle should be subsidized to make it competitive.

By the late 1970s, several individuals were convinced that early space development and easy access to space would be possible only if the cost of getting into orbit were brought down significantly and that the job could be done more cheaply by private enterprise. They tended to share the view that it was important to establish alternatives to government control over access to space and to make it easier for large numbers of people to go.

OTRAG

The first enterprise to test a private launch vehicle meant for commercial use emerged not in America but in West Germany. Aeronautical and propulsion engineer Lutz T. Kayser, who began pursuing his dream of a low-cost launcher in the 1960s, got funding from the German government for rocket motor tests until 1974. With a motor well developed, he and some colleagues put together a company called Orbital Transport and Rockets A.G. (OTRAG) to develop simple expendable launch vehicles that could place civilian payloads into orbit for a fee, particularly for less-developed countries that did not have their own launchers. The chairman of OTRAG's board was Kurt H. Debus, a long-time associate of Wernher von Braun.[18]

In 1976, OTRAG obtained a tract a third the size of Colorado in eastern Zaire for a near-equatorial launch site, a sort of private Cape Canaveral. The firm conducted some initial tests, including a May 1977 launch to low altitude of its smallest module. However, OTRAG became politically controversial as the Soviet Union and some other countries accused it of developing military rocketry, and the company was forced to close down its Zaire range in 1979. After looking for other locations, OTRAG began testing in the Libyan desert, claiming a successful launch to low altitude in March 1981. By that year, the company reportedly had raised $65 million from private investors attracted by tax benefits.

In September 1981, the press reported U.S. government concern about the possible use of commercial rockets for military purposes, particularly the proliferation of a nuclear weapon delivery capability to additional states. By the end of that year, OTRAG had pulled out of Libya and Kayser was out of the company. Frank Wukasch, who became president of OTRAG, said the firm would seek to avoid political controversy in the future by using existing launch ranges for tests. The company was hoping to market a series of sounding rockets as a first step.

As of June 1984, OTRAG was looking for a non-German, preferably American, business partner. Ten years after its founding, it still was not a commercial success.

The Back Yard Rocketeer

In July 1938, a young U.S. Navy midshipman named Robert Truax, a member of the American Rocket Society, visited London to inform the British Interplanetary Society about his rocket experiments. For the next two decades, Truax pursued his interest in rocketry within the military services, rising to head the U.S. Air Force space program from 1956 to 1958. As president of the American Astronautical Society, he generated a recommendation to President Eisenhower for a strong civilian space program. Retiring from the Navy in 1959, he joined Aerojet General the next year. There he formed a small group to study a cost-effective Earth-to-orbit transport vehicle.[19]

At the time, the U.S. Navy had a team headed by Captain John Draim working on the "Hydra" project, a study of launching rockets from the water. Truax heard about it and got involved in tests of rockets off the California coast that showed that water launch was feasible. By 1963, Truax's group had completed a study of the "Sea Dragon," a very large but simple rocket that would be launched from the water. However, according to Truax, NASA was not interested.

Leaving Aerojet General in the mid-1960s, Truax headed a recoverable launch vehicle study sponsored by the American Institute of Aeronautics and Astronautics that he believes was instrumental in convincing NASA to begin work on the Space Shuttle. He intruded on the public consciousness in September 1974 when he built the rocket for stuntman Evel Knievel's jump across the Snake River Canyon. Although that jump failed, Knievel asked Truax what they could do next. Truax says he told Knievel that for $1 million, he could make Knievel the world's first private astronaut. However, Knievel was not able to raise the money.

Impressed by the media interest in Knievel's jump (and by what he had been able to do in his own back yard) Truax, in 1976, embarked on a project to build a small rocket that would carry a volunteer on a suborbital flight (as of 1983, he reportedly had 4,000 volunteers).[20] The rocket, which has been ground-tested, is labeled Project Private Enterprise (it also has been called the "Volksrocket"). Truax plans to use the media event to attract support for an intermediate-sized launch vehicle called Excalibur. Eventually, the firm might become the operator of a spaceline.

As of April 1984, Truax had spent about half a million dollars on this project and estimated that it would cost $1 or $2 million more. Still the rugged individualist, Truax was continuing to work on the rocket in the garage of his California home. "My area," he says, "is getting to space as cheaply as possible."[21]

The Dreamer from St. Paul

If Robert Truax represents the older generation of private launch vehicle dreamers, Gary C. Hudson represents the younger. An award-winning science student in high school in St. Paul, Minnesota, Hudson went on to study physics, astronomy, and microbiology at the University of Minnesota, although he never received a degree. Hudson foresaw two areas of high technology with good commercial prospects: biotechnology and space. Betting that space would come earlier, Hudson left the university in 1971 and began a career as a writer, consultant, and lecturer on private sector approaches to space activity – at the age of 21. Unable to find financing for a private company of his own (his businessman father had told him never to work for anyone else), Hudson started a nonprofit organization called the Foundation Institute to promote the development of space technology. Inspired by the foundation in Isaac Asimov's science fiction trilogy of that name, this small, home-based institution published *Foundation Report*, which contained visionary articles about the future in space (it later became the *Commercial Space*

Report). Advertising in *Astronomy* magazine as late as 1978, the institute described itself as being on the frontier of the next industrial revolution.[22]

During the 1969-72 period, Hudson concluded that the primary hindrance to the commercialization of space was transportation and began working on designs for launch vehicles. He recalls being concerned that NASA was not sympathetic to the development of private launch services that could become competitors to the Space Shuttle. Meanwhile, older aerospace experts such as Robert Salkeld and Philip Bono were proposing their own alternatives to the Shuttle.

Hudson pursued his private vision for years without success. Only when he made contact with other space commerce visionaries did a critical mass develop behind such ideas.

Spaceport

Free market principles also were a driving force behind moves in the mid-1970s to develop offshore launch sites for commercial enterprises. A Harvard Business School graduate, Paul Siegler, founded the company Earth/Space in Palo Alto, California, in 1975 to develop "low cost ways of getting to and using space," including the possible use of equatorial launch facilities. Siegler played a leading role in organizing the American Astronautical Society's October 1977 conference in San Francisco, apparently the first to highlight private investment in space industrialization. His organization put out several issues of *Earth/Space News* but folded in the late 1970s.[23]

In 1976, Siegler invited another young Harvard graduate, Mark Frazier (self-described as a "Friedmanite"), to join the board of the company. To maximize cost-savings for private launch vehicles, Frazier suggested in October 1976 that a tax-free trade zone be established in conjunction with an international equatorial launch site. During the late 1970s, Frazier's free enterprise-oriented Sabre Foundation, which had been financed by conservative, free-market interests, developed the idea of an Earthport, a free trade zone that would include an international satellite launch facility. The name apparently was suggested by lawyer Arthur Dula, later an activist with the Houston-based Space Foundation. Hudson has suggested that the name may have been inspired by Arthur C. Clarke's vision of the "ports of Earth" in his 1968 book *The Promise of Space*; Frazier thinks the source was Robert A. Heinlein's *Starman Jones*.[24]

Initially, the headquarters of this project was in Santa Barbara, California, where Frazier started the World Space Center. Early donors included Arthur C. Clarke and Barbara Marx Hubbard. In addition to putting out *World Space News*, Frazier prepared a brochure on Earthport and mailed it to officials in all equatorial countries, the most logical sites for such an installation.[25] According to Frazier, he got back many positive responses. However, the project reportedly met resistance from the State Department and a lack of interest from aerospace companies.

The Sabre Foundation decided to take an indirect approach, spinning off Free Zone Authority Ltd. as a nonprofit consulting firm on free trade zone development with Frazier as its president. The aim, says Frazier, is to get technology-oriented zones established and to show that free trade zones can generate revenue for training and economic development. In the free trade zone destined to house a spaceport, revenues could be used to train third-world personnel in the techniques of remote sensing from space and to invest in promising space enterprises. Frazier believes the space-oriented freeport would build up an international constituency and give less-developed countries a stake in space.

The ultimate goal of this enterprise is to help self-sustaining space commerce get started and to get more people into space. The implied goal of breaking the great power monopoly on space launch facilities is reminiscent of Lutz Kayser's OTRAG. Meanwhile, a press report has suggested that Brazilian authorities are interested in developing an equatorial spaceport for all of Latin America.[26]

Houston Meets Silicon Valley

In 1976, successful Houston real estate developer David Hannah saw an article about Gerard K. O'Neill's space colony ideas in *Smithsonian* magazine.[27] Hannah decided that he wanted to do something practical to help get people into space. He tried to get the Carter administration interested but without success.[28]

In the fall of 1979, Hannah and Gary Hudson both were attending a fund-raiser in Houston for the then new Space Foundation, where they were introduced by Arthur Dula. Hudson concluded that Hannah was a potential investor in a private space transportation company, and the two began negotiating.

Early in 1980, Hudson moved to California's Silicon Valley to work with young English-born aerospace engineer Stan Kent, the founder of the organization Delta Vee. There he met three other young pro-space activists who were to play a role in the private launch vehicle industry: James Bennett, Philip Salin, and Gayle Pergamit.[29]

James Bennett had studied anthropology at the University of Michigan, the first home of FASST (he recalls that the organization's efforts to build support for high technology failed to excite most students). Bennett was more intrigued by Gerard O'Neill's space colony ideas and organized an L-5 Society chapter on campus.

Bennett moved to California and attended the October 1977 American Astronautical Society conference in San Francisco, where Christian Basler stirred interest by proposing a financing mechanism for space ventures called the staging company.[30] In 1978, Bennett got involved in the Earthport project and relocated to Santa Barbara, where he worked with Mark Frazier.

In 1980 he moved back to the San Francisco Bay area where he began working with Philip Salin, a UCLA economics student and Stanford MBA who had founded the Stanford Center for Space Development in 1978 (Salin recalls being influenced by Gerard O'Neill's ideas). Gayle Pergamit, another Stanford graduate and a co-founder of the SCSD, joined them. In the spring of 1980, the three decided to work with Gary Hudson in creating the first American commercial launch company, to be called Advanced Propulsion Technologies. Bennett continued the political and regulatory research started at Earthport; Pergamit researched the market for commercial launch vehicles; Salin mapped out a strategy for raising funds. However, disagreements between these three and Hudson regarding what it would take to succeed led to a split. Bennett, Salin, and Pergamit left to found Space Enterprise Consultants, which surveyed potential launch sites and did market analyses of space commerce. Hudson formed a company called GCH (his initials).

Failure and Success

In January 1981, Hudson came up with a design for a launch vehicle he called the Percheron. Hannah and others created a Percheron joint venture to finance the project. Presented with the opportunity of a lifetime at age 29, Hudson felt under pressure to do the job quickly and cheaply; Hannah's enthusiasm for an early demonstration flight, and the fact that the funding available for the venture was closer to $1 million than the $5 million Hudson had estimated would be necessary led him to cut corners. With hindsight, he recognizes that this was a mistake. But history will record that on August 5, 1981, Percheron blew up on a pad on Matagorda Island off the coast of Texas, apparently because of a frozen valve that had worked two days before.[31]

Hudson and Hannah ended their business relationship, and Hudson went on to found a new company called Pacific American Launch Services in Redwood City, California. His hope is to build a large, squat launch vehicle called the Phoenix as a competitor to the Space Shuttle. Hudson appears unfazed by the estimated cost of $100 million.

Meanwhile, the political and policy climate was changing. The Reagan administration had initiated a study of space policy in August 1981, about the time of the Percheron explosion. On July 4, 1982, when the fourth Shuttle mission landed at California's Edwards Air Force Base and the system was declared operational, the White House released a fact sheet on national space policy that listed as one of its basic goals "Expand private sector investment and involvement in United States space activities," and as one of its principles "the United States encourages domestic commercial exploitation of space capabilities, technology, and systems for national economic benefit." The fact sheet went on to state, "The United States government will provide a climate conducive to expanded private sector investment in space activities."[32]

David Hannah's Space Services Incorporated continued work on developing a commercial launch vehicle. Getting the assistance of retired NASA experts from the Johnson Space Flight Center near Houston and hiring aerospace contractors, Space Services moved toward acquiring or developing a rocket based on tried technology. The central issue was finding a reliable propulsion system. With the help of a network of pro-space contacts including O'Neill associate T. Stephen Cheston, the company approached the Air Force in the fall of 1981 about purchasing an extra second stage of a Minuteman missile. After the initial disbelief of its officials ("You want a what?", quotes Cheston), the Air Force referred the firm to NASA, which eventually agreed to help. Meanwhile, Space Services acquired the services of former astronaut Donald K. "Deke" Slayton, who became the company's president. Leading Space Services figures met with Presidential Science Adviser George Keyworth, who reportedly said that the White House liked the firm's entrepreneurial spirit even though the first rocket had blown up.[33]

This time the launch was a success. On September 9, 1982, Conestoga I lifted off Matagorda Island on a near-perfect suborbital flight.[34] The first successful test of a privately financed American launch vehicle, it was a symbolic turning point for space commercialization and the new entrepreneurs. However, it had been done with a rocket developed originally with government money and tested by an established firm, the Space Vector Corporation.

One of those present at the launch was Charles Chafer, the experienced young space activist who had been with Stephen Cheston's Institute for the Social Science Study of Space and who gave papers on space politics at O'Neill's Princeton conferences. Chafer became a vice-president of Space Services as well as its representative in Washington.[35] His wife Sallie, who also had been active with the ISSSS, announced the countdown for the successful launch.

As of 1984, Space Services was working toward a larger launch vehicle called the Conestoga II.[36] According to Deke Slayton, the company's motto is "Have rocket, will travel."[37]

The Stars Truck

Having made a thorough survey of private enterprise opportunities in space, Bennett, Salin, and Pergamit confirmed their opinion that transportation was the key to space industrialization. Their analyses showed a market opportunity for private launch vehicle companies. In March 1981, they and engineer Bevin McKinney formed a company known as Arc Technologies, betting they could outperform GCH (McKinney had been Director of the Space Now Society at the time of Space Day 2 in 1978).

Arc Technologies spent 1981 raising funds in Silicon Valley, which Bennett describes as "a large pool of capital with independent perspective and judgement, autonomous of Wall Street."[38] By early 1982 ARC Technologies was developing an engine for a suborbital test vehicle called the Dolphin, which was to be launched from the sea.

The company was boosted by an investment in February 1982 by former Apple Computer President Michael Scott, who became president of the new company that December. Arc Technologies was renamed Starstruck (Stars Truck) in May 1983. Salin and Pergamit left the company, but Bennett stayed on as vice-president for governmental affairs. In Washington, Starstruck was represented by Courtney Stadd, the young space activist who had been the National Space Institute's manager and the editor of its newsletter.

After three attempts to launch the Dolphin from the ocean near San Clemente Island were postponed because of technical problems, Starstruck enjoyed its first successful launch on August 3, 1984.[39] Although this launch attracted less attention than Space Service's Conestoga I, it was the first time Americans had launched a private space launch vehicle that they had developed entirely by themselves, and it was the first private launch from the water. However, the development of this vehicle had cost more than expected. As of 1985, the future of Starstruck was in some doubt.[40]

Deregulating Space

The negative side of the Space Services success in 1982 was that it had taken the firm six months and $250,000 in legal fees to get permission from federal agencies for the launch.[41] The company, which was looking at a possible Hawaiian launch site, made contact with Congressman Daniel Akaka of Hawaii, who had become co-chairman of the Congressional Space Caucus. Akaka aide Diana Hoyt, then holding together both the caucus and the Congressional Staff Space Group, drafted a bill called the Space Commerce Act to streamline the regulatory procedure, in part by designating a "lead agency" within the government that would provide a single point of contact. Akaka introduced the bill in December 1982. At about the same time, the administration responded to requests for a U.S. policy on private operation of expendable launch vehicles by commissioning an interagency study.

Akaka's bill was reintroduced in January 1983, and hearings were held in May. Meanwhile, there reportedly had been debate within the administration between those who feared private sector competition with the Shuttle and those who ideologically favored commercialization. Space commerce activists including Charles Chafer, Courtney Stadd, and Klaus Heiss lobbied vigorously. The outcome was that in May 1983 the White House issued a new policy favorable to private launch vehicle operators, which said that the U.S. government fully endorses and will facilitate commercial operations of expendable launch vehicles by the U.S. private sector. This policy applied to those vehicles previously developed for U.S. government use, as well as new space launch systems developed specifically for commercial applications.[42]

Shortly thereafter, NASA announced a worldwide effort to market Shuttle launch services. Meanwhile, the Akaka bill was being marked up in the House Science and Technology Committee. Hearings were scheduled for November 1983. Representatives of private launch vehicle companies such as Chafer and Stadd lobbied for their preferences for a "lead agency," which tended to lean toward the Department of Commerce. However, the White House announced just before the hearings that the Department of Transportation had been chosen as the lead agency. In February of 1984, Transportation Secretary Elizabeth Dole announced the creation in her own office of an Office of Commercial Space Transportation, headed by Jennifer (Jenna) Dorn, which was to expedite applications for launch permits. A Commercial Space Launch Act (a revised version of the Space Commerce Act) was signed by President Reagan in October 1984.[43]

Privatizing Government Launch Vehicles

Once the Shuttle became operational, NASA's expendable launch vehicles were to be phased out and their production lines shut down. However, some companies saw a commercial opportunity in taking over these established systems, and the Reagan White House was sympathetic to the privatization of these vehicles. Aerospace engineer David Grimes founded a company called Transpace Carriers Inc. in September 1982. In May 1984, the company signed an agreement with NASA to transfer that agency's Delta launch vehicle program to the firm, which describes itself as "the first operational privately owned spacecraft launch services company in the U.S."[44]

Established corporations also showed interest in privatizing expendable launch vehicles. General Dynamics submitted a proposal to take over the Atlas Centaur launch vehicle, and it looked as if Martin Marietta Corporation might market the launches of its own Titan vehicle.

Buy Your Own Shuttle

The idea of commercializing the Space Shuttle occurred to several people in the 1970s. One was an ex-Braniff Airline pilot named William A. Good, who founded a small company called Earth Space Transport Systems. "Why," he reportedly asked, "should we continue to allow NASA to make the decisions on who sends what, or whom, into orbit?" In 1979, Good proposed a Space Transportation Act that would establish

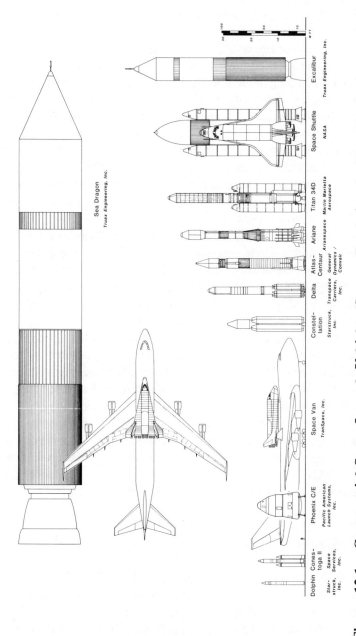

Figure 12.1 – Commercial Space Launch Vehicles (*Source:* Renderings by Tom Brosz, ed., *Commercial Space Report*, printed in *Space Calendar*, July 2-8, 1984, p. 3. Reprinted with permission.)

263

a commercial space transportation system in cooperation with other countries "to provide global access to outer space at the earliest practical date." In language reminiscent of OTRAG and Earthport, care and attention was to be given to providing such services to economically less developed countries.[45]

Another advocate was Klaus Heiss. In 1978, when the Carter administration cancelled funding for the fifth Shuttle orbiter, Heiss asked the Office of Management and Budget if they would agree to a private Shuttle purchase. According to Heiss, OMB said yes. Working closely with the Boeing Company, which had been studying the concept, Heiss, in 1979, started the Space Transportation Company (Spacetran) in a joint venture with the investment firm William Sword and Company, with the intention of buying the yet-unbuilt fifth Shuttle orbiter and making it a commercial venture. The company, which reportedly got a positive reaction from the Reagan administration, made proposals to NASA in 1982, but the government decided that no fifth orbiter was necessary. Spacetran decided in May 1982 that an expendable launch vehicle backup was needed and made a proposal to commercialize the Titan launch vehicle. In the view of Heiss, there could have been a commercial Titan in the spring of 1983 if NASA had not been opposed to competition for the Shuttle. In May 1983, Federal Express bought into Spacetran and created a new company called Fedex Spacetran.[46]

In 1983, aerospace executive Willard Rockwell stepped into the picture with a firm called Cyprus International, later Astrotech International, which approached NASA about purchasing a fifth orbiter (which Rockwell International manufactures). Astrotech has since established a subsidiary called Space Shuttle of America.[47]

In May 1984, a senior NASA official said chances are ten to one that the space agency will eventually turn over Shuttle operations to the private sector, beginning in two to three years.[48] Meanwhile, a firm called Third Milennium Inc. (formerly Transpace Inc.) proposed a mini-shuttle to be launched from the top of a Boeing 747 aircraft.[49]

Upper Stages

By design, the Space Shuttle carries payloads into a low Earth orbit; getting to higher orbits requires an upper stage. For its large satellites, the Air Force has developed the Inertial Upper Stage. NASA is adapting the Centaur upper stage used on expendable launch vehicles for use on the

Shuttle beginning in 1986. For small payloads, McDonnell Douglas Corporation provides the Payload Assist Module, developed and marketed under a 1976 agreement with NASA. Some satellite manufacturers are including propulsion units in their satellites.

A group of new space entrepreneurs believes that this array still leaves niches unfilled. During the 1980-81 academic year, Harvard Business School student David Thompson put together a team of students from the business and law schools to perform a six-month study for NASA on the prospects for space commercialization. Meanwhile, Thompson had learned that the Houston-based Space Foundation was offering fellowships for research on the same subject. After the NASA project was completed, Thompson and his colleagues applied for and won the Space Foundation award.

In early 1982, Thompson, Scott Webster, and Bruce Ferguson – none of whom were over 30 – saw a commercial opportunity in upper stages. They set up Orbital Sciences Corporation (originally Orbital Systems Corporation) to develop low-cost space transportation hardware. Texans Fred Alcorn and Sam Dunnam, both Space Foundation supporters, provided $300,000 in seed capital to finance the venture in its early phases.

Orbital Services signed a memorandum of understanding with NASA in December 1982, and in April 1983 the company signed an agreement with the space agency through which responsibility for the development of the Transfer Orbit Stage was transferred to OSC. NASA agreed not to develop a similar, competing vehicle of its own. Since then, OSC has outlined plans to develop an Apogee and Maneuvering Stage to serve a different class of payloads. In June 1984 the company announced the completion of a $63 million financial package, the largest private investment to that date in the space transportation industry.

OSC is aiming for an initial launch capability for its Transfer Orbit Stage in 1986.[50] In the best space enthusiast tradition, Thompson reportedly is seeking an assurance from NASA that he will be on that first flight.[51] His company is not just a source of income but a ticket into space.

SPACE MANUFACTURING

Space factories that would manufacture products for sale are central to the idea of space industrialization. The primary concept has been that of materials processing in space (MPS), taking advantage of the unique characteristics of the space environment, particularly microgravity. Although MPS experiments have been done since the Apollo days, many experts have remained skeptical about the commercial viability of space-based industry.[52]

The Space Shuttle helped bring the idea closer to reality. Tiny polystyrene spheres manufactured during the sixth Shuttle mission were sold by the National Bureau of Standards in 1985. Meeting two criteria for successful space manufacturing – they are expensive per unit and cannot be made as well on Earth – the spheres are the first commercial products made in space.[53]

More significant will be the first product made by a commercial firm. The Ortho division of Johnson and Johnson joined together with the McDonnell Douglas Corporation to build units that perform electrophoresis in the Shuttle orbiter, separating materials more efficiently than is possible on Earth. After successful experiments in 1983, which stirred widespread interest in space manufacturing, the companies hoped to begin production on the first flight of the Shuttle orbiter Discovery in the summer of 1984 (unfortunately, the electrophoresis products from that mission were contaminated).[54] The same mission also carried the first non-astronaut, McDonnell-Douglas employee Charles Walker (both the L-5 Society and the National Space Institute claimed him as their first member to go into space). NASA also has signed joint endeavor agreements with the small Florida-based firm Microgravity Associates to develop a process for the production in space of semiconductor materials. October 1984 saw the announcement of a ten-year agreement between NASA and the 3M Corporation for MPS work on the Shuttle orbiter and in the space station.[55]

If industries are to grow in space, they will need facilities more permanent than the Shuttle orbiter. Fairchild Space Company is developing an unmanned satellite called Leasecraft, which would provide an orbiting platform with control, power, and communications facilities to sustain various kinds of commercial operations in space.[56] Former NASA spacecraft designer Max Faget heads a company called Space Industries Inc., located near the Johnson Space Center outside Houston, which plans to build a pressurized orbiting module capable of providing a shirt-sleeved environment for technicians visiting automated manufacturing operations in space.[57] Like the Spacelab sometimes carried on the Shuttle, space factories will open up opportunities for more non-astronauts to go into space.

SPACE COMMERCIALIZATION GATHERS MOMENTUM

The space commercialization phenomenon visibly gathered momentum after the May 1983 White House announcement concerning expendable launch vehicles. That same month, the National Academy of Public Administration issued a report, done at the request of NASA, entitled *Encouraging Business Ventures in Space Technology*.[58] The report recommended that the administration declare and institutionalize a major commitment to the commercialization of space technology. Almost simultaneously, the National Chamber Foundation, which is affiliated with the U.S. Chamber of Commerce, initiated a study of how the environment for space commercialization can be improved. The principal researcher was Gregg Fawkes, himself a young entrepreneur. NASA established its own commercialization task force in June 1983.

In August, a group of aerospace executives met with President Reagan to discuss the commercialization of space, reportedly telling him that the White House must take a more aggressive and more publicly stated position that space commercialization would be good for the nation.[59] This was followed by the creation of a task force on the subject. In September, Cabinet Secretary Craig L. Fuller, who reportedly arranged the August meeting, was said to be writing a section on the commercialization of space for a "National Space Agenda."[60]

A symbolic culmination of this process came in the President's State of the Union address in January 1984 when he strongly endorsed space commercialization in connection with his space station initiative. In his radio address of January 28, 1984, the President said, "Obstacles to private sector space activities will be removed, and we'll take appropriate steps to spur private enterprise in space."[61] The White House issued a statement on July 20, 1984, concerning policy and legislative changes needed to achieve these goals and established a Cabinet Council on Commerce and Trade Working Group on Space Commercialization.[62] The Trade and Tariff Act of 1984 removed one impediment by permitting products manufactured in space by U.S. companies to enter the United States without duty.[63] In November, NASA announced the results of its own study of policy on space commercialization and outlined the functions of its Office of Commercial Programs.[64]

Meanwhile, the subject of space commercialization has become an industry itself. A Boston-based company called the Center for Space Policy Inc., staffed by young executives, provides consulting services for companies interested in the field. As of 1985, the newsletter *Space*

Business News appeared to be flourishing under its omnipresent editor Linda Billings and had been joined by *Space Commerce Bulletin, International Space Business Review, Commercial Space,* and others. Conferences on space commercialization are now a regular feature of the aerospace calendar.

THE FUTURE OF SPACE COMMERCIALIZATION

Space commercialization had become a legitimized field by the early 1980s and a growing one. Orbital Sciences Corporation President David Thompson has said that the number of companies in space enterprise went from 3 in 1980, when they invested about $10.5 million, to 25 in 1983, when their total investment was estimated at about $175 million.[65] *Aviation Week and Space Technology* and *Space Calendar* have published lists of companies involved in space commercialization; as many as 350 had expressed interest as of 1984.[66] The Center for Space Policy has projected the potential annual revenues to be generated by the end of the century by major category of space business as follows:[67]

Satellite Communications	$ 15 billion
Remote Sensing	$ 2 billion
Materials Processing in Space	$40 billion
Launch Services	$1 billion
On-orbit Services	$2 billion

One new organization betting on the future of space industrialization is the American Interstellar Society, a nonprofit membership organization seeking to build a financial base to support entrepreneurial space ventures. Like the Committee for the Future's Project Harvest Moon, the society believes that people who own a share of stock will have more of a proprietary stake in the development of space commercialization than those who are just onlookers.

Some observers, notably George Washington University space policy expert John M. Logsdon, have warned that space commercialization is in danger of being oversold.[68] Like the space phenomenon in general, it may not develop as quickly as some of its advocates would wish, and many of the new companies will fail. Both Jerry Grey and Charles Sheffield believe that true commercial activity will not get under way until the 1990s.[69] On the other hand, there also have been predictions that space commerce then will grow with unexpected rapidity, partly because there are fewer vested interests or legal and regulatory restrictions in space. "The most profitable thing we will be doing in 2000," says Gregg Fawkes, "cannot be predicted."[70]

Philosophically and politically, many of the new space entrepreneurs are closely linked with hard-line private enterprise ideology. Two of the most vocal exponents in Congress for space industrialization led by the private sector have been Congressmen Newt Gingrich of Georgia and Robert Walker of Pennsylvania, who also are leading spokesmen for the Conservative Opportunity Society.

However, the motives are not just economic. One finds in many of the space entrepreneurs that same streak of independence or libertarianism that appeared in some of the pro-space citizens groups, notably the L-5 Society and the American Society of Aerospace Pilots. They see in space commercialization an opportunity not just for profit but also for liberty.

Entrepreneurship also provides an alternative, possibly shorter route to career advancement in the space field than the traditional choices open to pro-space people, such as working one's way up through the bureaucracy of NASA or a large aerospace corporation or being active in a pro-space group. Some of the pro-space groups have provided avenues for individuals to get involved in space commerce; some of the new space entrepreneurs, such as GCH, Arc Technologies/Starstruck, and Space Services Incorporated, hired space activists. In a sense, the new, young space entrepreneurs are a subculture of the pro-space movement.

It would seem that some space advocates have found a new way of turning space as an avocation into space as a vocation. Commercialization is a short cut to the space dream.

Space commercialization could create a new and potentially influential set of economic interests, distinctively oriented toward space rather than being submerged in aerospace. This could strengthen the pro-space constituency. Many of the entrepreneurial companies are not contractors to NASA but independent actors whose lobbying in Washington is more against restrictions on their activities than for federal funding. As of 1984, lobbyists for the new launch vehicle companies were exchanging information in an informal network. With the support of the administration, the National Chamber of Commerce, and others, a loose coalition supporting private sector space development is emerging, far better funded than the citizens activists who played a role in legitimizing these ideas and bringing them to the attention of the public.

Space commercialization, like space militarization, is a potentially divisive issue for the pro-space movement. In the eyes of some, space commercialization tarnishes the "purity" of space. But the criticisms have been much less severe than in the case of space weapons. Many recognize that space commercialization could help to reduce the dependence of civilian space activities on politics and the annual budget process; it is an alternative to politically or militarily driven space projects (although it has become possible only because NASA and the Department of Defense funded the development of many of the technologies). Although it may be a slower route to the planets and the stars than a politically motivated space program like Apollo, space commercialization may be more lasting. Observes Robert L. Staehle of the World Space Foundation: "No frontier has ever prospered on government money alone."[71]

13

THE SPACE STATION DECISION

> The design, construction, and establishment of the manned space station should be completed by about 1970.
>
> Willy Ley, 1958[1]

> Tonight, I am directing NASA to develop a permanently manned space station and to do it within a decade.
>
> Ronald Reagan, 1984[2]

In January 1984, President Ronald Reagan told the nation that he was directing NASA to develop a permanently manned space station within a decade. Here was the next logical step in the realization of the classic agenda for manned spaceflight, which also would contribute to the more diverse agenda of space industrialization and colonization. It gave NASA and several of its centers a major new project, ensuring the continued existence of their skilled teams of space technology experts; without the station or a task of comparable magnitude, NASA would have been in line for a big layoff.[3] It gave the aerospace industry and aerospace professionals the prospect of new contracts and employment, a successor to the Space Shuttle program. The decision also gave pro-space Americans a new focal point for their support and another symbolic victory not long after the first flights of the Shuttle. Frustrated by the drought of the 1970s and stimulated by the first flights of the Shuttle, much of the pro-space constituency coalesced around the idea of a space station.

How did the space station decision come about? A study of the public record suggests that the principal lobby for the space station was NASA

271

itself, whose leaders conducted a long, patient campaign against skeptics and budget-cutters. NASA Administrator James Beggs in particular showed determination and political skill. Intermediaries in the White House played a significant role. The President's own interest in space, and its compatibility with the new pro-development, pro-private enterprise conservatism, were important factors. Friends in Congress provided support when it was needed. Aerospace companies and aerospace professional groups were interested and active participants in the process, although the decision appears to have taken much of a somewhat cynical aerospace community by surprise. Pro-space citizens groups had little influence on the decision itself but helped to create a favorable opinion climate and to provide a supporting chorus when the project first went to Congress.

A LONG HISTORY

The idea of a space station goes back a long way, having been featured in works of fiction in the 19th century, and in turn of the century writings by Konstantin Tsiolkowsky.[4] Hermann Oberth pointed out several potential uses of a space station in 1923.[5] In 1929, Hermann Noordung (real name Hermann Potocnic) proposed a toroidal design that was refined in new proposals by von Braun and others in the 1950s and was imprinted on our imaginations in revised form in the 1968 film *2001: A Space Odyssey*.[6] "Development of the space station," von Braun reportedly said, "is as inevitable as the rising of the Sun."[6] The station could be an observatory, a communications link, and a way station to the Moon and the planets. It was a central feature of the classic agenda for manned spaceflight and the beginning of permanent human presence in space.

Space station concepts have been under study by NASA and its contractors since the agency's earliest days.[7] The Institute of the Aeronautical Sciences sponsored a space station symposium in 1960. NASA's Langley Research Center established an office to study a manned orbiting research laboratory in 1963, at about the time that work began on the U.S. Air Force Manned Orbiting Laboratory. Planning for the Skylab short-duration space station began in 1967.[8] The Space Task Group's 1969 report to the President included a proposal for a continuously operating low altitude station with 6 to 12 occupants, to be followed by later stations in polar and synchronous orbits. A station and a shuttle were both part of NASA's vision of the American future in space.

The Nixon White House forced NASA to choose between a shuttle and a space station. According to former NASA Deputy Administrator Hans Mark, Wernher von Braun told a meeting of NASA officials at that time that the choice must be a shuttle, which was technically more difficult and which would lead naturally to the station. (Space writer James E. Oberg and others have noted that the Soviets made the opposite choice, first building a station and then a shuttle, and that the two great space powers will be converging again in the 1990s.[9]) The Nixon White House made it clear that NASA was to get a shuttle flying first; that policy was continued under the Ford and Carter administrations. However, NASA never gave up the space station idea. Throughout the 1970s, continued low-key planning efforts went on, particularly at the Marshall and Johnson Space Flight centers and in the Office of Space Flight at NASA headquarters. NASA released a booklet entitled *Space Station: Key to the Future* in the early 1970s, before agreement had been reached on the design of a shuttle.[10] Aerospace industry people and other aerospace professionals frequently presented ideas for space stations at conferences.

NASA got a temporary space station in Skylab, used in 1973-74. Skylab, which was basically an extension of Apollo technologies, burned up in the atmosphere in 1979, before the Space Shuttle went into operation. In 1975, NASA contractors conducted a study of the Manned Orbital Systems Concept, but the idea was not pursued "for budgetary reasons."[11] In that year, NASA released a booklet entitled *The Manned Orbital Facility: A User's Guide* to stimulate discussion about the uses of a space station; among the groups receiving it was FASST.[12] AIAA, an important forum for legitimizing new space program ideas, published a study favorable to space stations in the September 1975 issue of *Aeronautics and Astronautics*.[13] Meanwhile, the Soviet Union had begun putting up its Salyut space stations in 1971 and was continuing to use them as this book was written. Technologies for a space station clearly were within reach. However, the politics for a space station decision were not yet right.

In November 1976, *Aviation Week* reported that "NASA believes a program start for the first permanently manned space station could be part of a congressional budget request as early as fiscal 1979."[14] In May 1977, the United States and the Soviet Union announced they would discuss a possible international space station, but those talks fell victim to worsening U.S.-Soviet relations.[15] Subsequently, there was an increasing number of stories in *Aviation Week* about alleged Soviet development of a 12-man space station, possibly an invocation of the Soviet threat.[16]

By 1979, NASA's Johnson Space Center was studying a Space Operations Center, another euphemism for a space station. However, NASA then was dealing with cost overruns on the Shuttle and was not in a good position politically to seek a major new manned spaceflight project. *Aviation Week* reported in August 1980 that a consensus was growing in NASA "that the agency must pursue aggressively a permanently manned station as the next major U.S. manned space goal," with a projected new start for fiscal year 1984 or 1985. NASA official Rocco Petrone reportedly said, "The civilian space program today is in dire need of a bold central theme."[17] Two months later, NASA Administrator Robert Frosch reportedly said that the time was ripe to start seriously looking at a major new U.S. space initiative and that the Space Operations Center was a good example of what could be done. He added that NASA was then underutilized.[18] By 1980, NASA was hoping for a 1985 or 1986 flight of a science and applications platform, followed five years later by the Space Operations Center.[19] However, the Carter White House apparently never bought the idea of making the space station a new start. A Space Coalition poll reported in September 1980 that the aerospace industry was pessimistic about the prospects for a major new space initiative.[20]

CITIZENS RESPOND

Meanwhile, some pro-space citizens groups, notably the L-5 Society, had begun to focus on the space station as the next goal for their advocacy. The October 1976 *L-5 News* carried a major feature on the Manned Orbital Facility.[21] By 1980, an operational Space Shuttle was in sight, while space colonies and satellite solar power stations were beginning to fade farther into the future. The station provided a concrete goal, realizable within the lifetimes of citizen space enthusiasts and an organizing theme for their groups. "Such a project will almost certainly not happen," wrote Gerald W. Driggers, "if the L-5 Society does not turn its efforts toward causing it to happen."[22]

Most pro-space citizens groups saw the Reagan administration as an opportunity for a step forward. Once again, the L-5 Society and its affiliates tended to be out in front of other groups. The January 1981 *L-5 News* announced a letter-writing campaign urging support for the station, as well as for research on solar power satellites.[23] The Citizens Advisory Council on National Space Policy, in the report of its January 1981 meeting, recommended a low Earth orbit base, to be partially operational by the fall of 1988, which would serve as the hub of a "space industrial park" for private enterprise.[24] What was new was not the idea of a space station but using commercialization as a major argument for it. During the first flight of the Space Shuttle in April 1981 – a signal to aspire to a next step – the L-5 Society used its phone tree to press the Office of Science and Technology Policy to urge the President to adopt a permanently manned station as its next goal.[25]

Despite such campaigning, NASA was not given a major new mission during the Reagan administration's early years. One reason appears to have been a desire to see the Space Shuttle become operational before moving on to the next step. Once Shuttle development costs began to decline, a funding "wedge" would become available, assuming that NASA's total budget remained roughly the same.

THE SUCCESSFUL CAMPAIGN

At the beginning of the Reagan administration, NASA still was preoccupied with the Shuttle and with cuts in the planetary program. But the desire to start on the station remained strong. Ivan Bekey, head of advanced programs in NASA's office of Space Transportation Systems, was quoted as believing in March 1981 that a Reagan statement that "our next major step in space should be the establishment of a permanent manned presence in space" would suffice, leaving NASA to determine the details.[26]

The first clear political signal of NASA's aspirations under the Reagan administration came in May 1981 congressional hearings honoring the astronauts who flew on the first Shuttle mission, when acting NASA Administrator Alan Lovelace and astronauts John W. Young and Robert L. Crippen said that the station was the next logical step.[27] A few weeks later, NASA Administrator-designate James Beggs and Deputy Administrator-designate Hans Mark strongly supported the establishment of a permanently manned space station as the next U.S. space goal.[28] Beggs was a former vice-president of General Dynamics. Mark, a former director of NASA's Ames Research Center who had served as secretary of the Air Force in the Carter administration, attended the second meeting of the Citizens Advisory Committee on National Space Policy and was regarded by the L-5 Society as a friend in court.[29]

What followed was an excellent example of a sustained campaign by an agency's leadership to get White House approval for a major new program. NASA had learned lessons from its 1969-72 failure to win White House support of its grand agenda. This time the ground was laid carefully, and efforts were made to build alliances and create a consensus. Aerospace companies, although not driving the effort, were valuable allies in providing ideas and doing lobbying. Rockwell International, for example, presented a briefing in October 1981 that supported a near-term focus on an "Earth Support Base."[30] But the station did not come easy. Several sources report that Beggs and Mark and their colleagues in NASA made four unsuccessful attempts before they finally succeeded: (1) in the late 1981 preparation of the administration's fiscal year 1983 budget request and the President's January 1982 State of the Union message; (2) during the 1981-82 space policy study and the related preparations for the President's July 4, 1982 speech welcoming the return of the fourth Space Shuttle mission; (3) in the late 1982 preparations for the fiscal year 1984 budget; and (4) during the leadup to the President's October 1983 speech at NASA's 25th anniversary celebration. Each of these efforts provoked opposition.

In an interview published in the December 1981 *Omni*, Beggs said the station appealed to him as the next major goal for NASA.[31] However, despite public statements by Beggs and Mark that the station was the next logical step, the late 1981 effort made little headway. White House Office of Science and Technology Policy official Victor Reis reportedly expressed doubts about the need for a station in November.[32] In connection with this attempt, James Muncy and other space activists launched a campaign through *Omni* magazine to get people to write to Vice-President Bush about a "Skyport," in the hope of getting the President to include a space station in his January 1982 State of the Union address.[33] Although this effort produced a fairly significant volume of letters, it had no detectable effect on White House decision makers.

After this initial failure, NASA broadened its efforts. Space station studies under way at Marshall and Johnson Space Centers reportedly were to be combined into a formal NASA position for presentation to President Reagan in the fall of 1982, with the objective of getting a fiscal year 1984 new program start.[34] In March 1982, *Aviation Week* reported that Beggs wanted to give President Reagan a plan for U.S. space station/space platform development by that summer as the next large U.S. space goal. NASA was considering inviting Canada, the Europeans, and Japan to participate and briefed foreign governments. Presidential Science Adviser George Keyworth reportedly was opposed to the goal.[35] In its first issue of January 1982, *Aviation Week* carried Planetary Society Executive Director Louis D. Friedman's letter opposing a space station.[36] However, the momentum behind the station was building in other parts of the space advocacy.

In May 1982, the AIAA published its proposed space policy in *Astronautics and Aeronautics*, calling on the government to establish a commitment to a manned space facility in low Earth orbit.[37] That same month, NASA formed a space station task force, while commissioning space station studies by the Space Science Board, the Space Applications Board, and various contractors. The political visibility of this step suggests that it was done with White House knowledge, if not White House approval. *Aviation Week* reported at the time that NASA would request about $50 million for fiscal year 1984 to initiate work on the station. Beggs reportedly believed that he was making progress in convincing the Reagan administration that justification existed to initiate the program, although the Department of Defense was indicating no need for a station. By that autumn, the European Space Agency, Japan, Canada, France, and Italy reportedly had begun studies to define possible participation in the station.[38]

Meanwhile, the White House had initiated a study of U.S. space policy in August 1981. The results were released on July 4, 1982, the day the fourth Space Shuttle mission returned to Earth and the Shuttle was declared operational. Although the study did not lead directly to a major new program start, it hinted at the future. A White House fact sheet on national space policy stated this as one of the policies that was to govern the conduct of the civil space program: continue to explore the requirements, operational concepts, and technology associated with permanent space facilities.[39] Speaking at the welcoming ceremony for the returning Shuttle astronauts, President Reagan stated, "We must look aggressively to the future by demonstrating the potential of the shuttle and establishing a more permanent presence in space."[40]

The President's phrase was the outcome of a fairly sharp bureaucratic battle between NASA on the one hand and the Office of Management and Budget and the President's science advisor on the other. Early drafts of the speech reportedly referred specifically to "a permanent manned presence in space," but this was changed through the intervention of Science Adviser George Keyworth and Budget Director David Stockman. NASA's lobbying was said to have included direct appeals to the President and the Office of White House Counselor Edwin Meese. Keyworth was quoted as saying later that "it was improper to put that kind of pressure on the President." Another player who was a space station advocate was Air Force Colonel Gilbert Rye, then handling space issues on the National Security Council Staff, who is said to have explained the policy statement to the President.[41] After July 4, 1982, Beggs could argue that NASA had a charter to study the potential and technology of a permanent manned facility. But Keyworth argued that a clearly defined objective for a station and a grand vision of what might lie beyond it were needed.[42]

The new space policy established a Senior Interagency Group on Space, which later was tasked to look at options on the issue of a space station. Reportedly because of Rye, who was believed to have had the backing of then National Security Adviser William Clark, Keyworth's office and Stockman's office were given only observer status on this new committee. Rye also helped in the creation of the U.S. Air Force Space Command, which came into existence in September 1982.[43]

Outside groups had become increasingly active on the issue. According to *Science*, hundreds of letters and telegrams were sent directly to Keyworth's office, in what he called "a carefully orchestrated campaign." The Office of Science and Technology Policy staff counted 17 newspaper and magazine articles that predicted an announcement of the space station during the President's July 4 speech.[44] Meanwhile, NASA's number three official Philip E. Culbertson published a major article on NASA's space station plans in the September 1982 *Astronautics and Aeronautics*.[45]

Having won a partial victory, NASA then tried to get the station into the fiscal year 1984 budget, reportedly asking unsuccessfully for money for industry studies of a space station.[46] Keyworth told a House committee in February 1983 that, from the point of view of scientific research, U.S. development of a manned space station would be "an unfortunate step backwards."[47]

AIAA, which had been consistently supportive, published a major section on space station technologies in the March 1983 *Astronautics and Aeronautics*. In that issue, space writer David Dooling (based in Huntsville, Alabama, where the Marshall Spaceflight Center is located) reported that the Office of Management and Budget had directed a cutback in space station studies, possibly "to head off the formation of a space station lobby or constituency."[48] However, *Aviation Week* reported that same month that it was possible that a formal space station line item could appear in the fiscal 1985 budget and that data acquired by the Senior Interagency Group on Space in a study started in October 1982 could lead to an administration statement on the station within the next year.[49]

Aviation Week also reported that NASA's space station studies had matured to a point "where findings and decisions made over the next eight months could affect the character of the agency for the next 10-20 years." This drew a negative response from Planetary Society Executive Director Louis Friedman, who wrote that "now is hardly the time to embark on such decisions."[50] Opposition from some scientists was to become increasingly visible later.

THE YEAR OF DECISION

By mid-1983, there was growing evidence that the Reagan administration was moving toward a major new decision on space. Science Adviser George Keyworth, who had been a critic of the space station, startled many in July 1983 by calling for an ambitious new space initiative, such as a Moon base or a Mars landing.[51] Meanwhile, NASA contracted with AIAA to hold a major conference on the space station that month, possibly to give the station idea greater credibility. At that meeting, Administrator Beggs said that he expected White House approval within the next 6 to 12 months of a space station as the next U.S. initiative in space and predicted that the President would approve $200 million for advanced space station design work in the fiscal year 1985 budget.[52] House Science and Technology Committee chairman Don Fuqua reportedly said at the same meeting that the House and Senate committees would respond to a space station initiative by saying, "We've waited long enough."[53] However, the Office of Management and Budget continued to oppose the station. Meanwhile, Spacepac announced in a summer 1983 letter that "the space station is the key to our goals."[54]

One factor that helped to break the log jam was the idea of space commercialization and its adoption by a key White House insider. During early 1983, the Senior Interagency Group (Space) was discussing the commercialization of expendable launch vehicles and other space activities. Cabinet Secretary Craig Fuller, a former advertising executive who occupied a key position within the White House, reportedly dropped in on one of these discussions and, according to *Science* magazine, became so interested that he volunteered to write the group's report on the subject.[55] Fuller, a strong supporter of private enterprise, seems to have bought the ideas of space commercialization and space industrialization, which were compatible with the administration's ideology. Patiently supporting these concepts within the administration, he became a key ally of the NASA leadership, helping to bring the space station in through the commercialization door.[56]

NASA's appreciation for Fuller's role came out in public later. After Fuller had spoken at the second Huntsville conference on space industrialization in February 1984, third-ranking NASA official Philip Culbertson stood up and said, "I have never heard an official close to the President who knew so much about the space program. I am sure that some of the President's enthusiasm is due to you. On behalf of NASA, I thank you."[57]

NASA, which had set up a space commercialization task force in 1983, had been consulting closely with company executives interested in investing in space enterprises. Apparently with Fuller's help, arrangements were made for 11 of these corporate managers to brief White House officials on August 3, 1983, on the kind of policy changes needed to obtain the financing and space program stability necessary to stimulate the growth of space commerce. This group then had a meeting with the President, reportedly telling him that approval of a space station program would be the best way to introduce such support and stability while providing facilities in orbit that would stimulate space commercialization. According to *Science*, the President said, "I want a space station too. I have wanted one for a long time." However, he qualified the statement by commenting on the realities of federal budgeting. Participants in the meeting reportedly got the impression that the President was enthused by the prospects for commercial activity in space.[58] This also helped to broaden the industrial constituency for the space program.

Meanwhile, Senator and ex-astronaut John Glenn added a political dimension by announcing his ideas for the future of the U.S. space program, including a commitment to a manned space station. An interview with Glenn, then a candidate for the Democratic nomination for President, appeared in *Omni* in October 1983.[59]

In November 1983, the Space Shuttle carried the first Spacelab into orbit. This European-built module, designed for scientific and industrial experiments, was similar to some concepts for space station modules and showed that researchers could work in space.[60] By implication, it also broadened the range of those who could go into space well beyond an elite astronaut corps.

Following review by the Senior Interagency Group (Space), space station options were to be presented to the President as a prelude to White House decisions on the fiscal year 1985 budget. Reportedly, many influential members of the group lined up against the station, including the Office of Management and Budget, the State Department, the Defense Department, and the Joint Chiefs of Staff. The move of the Defense Department from a neutral position was said to reflect concern that the station would drain federal development money. Favoring the station were NASA, the Commerce Department, and the Arms Control and Disarmament Agency.[61] This disagreement continued into October, when *Science* reported a lack of consensus within the government on the space station.[62] NASA reportedly had requested initial space station funding of $235 million. Meanwhile, at House hearings in October 1983, former NASA administrators supported the space station.

The National Academy of Science's Space Science Board issued a report in September that stated, "The board sees no scientific need for this space station during the next 20 years."[63] This set off a new round of media stories about scientific criticism of the space station. However, the board's chairman, Thomas Donahue, was careful to point out later that the board had not opposed the station.[64] Less noticed by the media was the fact that the academy's Space Applications Board took a relatively favorable stand on the station.[65]

Meanwhile the President's political advisers reportedly were urging him to announce the space station with great fanfare in the fall to steal the thunder of Senator and ex-astronaut John Glenn, who was then running for the Democratic Presidential nomination. The occasion suggested was the October 19 celebration of the 25th anniversary of NASA. This did not happen, with Presidential Adviser Edwin Meese being quoted as saying that "NASA's birthday cake is not going to be in the shape of a space station."[66] Some of the language for the President's speech was written by space activist James Muncy, by then working in the Office of Science and Technology Policy and acting as a liaison between that office and "Conservative Opportunity Society" advocates in Congress, particularly Newt Gingrich. The speech, using Ben Bova's "High Road" terminology, urged NASA to be more visionary:

Right now we're putting together a National Space Strategy that will establish our priorities and guide and inspire our efforts in space for the next 25 years and beyond. . . . We're not just concerned about the next logical step in space. We're planning an entire road, a "high road" if you will, that will provide us a vision of limitless hope and opportunity, that will spotlight the incredible potential waiting to be used for the betterment of humankind. On this 25th anniversary, I would like to challenge you at NASA and the rest of America's space community: let us aim for goals that will carry us well into the next century.[67]

By November 1983, press stories were predicting that President Reagan would decide within the next few weeks to go ahead with a station.[68] Beggs reportedly was encouraged to place more emphasis on the Soviet threat in testimony presented before the Senate subcommittee dealing with space in mid-November.[69] At a space conference in Virginia late that month, National Security Council Staff member Colonel Gilbert Rye said that the President would receive space station options "very, very soon."[70]

All this encouraged pro-space citizens groups to be more active in supporting the space station initiative. However, the effort still was fragmented and uncertain at that stage. At the November 15 meeting of the National Coordinating Committee for Space, both David Webb and Leonard David said it was too late for the citizens groups to have a sizeable impact on the administration's decision. "The space movement is no longer setting the context," said Webb; "the question is how will we react to the President."[71]

The issue went before the Cabinet Council on Commerce and Trade and members of the National Security Council on December 1, 1983, when Beggs briefed the group on NASA's space station plan.[72] *Science* magazine reported that the options ranged from an Apollo-style crash program to doing nothing at all.[73] According to *Aerospace Daily*, defense community representatives indicated no requirement for a space station as well as concern for the potential cost of the program, while Office of Management and Budget officials expressed concern about the deficit situation.[74]

The crucial meeting was a Cabinet-level budget review. Sources conflict as to whether this was held on December 5 or December 16, or whether two meetings were held. It was at one of these meetings that Budget Director Stockman reportedly received a rebuff to his argument that the deficit could never be cut if the government continued to fund such exotic projects.[75] According to *Newsweek*, Attorney General

William French Smith responded, "I suspect the comptroller to King Ferdinand and Queen Isabella made the same pitch when Christopher Columbus came to court."[76] *Science* reports that it was President Reagan who mentioned Ferdinand, Isabella, and Columbus, at the meeting on December 1. According to *Science*, it was on December 5 that Stockman once again pressed his case against the station, but Reagan vetoed him and ordered that the station be planned for the budget.[77]

In any case, the outcome was that the NASA request for $235 million was reduced to $150 million but stayed in the President's budget request. NASA had won its biggest campaign in a decade. One signal of its confidence was a meeting between NASA and aerospace industry officials during the second week of December to discuss space station design issues.[78] European and Japanese space agencies then were actively studying participation in the station.[79]

By mid-December the word was out. The New York *Times* reported on December 14 that the Reagan administration appeared ready to commit itself to the station in its next budget.[80] The director of the Defense Advanced Research Projects Agency told the National Space Club the same day that "it appears that the President is going to make a decision in the not too distant future for NASA to proceed on a new manned space station."[81] By the end of December the chairmen of a full committee and two subcommittees of Congress – Don Fuqua and Harold Volkmer in the House and Slade Gorton in the Senate – had supported the station in letters to the President.[82]

Meanwhile, the citizens space groups had become active. On December 9, 1983, Spacepac alerted its members to write five senior administration officials in support of a space station and a lunar base.[83] Campaign for Space sent out "Space Station Alert" postcards on December 10.[84]

New National Coordinating Committee for Space (NCCS) Chairman David Webb began calling meetings of Washington representatives of pro-space organizations to discuss what they should do about the space station issue. It was agreed that the first step was to get an immediate input into the Cabinet to encourage the support of those already convinced and to stimulate people to be bold. Individual pro-space organizations launched letter-writing campaigns. Ben Bova sent a letter to National Space Institute members.[85] On January 12, Space Studies Institute Vice-President Gregg Maryniak encouraged members to support the station and a lunar base.[86] Other groups joined the effort, including the AIAA, *Omni*, Space Calendar, *Liftoff* magazine, Spaceweek,

Write Now!, and Star Fleet Command (a major group of "Star Trek" fans). On January 24, the L-5 Society initiated a campaign to support the President's plan.[87]

By early January, the NCCS group had decided that it would not have any significant influence on the decision itself, which appeared to have been made anyway, and shifted its efforts to supporting the decision once it was announced. As Muncy put it to the group, "we want to be a megaphone for what the President says."[88]

Two days before the State of the Union address, an NCCS drafting group led by Webb hammered out a statement of support, which was distributed on the day of the speech. Although time for coordinating with organizations around the country was short, the Washington group managed to get 15 pro-space organizations to sign. Here is the text of the press release:[89]

> The following member organizations of the National Coordinating Committee for Space, representing 100,000 members, supports President Reagan's anticipated announcement committing the nation to the establishment of a permanent manned space station.
>
> A space station is not a goal in itself, but an essential first step in regaining the United States' preeminence in space exploration and development. Funding for the station should be in addition to – not in place of – NASA's regular budget. There must not be a repeat of the 70's when the development of the space shuttle cut heavily into all our other space programs.
>
> A space station will enable the United States to compete more effectively with the Soviet Union and other nations that are so actively engaged in the exploration and industrialization of outer space. It will also enable our corporations and entrepreneurs to compete in the commercial development of this vital new frontier. In addition, it will provide a less expensive means for our continued, preeminent exploration of the solar system on behalf of all Mankind.
>
> A space station will establish the necessary launch base for our early return to the Moon and for the exploration of other planetary bodies. Such exploration will enable us to procure essential resources for continued development on Earth and in space.
>
> We commend the President for committing the United States to the establishment of a space station and the long-range development of outer space. We encourage the adoption of this program by the Congress.
>
> Endorsing Organizations:
> National Space Institute
> American Astronautical Society
> Space Studies Institute

L-5 Society
Spaceweek
Spacepac
Campaign for Space PAC
American Space Foundation
Piedmont Advocacy for Space
Chicago Society for Space Studies
OASIS/L-5
Students for the Exploration and Development of Space
Sunsat Energy Council
Maryland Space Futures Association
American Institute of Aeronautics and Astronautics

This is a high percentage of the major pro-space organizations, plus several minor ones. It was no small achievement for Webb to get them all to agree to the text, even though it reflected fairly well many of the views of the mainstream of the American space movement. However, there is one significant name missing from the list: the Planetary Society, which would have more than doubled the number of people represented by the signers. Although approached, it declined to sign because of real difficulties with the content of the statement.

This was the first important agreed text the NCCS had put out since 1981. However, there is no evidence that it ever reached senior levels in the administration or that it appeared in major media.

What may be of greater long-term interest is the fact that those citizen group representatives present managed to reach a consensus on a basic platform for space that included the space station, a lunar base, the SSEC core program, strong space sciences and applications programs, and in the longer term an asteroid probe, space industrialization, and a manned mission to Mars.

The President delivered his State of the Union message on January 25, 1984. To the surprise of most of his audience, space was one of his four main topics. And there was the space station initiative, clearly linked to industrialization and commercialization:[90]

A sparkling economy spurs initiative and ingenuity to create sunrise industries and make older ones more competitive. . . . Opportunities and jobs will multiply as we cross new thresholds of knowledge and reach deeper into the unknown. . . . We can follow our dreams to distant stars, living and working in space for peaceful economic and scientific gain. Tonight, I am directing NASA to develop a permanently manned space station and to do it within a decade. . . . A space station will permit quantum leaps in our research in science, communications, and in metals and life-saving medicines which can be manufactured only in space.

Space industrialization, a concept limited to technical and space enthusiast circles only five years before, had been endorsed at the highest level.

POST-DECISION LOBBYING

Once the decision was announced, pro-space groups shifted their efforts to supporting passage through the Congress of funding for the station for fiscal 1985. There was scattered but highly visible opposition from some scientists and a few other figures, such as aerospace workers leader William Winpisinger, who called the station "lunacy."[91] The L-5 Society again played the most active role; its Washington representative, Gary Oleson, went on half time from his job to lobby for the station. When it appeared that the House Authorization Committee might cut back funding for the space station, L-5 mobilized its phone tree and a write-in campaign and other pro-space organizations helped. Mark M. Hopkins wrote later that 115,000 messages had been sent in connection with the space station campaign, of which Spacepac and the L-5 Society paid for 75,000.[92] The committee eventually approved the full $150 million. Convinced that their effort made a difference, L-5 leaders cite a letter of appreciation to them from NASA Deputy Administrator Hans Mark.[93]

However, the battle was not over. By the time of the L-5 conference in April 1984, concern had shifted to the House Appropriations Committee, where some members favored cutting station funding and even making the station initially an unmanned platform. Spacepac leaders Mark Hopkins and David Brandt-Erichsen sent out an urgent letter calling on addressees to contact three swing members of the committee.[94] Strikingly, the L-5 Society and some other groups were particularly vehement in denouncing the idea that the station should initially be unmanned.

Criticisms again were heard from prominent scientists, including leaders of the Planetary Society, who revived old manned versus unmanned issues heard in debates since the 1960s. One tactic used by the pro-station forces was to head off opposition from some scientists by getting support from others. NASA formed a committee of scientists to study the scientific uses of the station.[95] In an inspired improvisation, space activists James Muncy, Gary Oleson, and Joe Hopkins created almost overnight an organization called Scientists for a Space Station, which put out a press release.[96] This group's most visible member was astronomer and former NASA official Robert Jastrow, who also was a supporter of the Strategic Defense Initiative (SDI). Another scientist critical of both SDI and the space station, Carl Sagan, was, of course, president of the Planetary Society, the only major pro-space group that declined to sign the NCCS statement.

As of 1985, it looked as if the station might turn out to be a mix of manned and unmanned elements. The committee of scientists convened by NASA eventually reported favorably on the scientific uses of the space station, a theme taken up in the pages of *Science* by two NASA scientists in December 1984.[97]

NASA, with a little help from its friends, got the requested $150 million, although part of it was set aside for studies of unmanned options. The President signed the authorization and appropriations bills in July 1984. NASA announced the formation of a space station program office shortly thereafter and began issuing requests for proposals, setting off the biggest scramble for space contracts since 1972. The agency was back on track with the classic agenda for manned spaceflight. However, this was just a step toward grander visions. NASA official Philip Culbertson told the National Space Club that the station itself is not a mission but an enabling capability.[98]

In November 1984, the Congressional Office of Technology Assessment (OTA) released its report *Civilian Space Stations and the U.S. Future in Space*.[99] Although the media tended to focus on the report's criticisms of NASA's approach to the space station, OTA was in fact supportive of a permanent human presence in space. The report included OTA's own proposals for future U.S. objectives in space, which drew on some ideas supported by the pro-space movement: a modest human presence on the Moon, further exploration of Mars and some asteroids, and taking at least hundreds of members of the general public into space every year. Even the OTA accepted some of the space vision.

The future of the space station was not ensured as of late 1984; it faced difficult budget tests in 1985 and 1986 as the proposed funding increased at a time when the Reagan administration and Congress were emphasizing the need to reduce the federal deficit. That fiscal crunch could be a test of support for manned spaceflight, and for the pro-space lobby. But optimism was surviving. Satellite solar power station inventor

Peter Glaser wrote in November 1984 that the station would open the door to U.S. industry in space.[100] NASA Administrator James Beggs observed that, starting in the early 1990s, Americans would be in space permanently.[101]

SUMMATION

Former NASA press spokesman Brian Duff describes the station initiative as "a staffed decision," in which the classic bureaucratic moves worked.[102] Although NASA was the principal lobbying force, it was critically dependent on receptive attitudes in the White House. For its advocacy to bear fruit, at least two things had to happen: the proposed program had to gain credibility both within the technical community and the interested public and the White House had to find it politically useful and ideologically compatible. Space station advocates were fortunate to have a President who found space development personally interesting. However, the steadiness of purpose of NASA's leaders also was critical. Beggs and his colleagues never wavered.

The space station decision seems to support the validity of a thesis discussed in the space weapons chapter: in general, major new technological ideas and projects cannot be rushed through the governmental and political system but require the building of a constituency and improved credibilty. Major top-down decisions such as Apollo can occur only in rare circumstances.

There is a significant difference in perception between outsiders and insiders as to the role of citizens groups in the space station decision. The L-5 Society and its offshoot, the Citizens Advisory Council on National Space Policy, claim to have had real influence on the President's decision and on the content of his State of the Union speech. However, two people who worked in the White House Office of Science and Technology Policy during the period say that the citizens groups had no significant influence.[103] What the space movement clearly had done was to help prepare the way with the interested public and to signal the existence of a constituency for space.

PART III
SUMMING UP

14

CONCLUSION

Today, at least, the space underground continues to work and dream without substantial support. The promises it commits to paper cannot be transposed into plastic and metal. The movement will live on; but its future is not under its control.

Richard Hutton, 1981[1]

Nothing great was ever achieved without enthusiasm.

Ralph Waldo Emerson, 1841[2]

You can make history happen.

Klaus Heiss, 1983[3]

There clearly was a major upsurge of organized interest in, and advocacy of, space activity between the mid-1970s and the early 1980s. The motivation to be associated with the space enterprise remained strong long after the first "Spaceflight Revolution" was achieved. People had responded to it in different ways, depending on the strength of that motivation, the skills they had to offer, and the opportunities they saw open to them.

Why did this sudden broadening of interest in, and enthusiasm for, spaceflight occur when it did? The downturn in the fortunes of the space program may explain the reactions of those with an economic or professional stake in it, but what about the others?

CAUSES

A major underlying factor may have been that the years 1977 to 1981 – the years of rapid growth in the new space advocacy – marked a turning point in the maturing of the spaceflight revolution and its acceptance into the American cultural and intellectual heritage. When William S. Bainbridge published his book in 1976, the outcome of space advocacy still appeared to him to be in some doubt. "Either spaceflight will be proven a successful revolution that opened the heavens to human use and habitation," he wrote, "or it will be proven an unsuccessful revolution that demonstrated in its failure the limits of technological advance."[4]

A year later, in 1977, when American space interest groups began to proliferate, 20 years had passed since Sputnik I. In the United States, spaceflight was passing out of its early adventurous years and pausing before entering a time of expanded, routine spaceflight operations. The first generation of spaceflight technology had achieved a plateau of technical maturity; the second was not yet in operation. Space concepts that once had seemed exotic, such as escaping a "gravity well" into "free space," were being digested and assimilated by nonspecialists. The use of space for a broad range of human purposes – the utilization of this largest and, to us, newest environment – was being accepted by a growing number of Americans.

This did not happen easily; there was cultural and intellectual resistance. The wave of reaction to spaceflight and other high technology ventures in the late 1960s and early 1970s receded slowly. Many critics in the 1970s still saw spaceflight as an aberration, a technological "stunt"; some regarded space as an alien and hostile environment, appropriate only for scientific investigation. Some Americans, including Walter Mondale, still seemed to associate space with heaven. The decline in public support for spaceflight reflected not only concern with other priorities but also a cultural lag in the acceptance of space as a place where humans belonged.

Advocacy was needed to help overcome this resistance to a paradigm shift, which some advocates believe to be of Copernican proportions. At its grandest, that advocacy made urgent and probably premature claims for space colonies, satellite solar power stations, and space-based anti-ballistic missile (ABM) systems. The debates over these proposals have reflected not only disagreement over questions of feasibility and cost but also an intellectual struggle over the full incorporation of the space environment into human affairs. In part, the new space advocacy reflected a gathering of forces in defense of a paradigm shift.

By the early 1980s, the balance appeared to have tilted toward the space advocates. One of the reasons was generational. In 1977, the first generation to have lived entirely in the Space Age began to enter its twenties; those who had been excited by the first space ventures while children were somewhat older and provided much of the leadership of the new space movement. Harrison Schmitt noted in 1985 that 80 million Americans (one third of the population) became intellectually aware during the Space Age.[5]

Through film and television, these younger generations enjoyed vicarious adventure and achievement. Planetary exploration stimulated them with imagery of other worlds. Works of science fiction and science fact with space themes found a ready audience among the young. Many wanted to get involved in the space enterprise. However, those who sought employment in the space field often met frustration after the downturn in NASA's fortunes.

These generations are the core of the new space advocacy. "You've got a lot of young people who grew up with the space program and now they're in responsible positions to do something about it," said Howard Gluckman in 1980. "We've believed in space all the time."[6]

The emergence of these first "space generations" into maturity coincided with a "participation revolution" in American society. During the 1960s and the 1970s, citizen activism became an increasingly widespread and important feature of American political life. Loomis and Cigler have described an explosion of group formation, notably including public interest or citizens groups that lobbied for causes not necessarily related to the occupations of their members.[7] There was a desire to achieve a feeling of efficacy through some sort of action, even if that were nothing more than taking part in a telephone or letter campaign. New information technologies promoted grass-roots lobbying. Affluence created a larger potential for group membership by lowering the relative cost of participation. Organizations became more numerous with increasing education. Issue-oriented groups did especially well; for a modest membership fee, people could make a statement without the burden of more direct involvement. The new reform and citizens groups depended heavily on the educated white middle class for their membership – exactly the kind of people who formed the bulk of the new space advocacy. There was a large pool of group organizers, who tended to be young, well-educated, middle-class people caught up in the movement for change and inspired by ideas or doctrine.

The conjunction of rising interest in space and citizen activism spurred organized space activism. In some cases, leading figures came from movements that had peaked or declined, such as the environmental movement and the anti-war movement; the role of space group leader defined some individual lives. Many pro-space people of younger generations shared the participatory vision in wanting to be part of the space enterprise, and to influence its course. Sometimes the appeal was direct. "Either you can be a spectator of Humankind's greatest adventure," wrote the American Society of Aerospace Pilots in May 1985, "or you can let your interest in spaceflight lead you to pioneering the next frontier."[8] Many were not content to see spaceflight remain an elite activity (as of January 1985, 249 people had gone into space).[9]

The late 1960s and early 1970s also saw attempts to assert greater democratic control over technology, in part by technology assessment. Space advocates wanted to democratize space technology not just by controlling it but also by participating in the activities that it makes possible. Many wanted to take space activity out of the exclusive control of large government and corporate bureaucracies, to decentralize it and spread participation. Planetary Society Executive Director Louis D. Friedman wrote in early 1985, "The Planetary Society democratizes space. It gives you the chance to take part in Humanity's most positive venture."[10] But such appeals still were largely to vicarious participation.

The most potent symbol of this "space populism" is the Space Shuttle, a major factor in reviving latent but frustrated interest in manned spaceflight. The Shuttle was not a cramped capsule but a truck to space, with a relatively roomy, liveable cab. The first launch of this vehicle in April 1981 implied not only an American return to space but also regular access by a larger number of people. To some space advocates, it was the symbolic beginning of the democratization of space.

In January 1979, *Space Age Review* editor Steve Durst had complained about the "exclusion" of most people from space. Three years later, NASA Administrator James Beggs announced his agency's intention to broaden the range of people going into space through what became the Spaceflight Participants Program. This stimulated Durst's Space Age Review Foundation (with the cosponsorship of Delta Vee and the Hypatia Cluster) to publish a "Space Shuttle Passenger Project," which summarized proposals for turning the Shuttle orbiter into a passenger liner that could give large numbers of people the space experience.[11] "It had become possible to envision space as the province of some sort of transport vehicle (even looking vaguely like a commercial transport) which could carry anyone instead of the exclusive domain of supermen in capsules," wrote Maxwell Hunter in 1984. "It is an absolute psychological triumph . . . in recasting the popular image of what space might be all about."[12] The Shuttle also is an enabling technology for space industrialization and, possibly, space colonization.

Landmarks in the emergence of the Shuttle – the first rollout in September 1976, the first drop test in August 1977, and the first flight in April 1981 – were major stimuli to the pro-space movement. President Reagan was on the mark when he said after the first flight that "the Space Shuttle did more than prove our technological abilities. It raised our expectations once more. It started us dreaming again."[13]

The new space advocacy also was inspired in part by a revival of technological optimism. New technologies can imply new power over the environment and the future. Nowhere is this more true than in spaceflight, a symbol of technological prowess that, with the assistance of mass media, has stirred visions of human power expanding throughout the solar system and beyond. Launch vehicle entrepreneur Gary Hudson expresses a belief widespread among space advocates when he rejects the idea that we are helpless to solve such problems as the threat of devastation by ballistic missiles; technology makes solutions possible. Satellite solar power stations, space colonies, and space-based missile defenses are proposed grand technological solutions to frustrating, large-scale problems. The generational connection with technological optimism is strong; surveys show that people under 35 are more inclined to think well of new technologies.[14]

One intriguing index of the times is interest in engineering education. In parallel with the downturn, it bottomed out in the early 1970s, but it then rebounded sharply later in the decade, roughly in parallel with the surge in space interest group formation (Figure 14.1).

Space missions also reflect collective achievement through the tools we have built. Anyone who has attended the launch of a manned space vehicle will recall the sense of audience participation, the shouts of encouragement to the rising rocket. There are associated physical sensations, such as the vibration and the rumbling noise, which cause the observer to feel the power of this technological instrument. If an automobile can be seen as an extension of individual power, a space launch can be seen as an extension of our collective power.

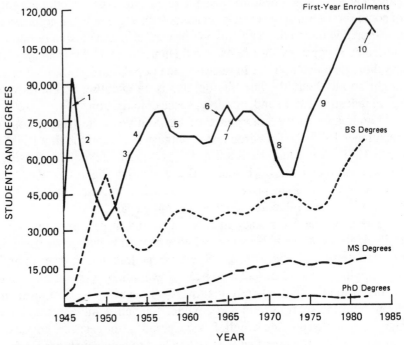

1   Returning WW II veterans
2   Diminishing veteran pool and expected surplus of engineers
3   Korean War and increasing R&D expenditures
4   Returning Korean war veterans
5   Aerospace program cutbacks and economic recession
6   Vietnam War and greater space expenditures
7   Increased student interest in social-program careers
8   Adverse student attitudes toward engineering, decreased space and
    defense expenditures, and lowered college attendance
9   Improved engineering job market, positive student attitudes toward
    engineering, and entry of nontraditional students (women, minorities, and
    foriegn nationals)
10 Diminishing 18-year-old pool

Figure 14.1 – Interest in Engineering Education (*Source:
National Research Council,* Engineering Education and Practice in the
United States [Washington, D.C.: National Academy Press, 1985].
Reprinted with permission.)

New technologies also can imply opportunities to reorder society, even to create technological utopias. In his 1985 book *Technological Utopianism in American Culture* Howard Segal described how industrialization stimulated some late nineteenth century and early twentieth century Americans to advocate specific designs for engineered human communities. Segal briefly mentions Gerard K. O'Neill as an example of a modern technological utopian, commenting that space colonies look something like suburbs in space.[15]

Perhaps the best analog to spaceflight is the technology of flight. In his 1983 book *The Winged Gospel*, Joseph J. Corn recounted how enthusiasm for the airplane between 1910 and 1950 was so strong that it became something akin to a secular religion for its adherents, and the airplane a kind of mechanical god that would usher in a millennium of peace and harmony.[16] Adherents to the cause, who consistently overestimated the ability of the airplane to cause social change, believed that aircraft would produce a democracy of the air, that it augured the expansion of freedom and the end of discrimination, and that it would not merely deter aggression but also bring humans closer together, thereby eliminating the conditions that cause wars. Corn placed this attitude within the context of American technological messianism and the age-old belief that flight was divine. But the winged gospel met its demise with the rise of strategic airpower, when Americans came to consider the airplane as an ambivalent, even malevolent agent.

The parallels with space advocacy are striking. Space enthusiasts at one time or another have promised greater abundance, enhanced freedom, and the elimination of war. To some of them, space technology had the potential to bring a utopian future. Such themes were particularly visible in the early writings of Gerard O'Neill and his followers. Corn wrote:

> If there is a latter-day religion of flight, it is to be found in the response of partisans of spaceflight. Many of their expectations for the space future recall the aviation gospel of the first half of this century. One of the most ardent and respected prophets of the space gospel is Gerard K. O'Neill. . . . In looking at his vision, it is helpful to know that Professor O'Neill for years has been a pilot, flying small planes and gliders.[17]

By 1984, ambivalent feelings about space technology were obvious, primarily because of the growing military uses of that technology.

The rise and decline of space utopianism reflects not only a recurrent pattern in American history but also the maturing of spaceflight as a technology and a human activity. By 1984 the uses of space had shifted toward the prosaic. But the dream of the space utopians remains alive in some parts of the new space advocacy and drives some hopes for the future.

Space rhetoric also is full of the imagery of the frontier, another powerful idea in American history. The frontier has promised Americans opportunities for exploration, adventure, success, new wealth, and a new start. With the loss of frontiers on Earth, space became the new frontier. "Space," wrote a *Spaceflight* editorialist in 1985, "was like America must have seemed to the Pilgrim fathers – huge, open, and free."[18] Gerard O'Neill made it the High Frontier, adding a transcendent element. Speaking to the 1979 Princeton Conference on Space Manufacturing, Freeman Dyson suggested a parallel between future space migrants and earlier colonists in the New World – the Pilgrims and the Mormons.[19]

In part, the American space movement is a revival of the idea of the frontier. Like the American frontier, it has encouraged visions of opportunity and adventure, and a proliferation of entrepreneurial schemes, many of which will fail. There also has been a strong association with the American West, both in the historic sense and in terms of current interest in space. Some of the social and political ideas associated with the American West fit well with the ethos of much of the new space advocacy; Joel Garreau has reported that the political culture of the West is characterized by optimism and by the absence of a sense of limits.[20] Science writer Dennis Overbye suggests that "space is the extension of the rovings of a restless people."[21]

Not everyone finds the frontier model attractive. "Those with a positive space program succumb to the old frontier illusion that an Eden of abundance and harmony awaits us in space," writes Daniel Deudney. "The urge to pick up and move to a new land when things start getting bad in the old country has taken on a new high-technology character."[22]

Many of the new space advocates are libertarians or other individualists who hope that space will offer new opportunities for freedom, liberty, and voluntarism. "A lot of people who want to go into space have trouble with authority," observes Carolyn Meinel.[23] Science fiction writer Robert A. Heinlein, one of the grandfathers of the new space movement, emphasizes voluntarism in his novels and abhors the atrophy of the will.[24] These tendencies may help explain the fractiousness of the new space movement.

This craving for freedom has a strong political element. "Space was the only area in our world left where governments could not control and coerce their citizens," wrote one student of aerospace engineering.[25] In late 1984, Gerard O'Neill commented that the Space Studies Institute "looks forward to an open future, in which the free choice of individuals, rather than the dictates of governments, will shape individual human destinies."[26] In 1983 and 1984, "Freeland" conferences were held in Southern California to discuss possible habitats (such as unclaimed islands and space colonies) outside the reach of "uncontrollable" national governments. Speakers at Freeland II included science fiction writer Poul Anderson, space entrepreneur Gary Hudson, and L-5 Society activist Conrad Schnelker.[27] Hudson has written that "space offers a political frontier, where people can live as they like, do as they like. . . . I think that technology is a great freer of human beings."[28] The relative lack of legal restrictions is one of the attractions of space both to potential colonists and to entrepreneurs and was a major reason for the L-5 Society's passionate opposition to the Moon treaty.

In their 1978 book *Space Trek* , Jerome Glenn and George Robinson wrote "Another facet of the emerging new perceptions of reality is that future Spacekind should be free from, and independent of, the political bonds of Earthkind." Their book closed with a proposed "Declaration of Independence" by space migrants.[29] This theme is especially congenial to some space activists with a conservative political orientation. "Our highest destiny," said James Muncy in April 1985, "is to spread free people throughout the Cosmos."[30]

The late 1970s and early 1980s also witnessed a revival of entrepreneurship in the American economy, typically associated with younger people and with high technology industries. This coincided, nicely with the ideological stance of the Reagan administration, which publicly endorsed the idea of space commercialization and encouraged entrepreneurial space ventures. It also reflected the ethos of some new pro-space groups, such as the American Society of Aerospace Pilots. Said James Muncy, "We want space to be a frontier for free enterprise."[31]

The sense of personal liberation also has an important physical dimension. "Your support of the L-5 Society is your declaration of independence from the restrictions of planetary surfaces," wrote the *L-5 News* in 1985, over a painting of a man and a woman moving ballet-like in space.[32] Many space advocates have on their walls a picture of astronaut Bruce McCandless floating free and untethered in space, an independent satellite of the Earth. A fundamental reason for pro-space activity is that many advocates want to have this experience themselves.

Another driving force was a revival of the old idea of alternate worlds. Some space advocates, notably in the Planetary Society, clearly wish to participate in a new age of discovery. For others, the appeal of new worlds goes farther. They want not only adventure but also alternatives to the familiarity and frustrations of the Earth. On new worlds, one might enjoy freedom from conventional restraints. New societies might be formed, independent of their ancestors, free of Earthly faults, offering opportunities for social experiment. Planetary exploration initially made the nearest potential alternate worlds (the Moon, Mars, and Venus) look less attractive than some had hoped, bringing the new worlds idea into temporary disrepute. But O'Neill's space colonies – artificial biospheres that could be anywhere in the solar system – revived it.

The resurgence of space activism also was part of a broader reaction to the cultural pessimism and hostility toward technology of the late 1960s and early 1970s, a revived belief in progress through science and technology. The questioning of the idea of progress, the dystopian view of the future, was unpalatable to many Americans. There also was concern that American power and economic competitiveness had declined. To many, the downturn in American space activity was a symbol of these negative times. For many space advocates, what was threatened was not only economic and professional interests but also a vision of the future.

The wish to escape the limits to growth, to believe in progress and in the strength and rightness of American culture, found a visible rallying point in the revival of American manned spaceflight in 1981. "Space," wrote *Aviation Week* three years later, "is in the good news department."[33] The concurrent revival of publicly expresssed patriotism, of a desire to strengthen American defenses and to match foreign economic and commercial competition, all were congruent with a revival of support for spaceflight. American faith in growth and technological progress were reasserted. The spring 1981 report of the Citizens Advisory Council on National Space Policy concluded, "The rediscovery of progress is a reasonable and feasible goal for the United States in the 1980s."[34] In part, the new space movement was a reflection of what *Time* magazine called "American Renewal."[35]

Since the earliest days of thinking about spaceflight, there have been transcendental elements in its advocacy, even hints of a secular, humanistic religion. In the *Spaceflight Revolution,* Bainbridge wrote that the initial urges that brought the spaceflight movement into existence were "non-economic, impractical, personal, and primitive desires." They could be described either in psychiatric or religious terms.[36] Like early aeronautical literature, space literature is liberally sprinkled with noneconomic justifications; both aircraft and spacecraft have been vehicles for transcendental aspirations, however ill defined they may be.

Clearly, such motivations were a factor in the revival of spaceflight advocacy in the United States. There were technological transcendentalists, who seemed to believe that whatever was scientifically and technologically possible would be done; Richard Hutton, in his 1981 book *The Cosmic Chase,* called them the "radical fringe" of the Space Underground.[37] There were others who saw a reformation of humanity in the space environment. Many space advocates are unembarrassed by the idea that they are driven by a dream.

Nowhere was the transcendental element seen better than in the space colonization movement. In a remarkable act of self-recognition, H. Keith Henson published an article entitled "Memes, L-5, and the Religion of the Space Colonies" in the September 1985 *L-5 News.*[38] Drawing on memetic theory, Henson suggested that the space colony concept is a meme with religious characteristics. Memes (information patterns that influence an organism to pass the meme on to other brains) lose their intense hold on people with the passage of time, especially when the promises of the meme are at great variance with reality. Henson sees the gradual replacement of human habitation with a general pro-space theme in the L-5 Society, and its loss of a clear goal, as byproducts of this process.

For many space advocates, there also is a sense of being part of a larger enterprise, of historic importance. Spaceweek's Dennis Stone put it well when he said, "We are not in it for ourselves."[39] Different advocates may define that historic enterprise in different terms. However, the underlying fact is that our generations are opening to humanity the largest of all environments, an act some compare to the emergence of sea life on to the land. The space "movement" is, in part, a symptom of that historic event.

THE STATE OF THE MOVEMENT IN 1984

As they approached the end of their first decade, the new space interest groups were becoming a mature phenomenon, a relatively permanent part of the American interest group scene. In her third report on the space movement, published early in 1985, Trudy E. Bell found that the number of space interest groups had stabilized at about 48. Their aggregate total memberships topped 300,000, and their aggregate budgets exceeded $30 million. The leadership of the groups had become increasingly professional. Bell speculated that the advent of corporate funding to various space interest groups might be an indication that the groups were finally being recognized as legitimate entities whose work was worthy of support.[40]

After a burst of organizational formation in the late 1970s and early 1980s, the space advocacy began to coalesce in the mid-1980s through increased cooperation and the weeding out of the less stable and more personalistic groups. There appeared to be movement toward the center. Relatively radical groups like the L-5 Society reconciled themselves increasingly with the classic agenda for manned spaceflight. Some older space interest groups, notably the American Astronautical Society, became more advocacy oriented and enjoyed an upturn in membership. The gap between professionals and enthusiasts narrowed; older and younger generations of spaceflight advocates, once separated by differences in cultural assumptions and style, seemed to draw closer after a period of standing off. There was some movement toward a consensus on a modified form of the classic agenda, with the Planetary Society being a major exception until 1984. Above all, there seemed to be a regained confidence in the future of spaceflight and a lessening of the sense of urgency that had prevailed a few years before.

Perhaps as a consequence, most space interest groups appeared to be near the top of their sine curves in 1984. The rate of organizational formation had declined sharply, and membership had leveled off or declined in most groups. Bell noted that the "flaky" groups were gone, but some initially credible organizations, such as the Institute for the Social Science Study of Space, also had vanished from the scene by 1984. The one major exception to this trend was the Young Astronaut Council, whose membership continued to climb through the end of the year. One of the staffers in that organization's Washington headquarters was Todd C. Hawley of Students for the Exploration and Development of Space, a successful migrant from citizens activism to a paying space-group position.

The space "movement" remained a diverse, eclectic, fragmented phenomenon, without a dominant leader or an agreed-on agenda. "The citizens groups are still too disorganized, and need to find more areas of agreement," observed the AIAA's Jerry Grey in 1983.[41] Coordinating mechanisms remained weak; the National Coordinating Committee on Space appeared to be inactive as of 1985. There remained the strains of differing priorities and the tension between grass-roots activism and professional credibility. Some citizens groups continued to try to play several roles instead of accepting a division of labor within the movement. The advocacy also seemed more divided by political partisanship than in the past; a new conservative space consensus, including suport for the Strategic Defense Initiative, was emerging by 1982, while liberal space advocates tended to emphasize international cooperation, including joint U.S.-Soviet exploration of Mars.

The fragmentation of the space movement probably was inevitable because of the board, inclusive, positive, and future-oriented nature of the space dream. "The big disadvantage of the space movement," said the National Space Institute's Mark R. Chartrand in 1983, "is that it is arguing for something abstract and future, instead of against something here and now."[42] Agrees the L-5 Society's Gary Oleson, "The positive nature of the space agenda is the most serious organizing problem."[43]

Despite these problems, the space advocacy was held together loosely by many interconnections — shared concerns, overlapping board memberships, frequent contact at conferences, shared media, and personal friendships. At some deeper level, it is held together by vague and often unarticulated ideas about the importance of the broad space enterprise, the sense that it is something new and of historic significance.

IS THIS A SOCIAL MOVEMENT?

In looking at the earlier spaceflight advocacy, Bainbridge found it to be outside normal science, and thus explicable only in terms of social processes that operate outside the conventional market mechanism – that is, a social movement.[44] In 1980, Trudy E. Bell found that "a space movement is in the air – a palpable excitement, almost a kind of euphoria in the proliferation of new groups, the discovery of one another's existence, the creation of new types of groups, and a sense of promise, direction, destiny."[45] Does the post-1972 space advocacy meet the criteria for social movements?

Certainly, the new space advocacy reflected a broad community of belief in the importance of space, held together less by shared economic interests than by shared ideas. It also demonstrated a rapid growth of organized activity, often by volunteers. However, this was a very fractious coalition, including individuals ranging from conservative aerospace engineers to former environmentalists and anti-war protesters. The advocacy also included very different types of organizations. Behavior ranged from conservative to radical. "You need some groups over the edge," says space writer James E. Oberg, "because they make the more moderate ones seem reasonable by comparison."[46] This phenomenon is not unusual in social movements, particularly in their early years.

Ralph Turner and Lewis Killian, in their book *Collective Behavior*, defined a social movement as "a collectivity acting with some continuity to promote or resist change in the society or group of which it is a part."[47] By this criterion, the results are mixed. Most of the older, established space groups behave more like traditional economic interest and professional groups, seeking to protect and advance the economic and professional interests of their members. Other space interest groups are basically enthusiast organizations, in which members want to communicate with others who share their interests. But the leaders of some of the newer pro-space organizations (particularly the L-5 Society and the Space Studies Institute) would argue that they and their supporters are in fact seeking significant change. They may indeed reflect a social movement.

Robert D. McWilliams argued in 1981 that space exploration supporters (particularly proponents of space industrialization) were rapidly evolving the organizational characteristics typical of social movements, such as common ideology, organized strategy and tactics, and the differentiation of roles and the distribution of power, and cited the L-5 Society as an example.[48] Bell's interviews suggested that within the citizen-support space-interest movement, the necessity for these additional factors was dawning.[49]

McWilliams also found that persons who wish to see more money spent on space exploration scored higher on indexes of social and moral liberalism than did those who opposed the space program; this fit with people involved in other social movements. They also scored higher on indexes of socioeconomic backgrounds, intellectualism, and attitudes toward organized religion. These indexes are comparable with those seen in the civil rights movement and the women's movement, for example. McWilliams concluded that "evidence such as this suggests strongly that the minority of Americans who wish to see more federal money allocated to space exploration are the sort of people who comprise social movements."[50]

Bell concluded in 1980 that "all of the signs indicate that by the mid-1980s an American space-interest movement could be as powerful as some of the major special interest movements existing today."[51] Two years later, however, she wrote that the spirits of leaders of space interest groups seemed notably dampened about whether or not there had been an evolution of a space interest movement since 1980; a good portion of the optimism expressed then had been based on the groups' discovery of each other.[52] By 1985, when she wrote an editorial on the subject in *Space World*, Bell had concluded that the space community was too fragmented and had too diverse an agenda to be called a movement.[53]

Perhaps it is most accurate to describe the new space advocacy as a set of overlapping social and political phenomena — partly a social movement and partly a set of interest groups, with the trend toward the latter. The L-5 Society had its roots in one kind of space-related social movement, having to do with space migration and the creation of new societies in space. The American Society of Aerospace Pilots emerged from another social ideology, concerned with free enterprise, voluntarism, and the re-creation of the American frontier in space. However, most other space interest organizations are not nearly so visionary and are more properly called interest groups.

The space "movement" also is relatively small by the standards of modern American social movements. Richard Hutton pointed out in 1981 that there were 20,000 environmental public interest groups, bringing in about $1.5 million a week from a constituency of about 4 million people.[54] This is more than ten times the membership – and more than twice the claimed budget – of the organized space advocacy in 1984. But the environmental movement is instructive in another way. It never coalesced into a single organization under a single leader, yet it often has been effective through its use of coalition politics.

WHAT KINDS OF INTEREST GROUPS?

If the new space advocacy is as much a set of interest groups as a social movement, what kinds of interest groups? Some, such as the Aerospace Industries Association and organizations of aerospace professionals and space scientists, have behaved in ways reasonably congruent with the economic model of interest group behavior. That model assumes an "economic man," a rational actor pursuing his own self-interest.[55] However, most members of citizens pro-space groups have no direct economic or professional stake in the space program. Despite this, many have devoted substantial personal time and effort (and, in some cases, money) to working for space. They seem to fit better with the theoretical revision of Olson's model offered by Terry Moe in 1980, in which he posited a "bounded rationality." Moe argued that citizens act not only on the basis of their economic interests but also on the basis of their values, their limited information, and their personal calculation of their influence. They form not only groups that offer "economic" and "selective" incentives but also groups that offer "purposive" incentives, such as ideological or moral satisfaction, or even solidarity.[56]

Political scientists Nathan C. Goldman and Michael Fulda have proposed a useful taxonomy of space interest groups.[57] The chart in Figure 14.2 is organized along two axes: degree of economic motivation and degree of purposive motivation.

According to Fulda and Goldman, examples of high purposive but low economic groups would include the Hypatia Cluster and Write Now!; high purposive and high economic, the National Space Club and

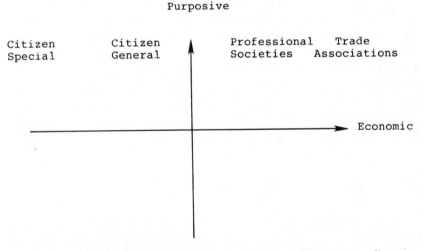

Figure 14.2 – Categorization of Space Interest Groups
(*Source:* Adapted from Nathan C. Goldman and Michael Fulda, "Space Interest Groups: Galaxies of Interests, Galaxies of Groups," to be published in volume 2 of the *Space Humanization Series*.)

the Universities Space Research Association; high economic but low purposive, the Aerospace Industries Association and the International Association of Machinists and Aerospace Workers; and low purposive and low economic, UFO groups and mystical cults. (There is, of course, some overlapping membership among space interest groups in different categories.)

Orenstein and Elder have observed that purposive groups are by nature unstable; they attract splinter groups and have difficulty in maintaining membership interest. Material benefit groups, once established, tend to be stable.[58] These principles are supported by the experience of many space interest groups. Yet both the National Space Institute and the L-5 Society, examples of new, high purposive groups, are now over ten years old.

The space advocacy also provides examples of both the "staff" and "grass-roots" models of American interest groups, with others along the spectrum between these extremes. The Institute for Security and Cooperation in Outer Space and the Institute for Space and Security Studies, for example, are pure staff organizations, with no real grass-roots base. Spaceweek, by contrast, is an extreme example of grass-roots decentralization.

There have been many motivations for forming space interest groups, such as protecting economic and professional interests, proselytizing for an idea, and seeking an application of knowledge and skills. However, many of the recent groups fit the entrepreneurial model of interest group formation described by Moe. The new space advocacy includes several groups founded by a single individual that often continued to revolve around that individual, with some collapsing when that individual went on to other pursuits (for example, Delta Vee). Howard Gluckman has observed that many space advocates who reacted to the downturn in the space program sought like-minded groups; when existing space interest organizations seemed inadequate, some of these people founded others.[59]

David Koch, who co-founded Spaceweek and who founded the American Society of Aerospace Pilots, and David C. Webb, active in the formation of Campaign for Space and U.S. Space 82, seem excellent examples of organizational entrepreneurs. James Muncy, who founded the Action Committee for Technology and Using Space for America, describes himself as a political entrepreneur.[60]

The multiplication of space interest groups also may reflect a democratic society's response to important new technologies; defaulting to a technological elite is not sufficient. Goldman and Fulda, addressing the space interest group example, have commented that the pluralist solution to technology may be more interest groups.[61]

Loomis and Cigler note the formal penetration of interest groups into the federal bureaucracy (advisory groups), the Congress (caucuses), and the Presidency (White House group representatives).[62] Parts of the space advocacy have achieved the first two of these. But the advocacy's purposes also have been advanced significantly by informal penetration of the administration, the Congress, and the Executive Office of the President. The space advocacy has allies (and sometimes formal members) within both the executive and legislative arms of the federal government. NASA, the United States Air Force, the Office of Science and Technology Policy, and congressional staffs have been "infiltrated" by people who share the space dream, who belong to that broad community of belief that underlies the pro-space phenomenon. Perhaps political science needs an interest group model that more formally recognizes that alliances across the rather artificial dividing line between government and interest group can be purposive in motivation and that such alliances are not limited to the classic "iron triangles" composed of federal agencies, their authorizing committees, and related economic interest groups.

HAS IT MADE A DIFFERENCE?

Writing in 1982, Bell concluded that space interest groups had failed in terms of their original expectations.[63] If we use the influencing of policy, budgets, and legislation as our main criterion, pro-space groups have not played a significant role in most major space policy and program decisions taken since 1972, except for the defeat of the Moon treaty. That lobbying effort succeeded in part because it was allied to more powerful interests, whose primary concern was the Law of the Sea. It also was directed against a proposal, not in support of a new program that would have cost the taxpayers large sums of money. The record is less impressive in other cases:

Despite claims that they influenced the outcome on the Jupiter Orbiter Probe, former House Committee Staffer James Wilson says flatly that the citizens groups did not play a major role in the decision.[64] Former NASA Associate Administrator for Legislative Affairs Joseph Allen agrees that JOP was not saved by such groups.[65]

Advocates of satellite solar power stations were unable to keep studies going, and none of the pro-SSPS legislation introduced into Congress as a result of their efforts after 1977 resulted in the appropriation of funds, despite lobbying by the L-5 Society and the SUNSAT Energy Council.

It has been claimed that the space movement saved the space program from extinction at the time of the Reagan administration's 1981 budget cuts.[66] While the scale of budget cuts may have been reduced, significant cuts were made and new starts were delayed. The Planetary Society's attempt to get funding for a Halley's comet mission failed completely. It is not clear that the continued existence of NASA was ever under more than rhetorical threat.

The coalition behind space commercialization included established figures, such as the aerospace executives who met with President Reagan in August 1983, as well as graduates from pro-space citizens groups. In the case of private launch vehicles, citizen groups graduates and other space advocates did achieve success in getting both administration and congressional endorsement. However, they suffered a defeat when the Department of Transportation rather than the Department of Commerce was chosen to be the "lead agency" for regulating launches and later on pricing policy for Space Shuttle launch services.

Although space advocates played a role in proposing space-based anti-ballistic missile systems, the President's 1983 decision was the result of defense-related arguments, rather than the borrowed ideas about space industrialization included by High Frontier.

Although some space interest groups had endorsed the space station for a long time, the President's decision to endorse it was largely the result of NASA lobbying, aerospace industry support, and the President's personal convictions.

All this suggests that major space program and policy decisions in the late 1970s and early 1980s continued to be made largely by the internal processes of government, supplemented by industry information and lobbying. Interviews show nearly unanimous agreement that the principal influences on such decisions are (1) the administration, particularly the President and his advisers, the Office of Management and Budget, NASA, and the Department of Defense, with the Office of Science and Technology Policy and the National Security Council staff sometimes playing important roles, and (2) aerospace companies. Congressional members and staff, aerospace professionals, and space scientists generally were considered intermittently influential. Pro-space citizens groups were seldom mentioned.

The L-5 Society's part-time lobbyist in Washington, Gary Oleson, agrees that the citizen group impact is not at the point of decision. "They can't go in and swing votes," he says.[67] His colleague from Spacepac, Gary Paiste, acknowledges that Congress has not had reason to take pro-space citizens groups seriously.[68] House Space Subcommittee Staff Director Darrell Branscome says that citizens groups rarely are the first on the doorstep; they follow up.[69]

John Loosbrock, the experienced public relations officer of the Aerospace Industries Association, noted in 1983 that 200,000 members is not much when distributed over 435 congressional districts; pro-space groups are spread thin.[70] Such groups may have access to individual members of Congress or state delegations (as the L-5 Society did with the Arizona delegation) but not across the board. Space writer David Dooling noted in 1984 that all of the space interest groups together then had a membership smaller than that of the Institute of Electrical and Electronic Engineers.

Pro-space citizens groups generally have even less impact on the administration; the September 1981 National Coordinating Committee for Space policy statement reportedly had no impact on policy (a possible exception is the Citizens Advisory Council on National Space Policy, some of whose members were well connected in the Reagan administration). Yet it was the President and his advisers who made the major decisions that are defining the next decade of space activity: the space station, the Strategic Defense Initiative, and space commercialization.

One would think that a natural point of contact for pro-space groups would be NASA itself. Although that agency has institutionalized channels for working with aerospace companies, aerospace professionals, and space scientists, as of 1984 it had no mechanism for regular liaison with pro-space citizens groups. Associate Administrator Patrick Templeton said in February of that year that the agency appreciates the support that citizens groups give but acknowledged that NASA does not have an official responsible for maintaining contact with them.[71] Such access as pro-space groups have enjoyed has tended to be through a few sympathetic individuals.

One is tempted to conclude that the belief of some pro-space citizens groups that they have had significant political influence confirms John

Kenneth Galbraith's observation that "resort to an instrument of power is widely confused in our time with an exercise of power."[72] Given the dependence of citizens groups on unpaid volunteers, one also could argue that the need to believe that one is having an effect on the policy process is inevitable. However, these conclusions would dismiss the modern pro-space advocacy too lightly. Although pro-space groups are not powerful in themselves, they play other important roles, indirectly influencing the process by which space decisions are made.

OTHER ROLES

It is clear that pro-space citizens groups do not rank with the American Dairy Association and the National Rifle Association as effective lobbying organizations. However, they do have influences of more subtle, indirect, and long-term kinds:

By their existence. Pro-space citizens groups are a visible demonstration of public interest in (and by implication support for) the civil space program. If they were not there, decision makers and opinion formers might be even more inclined than they sometimes are to see that program as an exercise in cynical self-interest by NASA and its contractors. The fact that nearly 300,000 Americans pay dues to pro-space organizations shows grass-roots, constituent support. They remind decision makers that space is popular.

By amplifying other space interests. Pro-space groups tend to reflect more than originate. "They are backstoppers, not initiators," says Courtney Stadd.[73] They are amplifiers for the concerns of NASA, the aerospace industry, aerospace professionals, and space scientists – and, more recently, space entrepreneurs – about the future of the space enterprise. Most of their speakers, much of the material they publish, and most of the quotes they use come from those sources. Yet it is notable that NASA has not made good use of most citizens pro-space groups to amplify its messages.

By circulating and legitimizing ideas. Most of the scientific and technological ideas in the spaceflight field come from individual professionals, whether they work in NASA, in aerospace companies, or in universities. Many of those ideas would remain confined to obscure journals and narrow interest groups if they were not picked up and spread around by nonexperts interested in space. Government

officials and company executives often are reluctant to endorse ideas
that seem "far out," and professionals and academics may find it
difficult to get a hearing for new concepts in respectable circles.
Space buffs in the media play an important role in helping to circulate
the more comprehensible ideas. (Walter Cronkite is a good example
from the recent past.) Space interest groups also provide a medium in
which those ideas can circulate, acquire adherents, and get broader
attention. Trudy Bell describes them as "a market-like testing ground
for astronautical concepts."[74] Pro-space groups clearly had
something to do with increasing the acceptability of space
industrialization and the idea of large numbers of people living
beyond the Earth. "There was a time when O'Neill could not get the
attention of NASA headquarters," said Universities Space Research
Association Executive Director David Cummings in 1983. "Now,
attitudes are shifting."[75] As of late 1984, the same process seemed to
be occurring in the case of extraterrestrial resources, as least partly
because pro-space groups had acted as amplifiers for the concepts of
space visionaries. "The idea of using lunar resources is gaining wide
acceptance in the aerospace community," wrote the Space Studies
Institute's Gregg Maryniak in early 1985.[76] Once far-out ideas are
being taken more seriously by decision makers, and pro-space
rhetoric has infiltrated into the language of those who would not have
used it five years before. Space interest groups have contributed to
that process.

Educating the public. Many citizens groups have the education of the
public about space as their primary aim. They not only encourage
membership and publication readership but also conduct public
programs and place materials in the media to reach wider audiences.
Over time, this helps to create a more favorable opinion environment
for future program decisions. Their biggest effect may be on younger
generations. "Citizens groups have not been successful with the
sophisticated elite," observes space law and policy expert Irwin
Pikus, "but with young people."[77]

Holding the community together. Most of the widespread support for
space activity is inchoate and has no outlet for expression.
Enthusiasts need rallying points: organizations, meetings, magazines.
The citizens groups bring together the pro-space constituency and
give it some coherence, a set of focal points for activism. Publications
and conferences play the major role in creating a nervous system for
the pro-space community. They encourage self-recognition by the

space advocacy. The discovery that many pro-space citizens organizations existed, in part the result of Trudy Bell's late 1980 and early 1981 articles, may have given space activists greater confidence and spurred their efforts.

Outlets for idealism. One is struck by the power of noneconomic motivations in the space advocacy. Other motivations, which could be described as idealistic or transcendental, clearly drive many of those involved. The citizens groups have provided useful outlets for this idealism, particularly among generations that suffer from high expectations and frustrated ambitions.

Alternatives to big organizations. For those who want to do something about space but are unwilling or unable to work their way up through a government or corporate bureaucracy, space interest groups can provide an alternative route. Trudy Bell comments that "Perhaps one third of the groups have evolved into an additional legitimate environment for the training and recognition of rising talent."[78] They provide a way for ordinary people to get involved and for leaders to get recognition. Space interest groups also can provide an alternate means of getting things done. Of particular note are those organizations that are doing technology development work themselves, such as the Space Studies Institute, the World Space Foundation, and the Independent Space Research Group.

Political and organizational training. Being in a pro-space group is a learning experience, not just about space but also about politics and organizations. Many enthusiasts have improved their political and organizational skills by working with citizens groups. They have been "incubated" before moving on to other walks of life, where they may rise to positions of influence and see their pro-space views have real impact. In a sense, the groups are transmission belts for activists.

Creating a network. Space enthusiasts establish a wide variety of contacts through their organizations and through conferences and informal meetings. There are extensive overlaps among groups, both at the leadership and ordinary member levels; former American Astronautical Society President Charles Sheffield, for example, was on the boards of both the L-5 Society and the National Space Institute in 1984 and also served as one of many links to the science fiction community. There are interconnections among pro-space citizens groups, professional groups, entrepreneurs, the aerospace industry, academia, and NASA; David Hannah, for example, was on the boards of the Space Studies Institute and the Institute for the Social

Science Study of Space as well as being the head of Space Services Incorporated. Many pro-space people stay in touch even after they "graduate" from the citizens groups, forming a network that already has begun to operate in the case of the new launch vehicle companies, both as an employment exchange and as a political communications channel (Trudy Bell calls it a "new boys network").[79] Space advocates also are filtering into a variety of institutions of American society outside the aerospace community. Space interest groups help to hold this network together.

Changing the political culture. In the short term, major space policy decisions are made primarily by the internal processes of government, marginally affected by industry and professional lobbying. In the longer term, however, space interest groups influence future events by helping to change the context for policy – the intellectual and political culture. They do this not by overnight conversion but by eroding assumptions about what can not be done.

Stimulating interest in science and technology. In a broader sense, space interest groups help to attract more people (particularly young ones) to science and technology. The space enterprise has higher visibility than any other technological activity. This clearly was recognized by the founders of Project Liftoff, whose main component is the Young Astronaut Program.

THE SECOND SPACEFLIGHT REVOLUTION

Bainbridge wrote in 1976:

My overall contention is that the next 20 to 50 years will be marked by a gradual upward coasting of space technology capabilities – a period of normal technological change. Somewhere soon after the turn of the century there is the real possibility of a Second Spaceflight Revolution, or at least very rapid progress of the normal type.[80]

The new space advocacy has a complex agenda. At the center of it, however, is the idea that many people, not just a few, can and should go into space to explore, to work, and to live. Many want the experience and the adventure of going into space. For some, the goal is an old one – to explore new worlds. For others, the goal is a space-based civilization, the extension of human activity of all kinds into a new environment, and the permanent incorporation of nearby space into the human realm. Gerard

O'Neill speaks of the attraction of establishing small-scale, manageable social units in space, reviving cultural diversity.[81] The L-5 Society states explicitly that it seeks "an independent society in space."[82] These people want a second spaceflight revolution: the permanent human occupation of space.

NASA officials have advocated this for some years, although in more cautious, bureaucratic language. In the mid-1970s, senior NASA official John Yardley stated that "NASA's principal long-term manned space goal is the achievement of permanent occupancy and limited self-sufficiency in space."[83] The Space Shuttle and the space station are the first, necessary steps, the first enabling technologies. The Campaign for Space wrote in 1984, "President Reagan's space station initiative represents the beginning of a new era for NASA, an era that will culminate in the permanent habitation of a totally new environment for Mankind."[84]

If the spaceflight movement of the 1920s and the 1930s laid the groundwork for our first, exploratory ventures into space, the new space advocacy is laying the groundwork for permanent human occupancy. The modern spaceflight movement is establishing the culture for a "Second Spaceflight Revolution." Its leaders are attempting to build the constituency for a paradigm shift that they believe is as profound as the Copernican revolution in astronomy.

Those who want a "Second Spaceflight Revolution" face a continuing challenge. Space involves big ideas, which take a long time to sell. The space movement has long time horizons; the bureaucracy, the Congress, and the industry do not. Those institutions are dominated by incrementalists not revolutionaries.

FAULT LINES

By 1984, old divisions within the space advocacy were becoming more visible, heightened by recent events. Familiar issues were revived, with some scientists, for example, criticizing manned spaceflight, and with many liberals criticizing the military uses of space technology.

Political scientist Nathan C. Goldman sees phases in the development of the new space advocacy. In the first phase of "mass consensus," we saw widespread backing for the naming of the Shuttle Orbiter Enterprise and for the Jupiter Orbiter Probe. In the second, we saw a breakdown of the coalition over the Moon treaty and the satellite solar power station, with a split between spaceflight advocates and environmentalists. In the third phase, new divisive issues arose, notably the space station and the Strategic Defense Initiative.[85]

These splits appear to reflect a deeper rift that goes back to the beginning of the Space Age. At the extremes, there are two identifiable schools of thought, one one each side of the fault line. Most people interested in space probably do not line up entirely on one side or the other. But a model divided along the following lines suggests the underlying strain:

Space Development	*Space Exploration*
Manned spaceflight	Unmanned spacecraft
Space commercialization	Space science
Aerospace industry	University researchers
Space defense	Space arms control
American pre-eminence	International cooperation
Outward migration	Public service
Limitless resources	Limits to growth

There are, of course, reasons of self-interest underlying this division. For example, NASA and the aerospace industry have very practical reasons to favor large, complex, long-duration projects, while individual scientists may be more advantaged by a diversity of smaller projects that enables more researchers to achieve individual recognition.

There also appear to be deeper reasons. Among those interested in space, there may be a difference in world view between those who study the environment (whose scientific training is to see celestial bodies as objects) and those who see them as fields for action, as worlds. There may be a difference in ethos between those who enjoy the pleasure of finding out, knowing, and telling others, plus the vicarious adventure of remote exploration, and those who enjoy the extension of individual and collective human power and influence through technology. And there may be a division between those who see space exploration and its findings as ends in themselves and those who see them as means to other ends.

Nathan Goldman comments that there are two paradigms of the future in space, but that the pro-space movement thinks of itself as the reflection of one paradigm.[86] Are these different world views permanent features? Or does one reflect the true future of space advocacy, while the other is a temporary, even reactionary phenomenon? We may need the perspective of another generation to tell.

THE HIDDEN AGENDA

Some experts comment that space policy is just the space dimension of other policies. However, this does not take into account a deeper motivation for the space advocacy, one whose roots reach far back before the present. Although not always stated explicitly, that motivation is shared by many of those involved in the new spaceflight movement. It is *expansion*.

In both science fiction and speculative fact writings about the future in space, the outward expansion of humanity is a recurrent theme. At the end of the 1936 film *Things to Come,* H. G. Wells has one of his characters express this in an extreme form:

> But for Man no rest and no ending. He must go on – conquest beyond conquest. This little planet and its winds and its ways, and all the laws of mind and matter that restrain him. Then the planets about him, and at last out across immensity to the stars. And when he has conquered all the deeps of space and all the mysteries of time – still he will be beginning.[87]

There may be an historic connection with cultural dynamism. "In human records," wrote the anthropologist J. D. Unwin, "there is no trace of a display of productive energy which was not preceded by a display of expansive energy."[88]

As pragmatic a man as NASA Administrator James Beggs has recognized the significance of this theme. "I believe that one of the indisputable truths of our time is that humanity is slowly but surely expanding into space," he said in 1985.[89]

Although few would wish to be so labeled, many space advocates are human expansionists. They wish to break out of the limits of the Earth and to extend human power and presence outward as far as it can reach. However impractical or uneconomic that may seem, it does much to explain the frequent hints of transcendental aspiration in the space literature. The division between expansionists and nonexpansionists is one of the elements underlying the schism in the space community. Yet by 1984, even the Planetary Society had begun to recognize the appeal of the expansionist motivation by endorsing a Moon base and manned missions to Mars and the asteroids.[90]

THE UNDERGROUND LIVES

Looking backward over the history of the astronautical movement, we find a kind of rough continuity reaching at least as far back as the nineteenth century. Originally, small groups of people who shared an insight and a passion formed a kind of friendly conspiracy, an open cabal advocating the use of the rocket to reach into space. Bainbridge categorized the revolutionaries as dreamers, practical visionaries, and implementers.[91] Periodically, someone catalyzed the spaceflight idea. Space societies rose and fell with changes in leaders and generations. People independently rediscovered the promise of space.

By the late 1960s and early 1970s, rocketry became a mature technology, reaching a symbolic climax as a means of exploration with the Apollo expeditions to the Moon. The rocket advocacy was no longer an underground.

To some extent, history is repeating itself. In the new space advocacy, we have seen dreamers and pragmatic visionaries; the implementers of the "Second Spaceflight Revolution" may be indentifiable in the 1990s. Some space advocates have made their peace with the classic agenda. But those with the greatest dream – massive human migration away from Earth – may be falling into the pattern of the first spaceflight revolution: first a proselytizing advocacy that meets frustration, then a military or political detour, then civil and private uses of the new technology.

In the 1970s we saw the elaboration of a new underground, seeking to democratize space, to establish a space-based economy, and to build permanent space settlements. This flowered in organizational form in the late 1970s and early 1980s, a resurgence of the space advocacy on an unprecedented scale. But much remained to be done.

In his 1981 book *The Cosmic Chase,* Richard Hutton wrote that the "space underground" was more like a loose federation than a tightly knit group of like-minded people.[93] That remained true four years later. But some things had changed. The underground had grown significantly; it had become more diversified, more open, more visible, more political. No longer just a friendly conspiracy, it aspired to be a social movement. By 1985, it was successfully co-opting a growing number of established figures who did not originally belong to it. Its organizations, while they may have conventional roles, also were arms of the space dream being turned into action. The organizations may change, but the dream goes on.

The conspiracy continues in an altered way, composed of overlapping subunits of people dedicated to exploring the cosmos, to expanding the human uses of space, to extending the human reach outward, and to spreading humanity beyond the Earth. The agenda has become more complex and diversified; there are several schools of thought as to what the next step should be, presenting the space advocacy with an organizational challenge.

The specifics, the tactics, and the justifications change, but the idea of humans going into space does not. Pursuers of the space dream have adjusted their appeal to the issues of the times, arguing that new ventures in space would be solutions to current social, economic, military, or political problems:

Long-range rockets in World War II and the Cold War
A race with the Soviet Union for international prestige in the late 1950s
    and 1960s
Practical, Earth-oriented applications in the early 1970s
Space industrialization in the late 1970s, including a response to the
    energy problem and the limits to growth
Space-based defenses and space commercialization in the 1980s

The space underground continued to live in 1985. Even with space activity revived, it continued to plan and advocate further steps, seeking to coalesce around new goals, searching for new justifications. What mattered to the space advocacy was not so much the solution of specific Earthly problems but the continuation of outward exploration and expansion by the human species.

By 1985, William Bainbridge's 1976 question about whether the spaceflight revolution would succeed or not had been answered. Despite predictions of its demise a decade ago, the American space program had survived and had been revitalized. The political climate for incrementally increasing American activity in space was generally good. Space activity, once the exclusive province of U.S. and Soviet government civil and military space organizations, had more diverse and widespread bases; space power was decentralizing, with more nations, government agencies, and independent companies becoming directly involved. Powerful new forces supporting increased activity in space were increasingly visible by 1984 in the form of defense and commercial interests, which seemed likely to create expanded, long-lasting constituencies. The means of access appeared to be diversifying, much in keeping with the ethos of a pro-space movement that wants more people to go.

Strikingly, there was virtually no organized opposition; only sniping about priorities from budget cutters and some scientists. As long ago as 1982, the National Space Institute's Leonard David had poked a little fun at the term "pro-space," pointing mockingly to "the invisible anti-space program invaders."[93] The space enterprise had become a permanent fact of our lives; "space" was here to stay.

15

EPILOGUE AND FUTURE

Not even the most clairvoyant space scientists and science fiction
writers can be sure what the shape and scope of our ventures into
space will be thirty years from now, but it seems safe to
speculate that we will be entering space with great regularity, that
large numbers of people will be making the journeys, and that the
term "astronaut" will have become a thing of the past.

Joseph P. Allen, 1984[1]

EPILOGUE

In late 1984 and 1985 we saw many events illustrating the vitality of
the American space enterprise and the relatively high profile of space
issues in public affairs. The Space Shuttle proved its versatility as a
multipurpose vehicle, enabling its crew to retrieve and repair satellites
while conducting scientific experiments and testing technologies for the
expanded use of the space environment.[2] Despite ciriticisms aimed at the
cost of shuttle missions and at some technical problems, the Shuttle
continued to be remarkably successful for a developmental vehicle until
the Challenger tragedy of January 28, 1986.

That searing disaster, which took the lives of five astronauts, a
commercial payload specialist, and a schoolteacher, jolted Americans into
a heightened awareness of human spaceflight. It brought with it a loss of
innocence about the dangers of space travel and temporarily challenged
the idea that ordinary citizens should go. It seemed certain to trigger a

broader debate about the future of the American space program, including a revival of the manned vs. unmanned issue. As of this writing, that debate held some potential to be a new threat to the spaceflight dream. At a minimum, Shuttle operations seemed likely to be suspended for months.

Yet the initial reaction to the loss of Challenger and the crew was remarkably positive. President Reagan certainly spoke for most Americans when he said that the manned space program must go forward. At the memorial service in Houston, he stated that both the Space Shuttle and the space station program would continue. There will be strong pressures to restore and perhaps expand America's national launch capability, which may well be more impressive after the accident than before. The American space program may be stronger because of Challenger's sacrifice. Sooner or later, the spaceflight participants program will be revived, once again giving ordinary Americans a more direct sense of participation. In the longer term, as we look back on the tragedy with some historical perspective, we may find that it had the effect of emotionally bonding many more Americans to the spaceflight dream.

The aerospace community's confidence in the future of America's civil space program had seemed to be reinvigorated by the events of 1984, including the space station decision, the lively interest in space commercialization, and Ronald Reagan's crushing victory in November. ("Four More Years: Aerospace Wins Big," reported *Aerospace America*.[3]) Funding for the Space Station survived challenges in Congress in 1984 and 1985, although at a lower level than NASA wanted. In the spring of 1985, the United States signed agreements with its European, Japanese, and Canadian partners for the design of the station.[4] Meanwhile, space science also was showing new life, with the Galileo mission to Jupiter scheduled for launch in May 1986 and the Hubble Space Telescope for August 1986. Work was underway on a Venus Radar Mapper and a Mars Observer.

Military activity in space continued to attract public attention. The first all-military Space Shuttle mission flew in January 1985, amid media controversy.[5] Under Secretary of the Air Force Edward Aldridge was to fly on the first military Shuttle mission launched from Vandenberg Air Force Base in California in 1986.[6] A unified Space Command went into operation at Colorado Springs in September 1985.[7] The United States conducted a third test of its new anti-satellite (ASAT) system against a still-functioning Air Force satellite in September 1985.[8]

Meanwhile, the Strategic Defense Initiative (SDI) continued as a major research program, despite annual efforts by opponents to cut its funding. There was a recurrent public policy debate about whether the SDI was or was not a bargaining chip in arms control talks with the Soviet Union, which continued to criticize the SDI. In September 1985, Daniel Graham of High Frontier announced the formation of a Coalition for the Strategic Defense Initiative.[9] Space arms control efforts achieved limited successes in restricting ASAT tests.[10]

Space commercialization remained a hot topic despite setbacks. Tiny latex spheres, the first products made in space, went on sale in June 1985.[11] By the spring of that year, the McDonnell Douglas/Ortho Pharmaceutical electrophoresis device was returning usable products to the Earth, although Ortho later pulled out of the project. Other successful experiments suggested that space industrialization may indeed have a future.[12] Space Industries Incorported reached agreement with NASA in August 1985 on a human-tended "space factory" module.[13] Efforts continued to identify and remove obstacles to space commercialization, including a report on the subject for the Aerospace Industries Association by space advocates David C. Webb and Courtney Stadd.[14] But commercialization suffered a setback in the insurance field after satellite losses and the Challenger disaster, bringing the future availability of insurance and launches for commercial space ventures into some doubt.

The future of the entrepreneurial launch vehicle companies also appeared to be in question by late 1985, although the loss of Challenger might open up some niches. After long and contentious debate, the U.S. government established a pricing policy for Space Shuttle launches that some argued would drive the small companies out of business.[15] The pricing issue stimulated arguments about whether the government should or should not subsidize early commercial ventures into this new environment.[16] Despite the decision, however, Space Services Incorporated pressed ahead with plans to launch cremated human remains into orbit.[17] Meanwhile, the outlook for the privatization of space systems remained clouded. Agreement was finally reached on the transfer of the LANDSAT system to a company called EOSAT, but the expected government subsidy may not be provided.[18] NASA appeared to have dropped the idea of private management of Space Shuttle operations.[19]

The idea of broader participation in spaceflight got a boost with the announcement that a schoolteacher would fly on the Space Shuttle in 1986. Over 10,000 teachers comepted for the honor.[20] Sadly, winner Christa McAulliffe died in the Challenger explosion. Senator Jake Garn of Utah flew on the Shuttle in the spring of 1985, and Congressman Bill Nelson of Florida flew in January 1986.[21] A journalist was to be the next citizen chosen for a ride on the Shuttle.[22] However, there still were complaints that NASA's astronauts were selected from too narrow a base.[23] Meanwhile, former Office of Technology Assessment official Thomas Rogers proposed another approach to democratizing space: homesteading in orbit, using modified Space Shuttle external tanks.[24]

The U.S. space enterprise remained a continuing media event during late 1984 and 1985. A television adaptation of James Michener's novel *Space,* an excellent four-part Public Broadcasting Service documentary on spaceflight, and a "Nova" program on "Space Women" were broadcast during the spring of 1985. Spectacular films shot from the Space Shuttle were shown at the National Air and Space Museum and at other locations. Advertisers continued to make heavy use of space themes, often using the Space Shuttle to symbolize achievement.

There were other signs of strong public interest. The Space Camp at the Alabama Space and Rocket Center remained hugely popular and added a course for adults.[25] Society Expeditions of Seattle distributed a prospectus for space tourism, with initial flights to cost $1 million on the Space Shuttle, or perhaps $50,000 on some future vehicle. The firm signed an agreement with launch vehicle entrepreneur Gary C. Hudson to develop a new launch vehicle by 1991.[26] Meanwhile, entrepreneurs proposed a "space resort" in California.[27]

THE NEW ADVOCACY

As for the new space advocacy, changes in the groups founded after 1972 suggested consolidation, professionalization, and a recognition of the central importance of manned spaceflight. Throughout much of 1984 and 1985, the L-5 Society and the National Space Institute discussed a merger of the two organizations, laying out a specific plan and polling their members on the subject. It seemed clear in both cases that an important motivation was declining membership.[28] The L-5 Society's H. Keith Henson warned of what happens when an activist organization merges with a stodgy one.[29] Meanwhile, the two organizations cooperated in supporting the passage of space station funding, with the L-5 Society and Spacepac doing most of the grass-roots lobbying.[30] An April 1985 Space Development Conference in Washington, D.C., was cosponsored by the L-5 Society, the National Space Institute, the American Astronautical

Society, the American Space Foundation, Spacepac, and the Students for the Exploration and Development of Space.

Of the new groups that appeared in late 1984 and 1985, the most significant may prove to be the Young Astronauts. In November 1983, columnist Jack Anderson had proposed to President Reagan that the space theme would be an effective means of stimulating interest in science and technology among young people and of improving the American educational effort in science and mathematics.[31] In June 1984, the White House announced "Project Liftoff," whose major component was the Young Astronaut Program. This was put together by the White House Office of Private Sector Initiatives, along with a number of corporations, professional societies, and nonprofit groups, including the National Space Institute. A Young Astronaut Council was inaugurated in Washington, D.C., on October 17, 1984.[32] Its Director of Operations was Todd C. Hawley, leader of Students for the Exploration and Development of Space. The Council began sending materials to schools all over the country. Advertisements even appeared on supermarket shopping bags. By April 1985, Young Astronaut memberhship had grown to about 50,000, making it the second largest space interest organization in the United States. Even before then, Trudy E. Bell had speculated that the Young Astronaut organization would be a major success story for the space interest movement.[33]

Other new groups of interest included the American Interstellar Society, which proposed to spread participation in space activity by putting its members' investments only into space enterprises, and Harrison Schmitt's New Worlds organization. Both were reminiscent of themes enunciated by the Committee for the Future 15 years earlier. Diana Hoyt of the Congressional Space Caucus became president of a new Washington-based group called Women in Aerospace, and Save the Apollo Launch Tower became the Apollo Society. The American Society of Aerospace Pilots replaced its entrepreneurial founder, David Koch, and Marcia S. Smith became the first woman president of the American Astronautical Society.

One of the most remarkable developments within the new space advocacy during 1984 was the endorsement of manned spaceflight by groups that had emphasized unmanned exploration. Impressed by an April 1984 *USA Today* poll that showed that 45 percent of Americans wanted to fly in the Space Shuttle, the World Space Foundation launched an "I Want to Go" campaign, featuring T-shirts and a proposed book on the subject.[34] More significant was the "conversion" of the Planetary

Society to the cause of manned spaceflight. The society commissioned Science Applications Incorporated to do studies of a lunar base and manned missions to Mars and the asteroids. The results, announced in November 1984, showed that such missions could be done more cheaply than critics had suggested.[35] The *Planetary Report* devoted its March/April 1985 issue to human exploration of the solar system. Meanwhile, the society continued its support of the search for extraterrestrial intelligence with the help of a $100,000 donation from "Close Encounters of the Third Kind" director Stephen Spielberg.[36] The society also supported a television special on Halley's comet in November 1985.

COLORADO – THE NEXT "SPACE STATE?"

Since the 1960s, California has been the leading American "space state." NASA sends more of its contract money to California than anywhere else and employs more contractors there. Not surprisingly, during the late 1970s and early 1980s California was the home of more space enthusiasts and more space interest groups than any other state. Texas, Florida, and Alabama also were important "space states."

As this book was being written, Colorado was emerging as an increasingly active locus for interest in space, primarily because of the location of the U.S. Space Command and the Consolidated Space Operations Center at Colorado Springs. Congressman Ken Kramer and Senator William Armstrong have given the Colorado space enterprise high political visibility. In 1983, Kramer and other interested citizens founded the United States Space Foundation, a Colorado Springs-based group that strongly supports space defense. Its leadership is heavily weighted with retired military officers. At its first annual conference, held in the fall of 1984, the agenda was dominated by the SDI.[37] Meanwhile, a different center of space interest was emerging at the University of Colorado at Boulder. Colorado may be destined to join California, Texas, Florida, and Alabama as a leading "space state."

THE SPREAD OF SPACEFLIGHT

The United States was not the only place where space activity was on the upswing. The Soviet Union continued to operate and resupply its

Salyut space stations. Its much-reported testing of its own spaceplane and space shuttle may lead to manned flights within the next decade. Eventually, the Soviet Union may move in the direction of its long-predicted Cosmograd, or "space city." In March 1985, Soviet scientists dazzled an American audience in Houston by unveiling plans for future planetary exploration, including a 1988 mission to Mars and its moons.[38] Soviet spacecraft have been launched to fly by Halley's comet in early 1986, after looping around Venus. Meanwhile, the Soviets continue to conduct a very active military space program and are reported to be working on space weapons.[39]

Space capabilities have continued to spread beyond the two superpowers. European Space Agency members and Japan have ambitious plans for future projects, including manned space vehicles.[40] They entered into cooperation with the United States on a space station largely to improve their own capabilities but may also have been encouraged by rising public interest in manned spaceflight, stimulated in part by the Space Shuttle. Both ESA and Japan launched probes toward Halley's comet in 1985.[41] The Peoples Republic of China has entered the commercial launch vehicle competition with its Long March 3.[42] During the 1984-85 period, new national space centers were established in the United Kingdom and Italy.[43] In these and other countries, space also is here to stay.

OUTLOOK FOR SPACE

As of 1985, it appeared that the U.S. space agenda for the next decade would be shaped largely by two major programs initiated by the Reagan administration: the space station and the SDI. Both seemed likely to require similar support technologies, such as new launch vehicles offering lower cost to orbit, whether those are heavy lift launch vehicles, shuttle-derived vehicles, or something else.[44] Both programs also may need orbital maneuvering vehicles and orbital transfer vehicles. Along with expanded experience with working in space and assembling large structures there, these should improve the prospects for space industrialization and, in the long run, space settlement. The Space Shuttle will remain the workhorse of the American space program for the near future. However, initial planning for a second-generation manned vehicle seemed well underway by 1985, and a request for a new start seems likely within a decade.

The space station has its critics. Some are opposed for budgetary reasons at a time of heightened concern about federal deficits; others want to use the station as a lever to get funding for their own programs (such as automation and robotics); others do not accept the need for a permanent human presence in space. The station satisfies the space developers but gives little to the space explorers. Like any major space program, the station will have its development problems and possibly minor crises over funding or international negotiations. However, a space station eventually will fly, at some time in the 1990s. When it does, U.S. participation in the "Second Spaceflight Revolution" will become a permanent fact. "Starting in the early 1990s," said James Beggs in 1984, "I believe there will always be Americans living and working in space."[45]

The first grand attempt to achieve this second spaceflight revolution – by Gerard K. O'Neill and his allies – effectively ended around 1980, just as the Space Shuttle was about to restore the classic agenda. But the classic agenda and the space colonization agenda merge in the permanent manned space station. Eventually, that station and its successors may achieve what science writer Walter Froelich predicts: "In time the space station may cause us to adjust our point of view, so that we will look on these Earth-orbiting regions as integral parts of the Earth, like newly found continents and oceans."[46]

The debate around the SDI is likely to be far more acrimonious. This initiative is a huge research program; it could prove to be the single most important technology driver in the United States during the next decade, even if space-based anti-ballistic missile systems are never deployed.[47] In that sense, it is the successor to the Manhattan Project and the Moon landing program.

Unlike the space station, however, SDI also is a defense and arms control issue. Opponents may continue to attack its funding and seek to block specific tests of SDI-related systems. This could overlap into some systems useful for the expansion of the civil space enterprise, such as compact energy sources in space.

As of 1985, many liberal opponents of the SDI were gathering around the idea of expanding U.S.-Soviet space cooperation, with several proposing a joint manned mission to Mars as an alternative to an arms competition in space. Among the prominent advocates were Senator Spark Matsunaga of Hawaii and the leaders of the Planetary Society. At a January 1985 symposium on space weapons at the National Academy of Sciences in Washington, D.C., Planetary Society President Carl Sagan called for a joint U.S.-Soviet landing on Mars by 2003.[48] Another symposium in July 1985, marking the tenth anniversary of the Apollo-Soyuz Test Project, was entitled "Steps to Mars." There Senator Matsunaga proposed an "International Space Year" beginning in 1992, the five hundredth anniversary of Columbus' first voyage to America.[49] However, any significant expansion of U.S.-Soviet space cooperation almost certainly would have to be preceded by an improvement in the political relationship between the two nations. This was recognized in the cautious assessment of U.S.-Soviet space cooperation issued by the Office of Technology Assessment in July 1985.[50]

In the near term, at least, NASA will remain the central pillar of the U.S. civil space enterprise. That agency can undertake missions that commercial firms would find unprofitable and that nonprofit organizations could not afford. The central course of U.S. civil space activity will continue to be determined by major program and budgetary decisions, regardless of what generalized policy statements or space policy acts may say. This will be true not only for "goal" projects such as a Moon base or a Mars landing, but also for major enabling technologies such as a human-rated orbital transfer vehicle. That means that the major decisions about space will be as much political as technical.

There is risk in this. Arthur Kantrowitz and others have pointed out the example of the Ming navy in China, which was conducting great, politically motivated voyages of exploration until bureaucrats cut off funding for this "impractical" activity.[51] Historian Walter A. McDougall has pointed out that massive state support of science and technology is a relatively new phenomenon, by implication raising the question of whether it is permanent.[52]

On the other hand, powerful sustaining forces for the space enterprise emerged more visibly during the first half of the 1980s: commerce and defense. Although still modest in scale as of 1985, space commercialization will in the longer run expand the economic constituency for space activity and reduce its dependence on federal appropriations, making it less subject to politically motivated interruptions. A space commercialization consortium was under discussion in 1985.[53] It would not be surprising to see the formation of a space industries association within the next ten years.

Defense interests, always important in the space field, seem likely to grow. Whatever may be done in the space arms control field, the defense establishments of the major powers will remain active in space. The existence of a Space Command could enhance the prospects for the eventual creation of a separate U.S. space force, possibly with dedicated manned military spacecraft such as the proposed trans-atmospheric vehicle. Although a particular military system might not be available or suited for civil uses, history suggests that sooner or later many of its technologies would be transferred to the civil and commercial sectors.

Older forces also will continue to affect the pace of government space programs. Competition from the Soviet Union – and increasingly from other nations and the European Space Agency – may spur greater U.S. efforts. Pride in the country and its achievements will remain an important factor. The use of space as an alternative to mlitary competition or even for some sort of international security system may have potential. However, the space agenda of the next decade will be dominated by defense and commercial interests if liberals do not come up with a persuasive, positive alternate agenda for space.

UPS AND DOWNS

There seem to be rough cycles in U.S. space history that may be repeated in the future. So far, each new administration since 1969 has had to be convinced that the broad space enterprise is worth not only expanding but continuing at the same level; each has been tempted to make the space program a target for budget-cutting because space spending (at least its civil side) is considered to be "discretionary." This suggests that the civil space program could face another trauma in January 1989, when a new administration takes office. If that new administration is a Democratic one, will the Democrats by then have internalized a positive space agenda?

There also seems to be a roughly once-a-decade cycle in the announcement of major civil space program decisions: Apollo in 1961, the Space Shuttle in 1972, the Space Station in 1984. This is both a political and a budgetary phenomenon; once the development costs of the current major project begin to turn down, it becomes possible to move toward the next one within an envelope of budgetary and political acceptability. This suggests that another major program decision will become politically feasible in the early or mid 1990s. If the classic agenda for manned spaceflight is followed, that decision will be for a Moon base. That also may be the time when a decision is made as to whether or not to deploy SDI systems in space, suggesting that the early and mid 1990s will be another time of intensified debate about the uses of space.

There also may be further up and down cycles in cultural attitudes toward the space enterprise, probably connected with broader questions such as attitudes toward technology, the military, and the business world. However, growing public and political acceptance of the permanence and utility of space activity may mean that the ups and downs of U.S. civil space activity will be less sharp in the future; the cyclical changes may be less dramatic. The crest of the mid 1960s and the trough of the mid 1970s may never be repeated.

Can we foresee provoking events that would disturb such a smooth continuity? International political and mlitary competition could produce such events, for example a sudden unilateral deployment of space-based weapons by the Soviet Union, although that seems both unlikely and qualitatively different from the broad challenge that Sputnik presented to the United States. There could be renewed concern about mineral and energy resources, with the space alternatives made more accessible through technological advance. Religious or social movements could have an impact. New ideas might appear on how to make other worlds more habitable. Astronomical discoveries, such as the detection of planets orbiting nearby stars, could spur interest. The detection of an extraterrestrial civilization could stimulate spaceflight activity.[54] Or humans could reach some new consensus about the purposes of going into space.

THE DEBATE OVER GOALS IN SPACE

In the years 1984 and 1985 there was an intensified debate in the United States over future U.S. goals in space, spurred primarily by the space station decision and the SDI. At the heart of one school of thought is NASA's own vision of what its goals should be, within political and funding constraints imposed by the White House and Congress. That vision comes close to being an updated version of the classic agenda for manned spaceflight and resembles the Space Task Group's grand plan of 1969.

Since the failure of that plan, NASA's formal setting of goals has been limtied to relatively near-term projects. However, these have been accompanied by frequent suggestions of the agency's longer-term aspirations. NASA's 1985 Long-Range Program Plan, for example, is concerned primarily with the implementation of major program decisions already made. In the manned spaceflight field, the only specific major goals are to develop a fully operational Space Transportation System (of which the Space Shuttle is the heart), and to establish a permanent manned presence in space (the Space Station). However, there are strong hints of the directions in which NASA would like to go.

In a section entitled "A Vision of the Space Era Beyond the Year 2000," the 1985 Program Plan foresaw one or more permanently manned space stations and an orbital transfer vehicle for manned sortie flights to at least geosynchronous earth orbit. This OTV "should be able to provide the basis for transportation for longer flights to establish a lunar base." Noting that "For the United States to exercise space leadership and pursue the long-term purposes of improving the well-being of Humankind in the Earth and space environments beyond the year 2000, long-range goals will be required," the plan foresaw routine use of the lunar surface for planetary geoscience studies, solar monitoring, astronomical surveys, and possibly extraction of resources, a well as routine materials processing in space. The plan also noted the possibility of a manned mission to Mars. Mixing the language of the new space advocacy with political rhetoric, the plan said the following:

> Pioneering in space by building a permanent settlement on that vast frontier also has meaning with respect to the President's third and fourth goals: "Strengthening our community of shared values," and "a lasting and meaningful peace." A community in space, with U.S. personnel working and functioning there, will reflect our shared values of a democracy in action for freedom and peace.[55]

NASA Administrator James Beggs has spoken often of NASA's visions of the U.S. future in space. Speaking at the U.S. Naval Academy in March 1984, he said the agency's future could include a manned space station in lunar orbit within about 20 years, a colony on the lunar surface by about 2010, initial construction of a manned station on Mars by about 2030, and a healthy and growing colony on Mars by 2040.[56] Addressing the Aero Club in New York in October 1984, he reportedly pointed to the possibility of returning to the Moon and traveling to near-Earth asteroids to mine their resources and of the establishment of a manned base on Mars to explore and investigate the use of "this most habitable and resource-rich planet as a staging base for further exploration of space."[57] After interviewing NASA officials and space scientists, *Washington Post* science writer Thomas O'Toole predicted a space station in 1992, the Shuttle taking its first tourists around Earth soon thereafter, a Moon colony underway by 2010, and Martian colonists setting sail by 2035.[58]

Presidential Science Advisor George Keyworth also has spoken in visionary terms about the future in space. Addressing a June 1984 conference sponsored by the Conservative Opportunity Society, Keyworth saw a manned space station as "a doorway to exploring and developing the solar system." Sounding very much like Gerard O'Neill, he said "We can go back to the Moon, not as a circus stunt, but to establish a permanent manned lunar base to do research and to mine lunar resources for use in building large structures in space." Keyworth foresaw U.S. missions to Mars, first automated, then human. In short, he said, "we can expand Mankind's presence into the cosmos."[59]

Thomas O. Paine, the principal advocate of the grand plan of 1969, said in late 1984 that the United States should take a bolder, broader, longer range look at a spaceport in low Earth orbit and the Moon and Mars as the focal points of twenty-first century space operations. Paine foresaw industrial processing using lunar and Martian resources. Low cost transport to low Earth orbit and an effective spaceport are the first milestones on the future "high road" to the Moon and Mars, he said. Paine thought that the economic, human, and technical resources required for the settlement of the Moon and Mars in the next century would be broadly available, and that the total investment to occupy and develop new worlds would represent about a tenth of a percent of the twenty-first century's gross world product.[60]

Not everyone agrees with these expansive visions of the future in space. During 1984, the Office of Technology Assessment, some space scientists, and leading figures in the Planetary Society were particularly visible in criticizing the approach of NASA and the Reagan administration.

In its November 1984 report *Civilian Space Stations and the U.S. Future in Space,* the OTA argued that "because the Nation does not have clearly formulated long-range goals and objectives for its civilian space activities, proceeding to realize the present NASA space station concept is not likely to result in the facility most appropriate for advancing U.S. interests in the second quarter-century of the Space Age."[61] In general, said the OTA report, the choices of space "infrastructure" should not be made without prior agreement on the future direction of the civilian space activities of the United States." The most effective way to determine our direction in space, according to the report, "would be a national discussion of, and eventual agreement on, a set of long-range goals which the United States expects its civilian space activities to address." NASA Administrator Beggs responded by saying that NASA's major goals must be politically acceptable. The highly visible goals, he said, provide a focus around which NASA can develop its capabilities.[62]

The OTA report proposed national goals for discussion that emphasize democratic control of, and greater direct involvement with, space technology, but that end with an endorsement of the expansionist vision:[63]

To increase the efficiency of space activities and reduce their net cost to the general public

To involve the public directly in space activities, both on earth and in space

To derive scientific, economic, social, and political benefits

To increase international cooperation and collaboration in and regarding space

To study and explore the Earth, the solar system, and the greater physical universe

To spread life, in a responsible fashion, throughout the solar system.

The OTA report proposed objectives including a transportation service to the Moon and the establishment of a modest human presence there, and bringing at least hundreds of members of the general public per year into space for short visits.

In its 1982 report on *Civilian Space Policy and Applications,* OTA had recommended reestablishing a mechanism similar to the old National Aeronautics and Space Council.[64] In 1984, Congress authorized the creation of a National Commission on Space, and in 1985 the administration accepted the challenge. Addressing the National Space Club in March, President Reagan announced his appointees to the commission, which was to propose goals for the American civil space program by March 1986.[65] The chairman was Thomas Paine, who had been NASA's most visionary administrator. The new commission included representatives of both the older space advocacy and its newer successor: Paine, space colonization and industrialization advocate Gerard K. O'Neill, and David C. Webb, chairman of the National Coordinating Committee on Space. Other members were planetary scientists Laurel Wilkening, physicist Luis Alvarez, space scientist Paul

J. Coleman, physicist and astronomer George Field, ITT Executive Charles M. Herzfeld, Massachusetts Institute of Technology professor Jack L. Kerrebrock, former astronaut Neil Armstrong, astronaut Kathryn Sullivan, former U.N. Ambassador Jeanne Kirkpatrick, and three retired military officers of general rank: Bernard Schriever and Charles (Chuck) Yeager of the Air Force and William H. Fitch of the Marine Corps.

The three principal staff members of the commission were all representative of the post-Apollo space interest phenomenon: Library of Congress space expert Marcia S. Smith, IEEE Space Subcommittee chairman Theodore R. Simpson, and National Space Institute staffer Leonard W. David. The new space advocacy was drawn into the commission effort when David Webb arranged to have the L-5 Society, the National Space Institute, and the American Space Foundation conduct surveys of their members.[66]

Perhaps the most striking thing about the makeup of the Commisison is that one of its members was Gerard O'Neill. The visionary of the 1970s had been accepted by the establishment of the 1980s. It also is significant that the major antagonists in the new debate about U.S. goals in space agree on the need for goals broader than the space station, which most see as a stepping stone to other things. Many of those debating the issue have absorbed and used language introduced and spread by the new space advocacy. Most notably, many have accepted the idea of colonizing other worlds and possibly space itself, of sending humans into space to stay.

AND THE NEXT STEP AFTER THAT ...

Although some space advocates call for commitment to their entire vision as a package, coalitions form more realistically around one goal at a time. As soon as the U.S. government decides to take one major step, U.S. space advocates move on to the next, trying to build a constituency for it, to sell the idea over a period of years, and to make the eventual decision seem logical or even inevitable.

Even before the space station decision, advocacies were becoming visible behind two further steps in the classic agenda for manned spaceflight: a Moon base and a manned mission to Mars. The first is openly led by mid-level NASA officials, with plenty of help from industry and other advocates. The second, emerging from a "Mars

Underground," is more in the tradition of the space dreamers of the 1920s and 1930s, an informal coalition of experts and enthusiasts with little institutional or economic interest in the outcome.

The "Return to the Moon" lobby became visible during the debate over the space station decision, when some individuals found the project too limited and uninspiring and called for a "high option" of a lunar base. Presidential Science Adviser George Keyworth was a prominent member of this group. Former Apollo astronaut Edwin E. (Buzz) Aldrin and former NASA Associate Administrator George Mueller were lobbying openly for such a decision in late 1983 and early 1984.[67]

In fact, the agitation for a return to the Moon had been gaining momentum for at least two years before the space station decision, due in considerable measure to the efforts of Johnson Space Center officials Michael B. Duke and Wendell W. Mendell.[68] In October 1983, *Aviation Week* reported that Johnson Space Center lunar and planetary scientists were proposing a small base on the Moon by the late 1990s, preceded by an unmanned lunar polar orbiter in the early 1990s.[69] Duke and Mendell obtained a grant to conduct a workshop at Los Alamos National Laboratory in April 1984, with about 50 participants from within and outside the government. The resulting report, issued in September, provided the basis for a symposium held in Washington, D.C., in October 1984. The AIAA's magazine *Aerospace America,* playing a familiar role, provided a forum for Duke and Mendell in its October 1984 issue.[70]

At the symposium, entitled "Lunar Bases and Space Activities of the 21st Century," NASA Administrator James Beggs came close to endorsing the lunar base when he said, "I believe it is likely that before the first decade of the next century is out, we will, indeed, return to the Moon." Keyworth told the same symposium that "a lunar base is only one of the more obvious next steps." Duke pointed out at the symposium that most of the equipment needed for the project is already in existence or planned; the exception is that key piece of space technology, the orbital transfer vehicle. *Science* magazine, reporting on the symposium, described it as a trial balloon, an attempt by lunar base proponents to get the attention of the administration, the media, and NASA headquarters and judged that they were probably successful. "It remains to be seen," concluded the *Science* report, "if they can build a compelling political case for the Moon."[71] By January 1985, Edward Teller reportedly was pushing for a lunar colony within ten years.[72] A letter published in *Aerospace America* in February 1985 suggested selling the Moon by offering subscriptions to the common budget for its development, bringing back memories of the Committee for the Future's Project Harvest Moon.[73]

The Moon base symposium helped bring together the advocates of the classic agenda and the space developers with a project both could support. NASA and the aerospace industry should be comfortable with the idea, having studied it in the 1960s. Once the Space Station is firmly on track, the Moon base could provide a new rallying point for the American space advocacy. Familiar arguments are likely to be heard pro and con, with manned spaceflight advocates and space developers strongly for the Moon base, while those favoring pure scientific research using unmanned vehicles may be against it. There also could be a debate as to whether a lunar base should be American or international. Turning points for the Moon base advocacy will be decisions to build a powerful, human-rated orbital transfer vehicle and to send a lunar polar orbiter to do a complete survey of the lunar surface.

Judging by past experience, Moon base advocates will need several years to build a constituency for the project and to make the idea more familiar to the interested public through frequent exposure in the media. There the citizens' pro-space groups could be helpful. NASA may repeat the space station pattern, forming a working group and doing intensive studies to prepare for its own formal advocacy of the project. The World Future Society has predicted a permanent U.S. base on the Moon by 2007.[74]

THE MARS UNDERGROUND

Another school of thought is focused by the old idea of a manned mission to Mars. First formulated in specific planning terms by Wernher von Braun in the 1950s, the Mars mission was part of the 1969 grand plan.[75] However, NASA studies ceased after 1971.

The Mars mission was given a subtle intellectual boost in the 1970s by Mariner 9 and the Viking missions, which revealed the Red Planet's diverse and interesting surface in great detail, and which suggested that frozen water may be abundant on Mars.[76] Although confirmation that the Martian atmosphere is very thin and cold was initially discouraging, would-be colonizers found hope in the science fiction idea of "terraforming," altering a planet (particularly its atmosphere) to make it more Earth-like so that humans could live on it. Carl Sagan had suggested an approach to terraforming Venus in 1961 and to terraforming Mars in 1973.[77] In 1976, NASA published a booklet entitled *On the Habitability of Mars: An Approach to Planetary Ecosynthesis,* prepared by NASA's Ames Research Center in California and edited by M. M. Averner and R. D. McElroy. The study, which McElroy called the beginnings of planetary engineering, concluded that there was no fundamental insuperable limitation on the ability of Mars to support a terrestrial ecology.[78]

In 1978, Christopher P. McKay, a graduate student at the University of Colorado, led an informal seminar inspired by the NASA study (this was at about the time that a citizens pro-space movement was starting to diversify rapidly). The seminar developed into an ongoing effort that attracted other interested people. In March 1979, Penelope Boston and space writer James E. Oberg sponsored an informal session at the annual Lunar and Planetary Science Conference near Houston. In the spring of 1980, Leonard David of the National Space Institute suggested to McKay that the time was right for a conference to study the near-future possibilities of exploring and colonizing Mars. Stan Kent of the Viking Fund lent his support, and a decision was made to hold a conference entitled "The Case for Mars" at Boulder, Colorado. Hearing about the conference, mostly by word of mouth, speakers volunteered.[79] A Mars underground began to form.

The conference, sponsored by the University of Colorado Space Interest Group, the Viking Fund, and the Rocky Mountain sections of the American Astronautical Society and the AIAA, was held at the University of Colorado from April 29 to May 2, 1981 – just after the first Space Shuttle mission. Since no formal study of a manned Mars landing had been done since 1971, it was essentially a brainstorming session. However, it concluded that a manned Mars mission was a viable option for our space program. The conference brought the participants into direct contact; as Carol Stoker puts it, "We wanted to bring people out of the closet."[80] It also helped bring the Mars idea back into the view of the interested public.

The second case for a Mars conference, cosponsored by the Boulder Center for Science and Policy and the University of Colorado Space Interest Group, was held July 10 through 14, 1984. By this time, the Planetary Society had established a Mars Institute, an informal network of experts headed by Chris McKay that is dedicated to studying the Red Planet. The Case for Mars II, whose purpose was to explore the potential for the colonization of Mars, drew many of the biggest names from the space advocacy and got a fair amount of media attention. A specific planning model for the Mars expedition, involving 15 people in three vehicles, was proposed.[81]

Planetary Society leaders and others, such as Senator Spark Matsunaga, have seen political potential in the idea of an international manned mission to Mars. In January 1985, Matsunaga introduced a resolution proposing cooperative efforts in space that could lead to a joint U.S.-Soviet mission.[82] Carl Sagan appears to believe that a political motivation is the only convincing one for funding such a mission. Others have used the traditional tactic of pointing to alleged Soviet preparations for such a mission as a motivation for an U.S. response.[83]

The Mars option also is an alternative to the Moon base/cislunar (Earth-Moon) space design favored by space developers. The "Mars lobby" was challenged forcefully by L-5 Society activist K. Eric Drexler in an October 1984 *L-5 News* article entitled "The Case Against Mars." Drexler suggested that a Mars mission would divert public attention and support from more utilitarian projects such as a lunar base and asteroid mining. Seeing two schools of thought about space, he wrote, "One way promises to open space as a true frontier with expanding opportunities and room for freedom. The other way promises to bring a space program centered on a political stunt, with some hope of setting up an open-ended charity."[84]

By 1985, a new fissure seemed to have merged within the U.S. space interest movement, reopening the gap between the space developers and the space explorers and between different schools of thought about the political role of space activity. That split also may reflect a difference between the official, NASA view of the next step in space and a more visionary "High Option." However, the American Space Foundation, which had appeared to support the classic NASA agenda, added a Mars Mission to its list of objectives in early 1985.[85]

THE ULTIMATE TRIP

There also is a small but persistent lobby for the ultimate trip: interstellar flight. Long a feature of science fiction, travel to other stars was taken up in a serious way by British space advocate Leslie Shepherd in an article published in 1950.[86] Carl Sagan, in a 1966 book coauthored with Soviet astronomer Iosif Shklovskii, argued that flight to nearby stars is a feasible objective for humanity.[87] In 1973, the British Interplanetary

Society began a study of how an interstellar probe could be built with foreseeable technology. Published in 1978, *Project Daedalus* concluded that interstellar flight, although expensive and difficult, could be accomplished by a civilization only slightly in advance of our own.[88]

In the United States, the leading spokesman for the interstellar flight lobby has been Hughes Research Laboratory physicist and science fiction writer Robert L. Forward, who as long ago as 1975 presented a paper to the space subcommittee of the House of Representatives entitled "A National Space Program for Interstellar Exploration."[89] He and aerospace engineer Eugene Mallove also edited a bibliography on interstellar flight and communication.[90] Forward, who finds U.S. aerospace technical societies too conservative, believes in selling the "pizzaz" of space exploration.[91] Mallove also is a proponent of colonizing Mars.[92]

This small interstellar flight conspiracy, like the interplanetary lobby of the 1930s, is international. Its principal medium of communication is the quarterly "Interstellar Studies" issues of the *Journal of the British Interplanetary Society*. That journal has long been an outlet for ideas too visionary to be accepted by conservative U.S. technical journals.

This apparently far-out idea was given a powerful boost during 1984 by space science discoveries that suggested that planets may be common attendants of other stars. The Infrared Astronomical Satellite revealed evidence that small objects were orbiting some nearby suns, and astronomers later succeeded in producing an image of a circumstellar disk.[93] These discoveries provide new targets for exploration and possibly new alternate worlds to fire the imaginations of future generations.

THE OUTLOOK FOR THE SPACE MOVEMENT

In 1976, Bainbridge wrote that "the next two decades will be a period of consolidation for the spaceflight movement."[94] This may have underestimated the reaction of the latent pro-space constituency to events such as the dramatic turndown in NASA's fortunes. The organized part of that constituency grew rapidly after the publication of Bainbridge's book.

It now seems likely that the formal organized expression of the U.S. space advocacy will shrink with success. The surge of growth seen in the late 1970s and early 1980s has not continued at the same pace; the movement appears to be near the top of its sine curve for this generation. Formal memberships are likely to decline as the sense of emergency passes. The number of lasting groups will shrink before stabilizing. If the environmental movement is any example, the strongest space interest organizations will survive as semi-permanent features of the interest group landscape. The surviving larger organizations are likely to become more professional in their operations, and less strident than some have been in the past. Many of the smaller groups will disappear as their leaders move on to other things; others will merge with larger organizations.

Space interest groups may find it easier to work together as they mature, as their entrepreneurial founders are replaced, and as proliferation ceases and consolidations occur. As the insecurity of new organizations fades, coalitions may become more feasible. However, it still will take sustained effort and good will to define an agreed agenda for the future in space beyond the Space Shuttle and the core program of the Solar System Exploration Committee. Old issues such as the manned vs. unmanned dispute may not go away. Newer ones, such as the question of weapons in space or the international legal regime for the uses of space, may become more divisive.

The surveys reported in Chapter 7 suggest that there remains enormous potential for organizing the space interest constituency, as was not done for Apollo. That constituency is very diverse and overlaps with other interests. Ideally, different sections of the organized space advocacy could appeal to different sectors of that constituency, and all could take advantage of more sophisticated audience analysis and marketing techniques. Each organization might concentrate on its particular skills – educating the public, grass-roots lobbying, direct mail, fund raising, research and development – instead of trying to perform all roles. However, if the organized advocacy wants to be politically effective, it must do a better job of coalition-building, outside the space interest field as well as within it.

Temporarily, at least, the new space movement may fall into the category of a post-success lobby. If the participation revolution fades, much of the organized space advocacy will fade with it. In the longer term, more members of younger generations may be able to turn space as an avocation into space as a vocation, turning them from citizens activism to conventional interest group activity.

Space remains a symbol of the use of advanced technology to better the human prospect, while stimulating us with exploration, discovery, and achievement. This suggests that spaceflight will flourish in times of cultural confidence and optimism and that support for spaceflight will tend to decline during periods of cultural self-doubt and pessimism. In the case of the United States, we seem to have gone through one cycle already. However, there also is a reactive quality to the new organized space advocacy; it is in part a response to a downturn in the prospects for space activity. Says Benjamin Bova, "There is an inverse ratio between the vigor of space groups and the success of the civil space program."[95] If and when a major downturn in the U.S. civil space program occurs again, the space advocacy may resurge to keep the dream alive.

Whatever organizational forms this advocacy may take, whatever its paid membership may be, the friendly conspiracy that underlies it will continue to exist, its members keeping alive what they believe is a revolution. The advocacy will continue to call for the next steps in the space dream, encouraging decision makers to be bold. This advocacy is different from others, in that it has in front of itself endless opportunities.

The positive, expansive, even transcendental nature of the spaceflight dream may be a tactical disadvantage. But it is a strategic advantage in providing an endless series of outward goals. Robert H. Goddard said it over 50 years ago: "There can be no thought of finishing, for aiming at the stars, both literally and figuratively, is a problem to occupy generations, so that no matter how much progress one makes, there is always the thrill of just beginning."[96]

Appendix A
American Space Interest Groups

This Appendix lists significant American space interest groups active as of 1984. The categories are those used in Chapter 8, The Space Group Boom. The memberhship figures are taken from Trudy E. Bell's monograph *Upward: Status Report and Directory of the American Space Interest Movement, 1984-1985*. A few of the groups included in Bell's directory are not included here because of my doubts about their significance or their relevance to the pro-space cause.

Name and Address of Group	Date Founded	Membership
Educational, Mixed Purpose, and Other		
American Space Foundation for Education, Inc.	1984	Section 7
Suite 550		
214 Massachusetts Avenue, N.E.		
Washington, DC 20002		
Hypatia Cluster	1981	85
Suite 200		
1724 Sacramento Street		
San Francisco, CA 94109		
L-5 Society	1975	9,500
1060 E. Elm		
Tucson, AZ 85719		
National Space Club	1957	1,200
Suite 300		
655 15th Street, N.W.		
Washington, DC 20005		

Name and Address of Group	Date Founded	Membership
National Space Institute West Wing Suite 203 600 Maryland Avenue, S.W. Washington, DC 20024	1974	10,000
The Planetary Society 110 S. Euclid Avenue Pasadena, CA 91001	1979	130,000
Spaceweek (National Headquarters) P.O. Box 58172 Houston, TX 77258	1980	100 local directors
Students for the Exploration and Development of Space 800 21st Street, N.W. Washington, DC 20052	1980	4,000
United States Space Education Association 746 Turnpike Road Elizabethtown, PA 17022	1983	1,000
United States Space Foundation P.O. Box 1838 Colorado Springs, CO 80901	1983	700
Young Astronaut Council Suite 950 1015 15th Street, N.W. Washington, DC 20005	1984	3,800
Economic Interest Groups Aerospace Industries Association of America, Inc. 1725 De Sales Street, N.W. Washington, DC 20036	1917	50 firms

Name and Address of Group	Date Founded	Membership
The GEOSAT Committee Suite 209 153 Kearny Street San Francisco, CA 94108	1976	90 firms
Nonprofit Interests Public Service Satellite Consortium Suite 907 1660 L Street, N.W. Washington, DC 20036	1975	100 orgs.
Space Science Working Group, Association of American Universities Suite 730 1 Dupont Circle, N.W. Washington, DC 20036	1982	20 univs.
Universities Space Research Association 311 American City Building Columbia, MD 21044	1969	55 univs.
Professional Organizations and Societies Aerospace Education Association of America 1910 Association Drive Reston, VA 22091	1976	10,000
Aerospace and Electronics Systems Society of the Institute of Electrical and Electronic Engineers 345 East 47th Street New York, NY 10017	1951	6,900
American Astronautical Society 6060 Tower Court Alexandria, VA 22304	1954	1,000

Name and Address of Group	Date Founded	Membership
American Institute of Aeronautics and Astronautics 1633 Broadway New York, NY 10019	1963	37,000
American Society of Aerospace Pilots 946 S.E. 8th Street Grants Pass, OR 97526	1981	1,300
Aviation/Space Writers Association Suite 311 1725 N Street, N.W. Washington, DC 20006	1938	1,300
Society of Satellite Professionals P.O. Box 19047 Washington, DC 20036	1983	550

Funding/Research Organizations

California Space Institute Building T-31, Mail Code A-030 University of California, San Diego La Jolla, CA 92093	1979	—
The Space Foundation P.O. Box 58501 Houston, TX 77058	1979	28 firms, 100 indivs.

Do It Yourself: Technology Development and Research

Radio Amateur Satellite Corporation (AMSAT) P.O. Box 27 Washington, DC 20044	1969	5,000
Independent Space Research Group P.O. Box 1246 Troy, NY 12180	1980	1,000

Name and Address of Group	Date Founded	Membership
Space Studies Institute 385 Rosedale Road P.O. Box 82 Princeton, NJ 08540	1977	5,000
World Space Foundation P.O. Box Y South Pasadena, CA 91030	1979	not discl.
Political Organizations American Space Foundation Suite 550 214 Massachusetts Avenue, N.E. Washington, DC 20002	1981	22,000
American Space Foundation Candidates Committee above	1984	above
American Space Frontier Committee P.O. Box 1984 Merrifield, VA 22116	1983	10,000 contrib.
Campaign for Space P.O. Box 1526 Bainbridge, GA 31717	1980	1,500 mail list
Congressional Space Caucus Rep. Daniel Akaka 2301 Rayburn House Office Building Washington, DC 20515 Rep. Herbert Bateman 1518 Longworth House Office Building Washington, DC 20515	1981	169

Name and Address of Group	Date Founded	Membership
Congress Staff Space Group 322 House Annex No. 1 Washington, DC 20515	1981	230
Spacepac Suite S 2801B Ocean Park Boulevard Santa Monica, CA 9405	1980	2,000 contrib. 18,000 on phone tree
Write Now! P.O. Box 36851 Los Angeles, CA 90036	1980	—
Space Defense/Space Arms Control High Frontier, Inc. Suite 1000 1010 Vermont Avenue, N.W. Washington, DC 20005	1981	40,000 subscrib.
Institute for Security and Cooperation in Outer Space Suite 102A 201 Massachsuetts Avenue, N.E. Washington, D.C. 20002	1983	—
Institute for Space and Security Studies 7720 Mary Cassatt Drive Potomac, MD 20854	1983	800 mail list
Progressive Space Forum 1724 Sacramento Street, Number 9 San Francisco, CA 94109	1979	250
World Security Council Suite 275 World Trade Center San Francisco, CA 94111	1980	328

Name and Address of Group	Date Founded	Membership
Unbrella Organizations		
National Coordinating Committee for Space No. 807 4500 S. Four Mile Run Drive Arlington, VA 22204	1979	18 groups

Appendix B
The British Interplanetary Society

The British Interplanetary Society (BIS), founded in 1933, is the oldest significant pro-spaceflight organization still existing under its original name. The first important group, the *Verein fur Raumschiffahrt,* went out of business in the 1930s. The second, the American Interplanetary Society, became the American Rocket Society in 1934 and merged with the Institute for the Aeronautical Sciences in 1963.

The BIS, being the principal space interest group in another English-speaking country, provides a useful standard of comparison for American space interest groups. Although there are some similarities, the differences are striking.

THE THIRTIES

In the early 1930s, a young Cheshire man named Philip E. Cleator, who had long held an interest in the possibility of space travel and who was an active reader of science fiction, discovered the existence of the American Interplanetary Society. Throughout 1933, he struggled to create a British counterpart, finally bringing six people together for a founding meeting on October 13, 1933. As in the case of the American Interplanetary Society, many of the early members of the BIS were science fiction writers or fans. Most were in their twenties; Arthur C. Clarke joined age 19. Cleator later described the prewar society as a small group of enthusiasts and cranks. The first issue of the *Journal of the British Interplanetary Society* was published in 1934.[1]

The raison d'être of the BIS was "to achieve the conquest of space and thence interplanetary travel."[2] Unlike its U.S. cousin, however, the BIS was severely restricted in its ability to do technology work by the Explosives Act of 1875, which was interpreted by British authorities to exclude private rocket experiments. As a result, the society became concerned primarily with theoretical and design work and with spreading the spaceflight message. However, historian Frank H. Winter notes that there was no immediate groundswell of public interest in Great Britain, nor was there a British champion of spaceflight with the stature of a Konstantin Tsiolkowsky or a Hermann Oberth.[3]

In 1936, Cleator published a book entitled *Rockets Through Space,* which introduced the public to basic spaceflight ideas and attracted some public notice.[4] In the same year, the film *Things to Come* also inspired a flurry of interest in spaceflight. Despite all this, the society's prewar membership peaked at a little over 100.[5]

The BIS had no success in persuading the British government of the utility of rockets. In a now classic letter, an Under Secretary of State in the Air Ministry wrote the following:

> We follow with the interest the work that is being done in other countries on jet propulsion, but scientific investigation into the possibilities has given no indication that this method can be a serious competitor to the airscrew-engine combination. We do not consider that we should be justified in spending any time or money on it ourselves.[6]

In 1937, the London branch of the BIS took over the leadership of the organization from the Liverpool branch. Meanwhile, the society had embarked on the design of a spaceship that could carry three people to the Moon; the results were published in 1939.[7] Many of the basic ideas of the BIS Moonship study were applied 30 years later in the successful Apollo landings. However, because of the beginnings of World War II in Europe, the BIS ceased activities in late 1939.

REVIVAL

The BIS was revived during the second half of 1945 and was incorporated in December of that year with a membership of 280. Its membership was boosted by the merger into the BIS of other British groups that had continued to function during the war.[8] The *Journal* resumed publication and became a respected international medium for spaceflight ideas. The Moonship was redesigned. The example of the BIS inspired the formation of the American Astronautical Society in 1954.

The postwar BIS became more professional in its membership, although it still was more of an amateur society than the American Rocket Society. Membership rose to 2,900 by 1953.[9] In 1956, the *Journal of the British Interplanetary Society* was joined by the less technical magazine *Spaceflight,* noted for its excellent coverage of international space events.

Continuing its role as an advanced planning department for the spaceflight movement, the BIS did work on an "aerospace transporter," which presaged the U.S. Space Shuttle.

Sociologists William S. Bainbridge comments that the BIS was struggling to become a professional technical space organization in a nation that had little use for one.[10] BIS activists tried to create a major United Kingdom rocket industry; in 1960 they urged a U.K. space program based on European collaboration, with the Blue Streak rocket as the first stage of a satellite launcher, and outlined a national program including the development of the Black Knight as a second stage. Their advocacy contributed to the formation of the European Launcher Development Organization.[11] However, their aspirations for a national program suffered failure in 1973 when the independent British launcher program was cancelled.

BIS activists also worked long and hard to persuade the British Post Office of the merits of satellite communications, an idea stemming largely from the work of early BIS member Arthur C. Clarke, but progress was slow. (The Society's longtime secretary, L. J. Carter, commented in 1984 that the BIS did not intend that space communications be the *only* area of British space activity.[12]) Increasingly, the United Kingdom conducted its space efforts through European multinational organizations, today the European Space Agency. Not until 1985 did the United Kingdom begin to create its own official space organization.[13]

In October 1973, at the fortieth anniversary of its founding, the BIS rededicated itself to the spaceflight revolution by announcing plans for studies of interstellar flight and communication. BIS members launched a study of the feasibility of an unmanned interstellar probe called Project Daedalus. The results, published in 1978, showed that such a vehicle could be built by a civilization only slightly in advance of our own, although it would be expensive and difficult.[14] As the Moonship study presaged Project Apollo, so this effort may presage a real interstellar probe in the future.

By 1984, BIS membership was maintaining a relatively constant level of about 3,500. Roughly one third of the members came from outside the United Kingdom, with most of those being Americans. The society has acquired a larger headquarters building near Lambeth Palace in London. The BIS holds a large annual meeting every October at the seaside resort of Brighton and smaller lecture and film programs at its London headquarters.

THE BIS AND ITS AMERICAN COUSINS

In his 1976 book *The Spaceflight Revolution,* Bainbridge observed that, unlike the American Institute of Aeronautics and Astronautics, the BIS had not been absorbed into the aerospace industry, in great measure because of the weakness of official support for spaceflight in the United Kingdom. It had remained independent. While the AIAA concerned itself with projects showing the promise of quick financial gain, the BIS kept the long view. Bainbridge concluded that the two organizations complemented each other by playing different roles.[15]

The contemporary BIS does not try to play all the roles U.S. space interest groups have explored, particularly the newer political ones such as lobbying aggressively or forming political action committees. "The BIS has not got into political activism" says Society Secretary L. J. Carter, noting that political activism can cause the more sober space-oriented person to back away.[16] This has enabled the BIS to avoid a split between the technical side and the enthusiast side of its membership.

BIS activists have sought to use modest means such as letters and information packets to persuade government officials, members of parliament, and opinion molders of the need for greater efforts in the spaceflight field. However, opposition remains strong. Carter recognizes that such mundane matters as politics are now of real concern. In his view, the most exciting thing now is the United States space station; the BIS would like to see major British involvement with the project.[17]

The BIS also recognizes that a new generation of people interested in space has made its presence felt in the United States. "This is creeping up on us," says Carter, who observes that "we need to transmit across generations."[18] The BIS also shares the U.S. space advocacy's hope of broader participation in spaceflight. Advocating more European activity in space, a *Spaceflight* editorialist wrote, "The time has surely come when we should strive for a broader participation in activity at the space frontier."[19]

The BIS faces a watershed in the process of rejuvenation. "It is becoming more practical, and is gearing itself into the system," says Carter.[20] In particular, this involves awareness of competing claims from nonspaceflight interests. The society, which follows the work of U.S. space interest groups, understands that it must learn new techniques.

In the end, what is striking about the BIS is its staying power. "There have been many competitors to the BIS," comments Carter.[21] But they have faded, and the BIS has survived. Unlike the U.S. case, in which the spaceflight movement has flowered into a diversity of organizations, the BIS has maintained a central, preeminent position in the field within its national culture for over 50 years.

"Progress is due to small groups with vision," Carter concludes. "Small groups are the way to make the greatest progress in the shortest time."[22]

Notes

OPPOSITE TITLE PAGE

1. Arthur C. Clarke, *The Promise of Space* (New York: Harper & Row, 1968), p. 314.

2. Carl Sagan, *The Cosmic Connection: An Extraterrestrial Perspective* (New York: Dell, 1975), p. 227.

3. Joseph P. Allen, *Entering Space* (New York: Stewart, Tabori, & Chang, 1984), p. 218.

4. *Space Industrialization – Final Paper* (Los Angeles: Rockwell International Space Division, April 1978), p. 20.

5. Frederick J. Turner, "The Significance of the Frontier in American History," in *Annual Report of the American Historical Association for the Year 1893* (Washington, D.C.: U.S. Government Printing Office, 1894), pp. 197-227, 227.

6. Jerry E. Pournelle, as quoted by the L-5 Society in its announcement of the Third Annual Conference on Space Development, held in April 1984.

7. Attributed to *Omni* magazine.

PREFACE

1. Sir Herman Bondi, in his review of Geoffrey Pardoe, *The Future for Space Technology,* in *Space Policy* 1 (February 1985):96.

2. See Wernher von Braun (with J. Kaplan et al.), *Across the Space Frontier,* ed. C. Ryan (New York: Viking Press, 195); Arthur C. Clarke, *The Exploration of Space* (New York: Harper & Brothers, 1951); Willy Ley (with paintings by Chesley Bonestell), *The Conquest of Space* (New York: Viking Press, 1949).

3. Interview with David C. Webb, October 13, 1983, in "Pro-Space: Interviews with the Space Advocacy," unpublished manuscript (see note 7).

4. See, among others, William S. Bainbridge, *The Spaceflight Revolution* (New York: John Wiley & Sons, 1976) and Frank H. Winter, *Prelude to the Space Age: The Rocket Societies, 1924-1940* (Washington, D.C.: Smithsonian Institute Press, 1983).

5. For the 1980 survey, see Trudy E. Bell, "American Space-Interest Groups," *Star and Sky* 3 (September 1980):53-60. A revised version appeared as "Space Activism," *Omni* 3 (February 1981):50-54, 90-94. Bell's 1982 survey, entitled "From Little Acorns . . . : American Space Interest Groups, 1980-1982," was to have been published in volume 2 of *The Space Humanization Series* (published by the Institute for the Social Science Study of Space), which had not appeared as of early 1986. Results of her 1984 survey appeared in *Upward: Status Report and Directory of the*

American Space Interest Movement, 1984-1985, published in 1985 and available for $20 from the author at 11 Riverside Drive, #13GW, New York, NY 10023.

6. Joseph J. Corn, *The Winged Gospel: America's Romance with Aviation, 1900-1950* (New York: Oxford University Press, 1983).

7. Edited versions of these interviews are in an unpublished manuscript entitled "Pro-Space: Interviews with the Space Advocacy." That project was supported by a grant from the Una Chapman Cox Foundation. Unless otherwise indicated, all interviews referred to in the notes to this book are in this manuscript.

CHAPTER 1

1. T. E. Lawrence, as quoted in Arthur C. Clarke, *The Challenge of the Spaceship* (New York: Curtis, 1953), p. 26.

2. John M. Logsdon, *The Decision to Go to the Moon: Project Apollo and the National Interest* (Chicago: University of Chicago Press, 1970), p. 130.

3. This story is recounted more fully in Arthur C. Clarke, ed., *The Coming of the Space Age* (New York: Meredith Press, 1967), pp. 106-7.

4. See Jules Verne, *From the Earth to the Moon* and *Around the Moon* (Philadelphia and New York: J. B. Lippincott, 1962); Kurd Lasswitz, *Two Planets,* trans. Hans H. Rudnick (New York: Popular Library, 1971).

5. See especially Percival Lowell, *Mars and Its Canals* (New York: Macmillan, 1906); Percival Lowell, *Mars as the Abode of Life* (New York: Macmillan, 1908); and Edgar Rice Burroughs, *John Carter of Mars* (New York: Ballantine Books, 1970).

6. Konstantin Tsiolkowsky's first important work was *Dreams of Earth and Heaven* (Moscow, 1895).

7. Hermann Oberth's first important work was *The Rocket into Interplanetary Space* (Munich: R. Oldenbourg, 1923). The film was *Frau im Mond,* which premiered in October 1929.

8. On these early societies, see Frank H. Winter, *Prelude to the Space Age: The Rocket Societies, 1924-1940* (Washington, D.C.: Smithsonian Institution Press, 1983) and William S. Bainbridge, *The Spaceflight Revolution* (New York: John Wiley & Sons, 1976).

9. Bainbridge, *Spaceflight Revolution,* p. 127.

10. G. Edward Pendray, *The Coming Age of Rocket Power* (New York: Harper & Brothers, 1945), p. 227.

11. See Clayton R. Koppes, *JPL and the American Space Program: A History of the Jet Propulsion Laboratory* (New Haven: Yale University Press, 1982).

12. Winter, *Prelude to Space Age,* p. 15.

13. Walter A. MacDougall, "Technocracy and Statecraft in the Space Age – Toward the History of a Saltation," *American Historical Review* 87 (1982):1010-40, 1011.

14. Bainbridge, *Spaceflight Revolution,* pp. 45-85. It should be noted that Winter has questioned this thesis.

15. See Frederick I. Ordway III and Mitchell R. Sharpe, *The Rocket Team* (New York: Crowell, 1979).

16. Koppes, *JPL and American Space Program,* pp. 40-41.

17. See Homer E. Newell, *Beyond the Atmosphere: Early Years of Space Science* (Washington, D.C.: National Aeronautics and Space Administration, 1980), pp. 46-49 and Jay Holmes, *America on the Moon: The Enterprise of the Sixties* (Philadelphia: J. B. Lippincott, 1962), pp. 46-47.

18. Holmes, *America on the Moon*, p. 74.

19. See, for example, J. N. Leonard, *Flight into Space* (New York: Signet, 1954), p. 51.

20. See James Gunn, *Alternate Worlds: The Illustrated History of Science Fiction* (New York: Visual Library, 1975), pp. 225-26.

21. Konstantin Tsiolkowsky's "fourteen points" are listed in a footnote in Nicholas Daniloff, *The Kremlin and the Cosmos* (New York: Knopf, 1972), p. 20.

22. Interview with Gary Paiste, January 9, 1984.

23. Interview with Thomas O. Paine, February 9, 1984.

24. Tom Wolfe, *The Right Stuff* (New York: Farrar, Straus, & Giroux, 1979).

25. Mary A. Holman, *The Political Economy of the Space Program* (Palo Alto, Calif.: Pacific Books, 1974), p. 247.

26. Logsdon, *Decision to Go to the Moon*, pp. 64-130.

27. Holmes, *America on the Moon*, p. 32.

28. Their advocacy is briefly described in Kenneth Gatland, et al., *The Illustrated Encyclopedia of Space Technology: A Comprehensive History of Space Exploration*, (New York: Salamander Books, 1981), pp. 156-57.

29. Eugene M. Emme, ed., *Twenty-Five Years of the American Astronautical Society: Historical Reflections and Projections* (San Diego: Univelt, 1980 [AAS History Series, volume 2]), pp. 165-69.

30. Interview with David Wilkinson, October 18, 1983.

31. Interview with G. Harry Stine, April 5, 1984.

32. Interview with Jerry E. Pournelle, February 8, 1984. According to the *Encyclopedia of Associations*, the SAST still had 30 members in 1984.

33. Speech by Congressman Daniel Akaka to the Federal Bar Association, November 3, 1983. The figure comes from NASA.

34. Holman, *Political Economy of Space Program*, p. 191.

35. As quoted in Amitai Etzioni, *The Moondoggle: Domestic and International Implciations of the Space Race* (Garden City, N.Y.: Doubleday, 1964), p. 93.

36. Hugh Dryden, as quoted in Edwin Diamond, *The Rise and Fall of the Space Age* (Garden City, N.Y.: Doubleday, 1964), p. 71.

37. Bruce Mazlish, ed. *The Railroad and the Space Program: An Exploration in Historical Analogy* (Cambridge, Mass.: M.I.T. Press, 1965).

38. Abraham Ribcoff, quoted in *Congressional Quarterly*, July 25, 1969, p. 1311 and in *The Futurist*, October 1969, p. 123.

39. Koppes, *JPL and American Space Program*, p. 94.

40. Neil Ruzic, *Where the Winds Sleep: Man's Future on the Moon* (Garden City, N.Y.: Doubleday, 1970).

41. Etzioni, *Moondoggle*.

42. The incident is described in some detail in David Baker, *The History of Manned Spaceflight* (New York: Crown, 1981), pp. 273-91.

43. Interview with Thomas O. Pæaine.

44. Quoted in Erlend A. Kennan and Edmund H. Harvey, Jr., *Mission to the*

Moon (New York: Morrow, 1969), p. xvii.

45. James Michener, *Space* (New York: Random House, 1982), p. 414.

46. See Logsdon, *Decision to Go to the Moon*, p. 140.

47. Quoted in Kennan and Harvey, *Mission to the Moon*, pp. 306-7.

48. Forward by Ralph E. Lapp in Kennan and Harvey, *Mission to the Moon*, p. xiv.

49. Ibid., p. 306.

50. *The Post-Apollo Space Program – Directions for the Future: Space Task Group Report to the President, September 1969* (Washington, D.C.: National Aeronautics and Space Administration, 1969).

51. Interview with Hans Mark, August 21, 1984.

52. John M. Logsdon, "The Space Shuttle Decision: Technology and Political Choice," *Journal of Contemporary Business*, 7 (Summer 1978):13-30, 14.

53. Interview with Benjamin Bova, December 5, 1983.

54. Speech by Newt Gingrich at the National Space Club's Goddard Dinner, March 16, 1984.

55. See Leland F. Belew, ed., *Skylab: Our First Space Station* (Washington, D.C.: National Aeronautics and Space Administration, 1977); and Henry S. F. Cooper, *A House in Space* (New York: Holt, Rinehart, & Winston, 1976).

56. A Soviet view of the project was presented in English by Lev Lebedev and Alexander Romanov, *Rendezvous in Space: Soyuz-Apollo* (Moscow: Progress, 1979).

57. Interview with Joseph P. Allen, March 14, 1984.

58. For one overview of these problems, see R. Jeffrey Smith, "Uncertainties Mark Space Program of the 1980's," *Science* 206 (1979):1284-86.

CHAPTER 2

1. Mary A. Holman, *The Political Economy of the Space Program* (Palo Alto, Calif.: Pacific Books, 1974), p. 3.

2. Erlend A. Kennan and Edmund H. Harvey, Jr., *Mission to the Moon* (New York: William Morrow, 1969), p. 315.

3. Amitai Etzioni, *The Moondoggle: Domestic and International Implications of the Space Race* (Garden City, N.Y.: Doubleday, 1964), p. 43.

4. See, for example, Edwin P. Hoyt, *The Space Dealers: A Hard Look at the Role of American Business in Our Space Effort* (New York: John Day, 1971).

5. Interview with Benjamin Bova, December 5, 1983.

6. A classic presentation is Mancur Olson Jr., *The Logic of Collective Action* (Cambridge, Mass.: Harvard University Press, 1965).

7. Interview with Joseph P. Allen, March 14, 1985.

8. Interview with Klaus Heiss, November 22, 1983.

9. Interview with Thomas O. Paine, February 9, 1984.

10. These centers are described in "A Visitor's Guide to NASA," *Sky and Telescope*, February 1984, pp. 102-5.

11. For the early politics of the U.S. space program, see Walter A. MacDougall, *... The Heavens and the Earth: A Political History of the Space Age* (New York: Basic Books, 1985).

12. See, for example, *Future Space Programs 1975,* hearings before the Subcommittee on Space Science and Applications of the Committee on Science and Technology, U.S. House of Representatives, July 22, 23, 24, and 30, 1975 (Washington, D.C.: U.S. Government Printing Office, 1975).

13. The early history of the House Committee is related in Ken Hechler, *The Endless Space Frontier: A History of the House Committee on Science and Astronautics, 1959-1978* (San Diego: Univelt, 1982 [AAS History Series, volume 4]).

14. Interview with Frank Moss, January 27, 1984.

15. Information on this subject is provided by the Aerospace Industries Association in its quarterly publication *Aerospace* and in its annual *Aerospace Facts and Figures,* published by *Aviation Week and Space Technology.*

16. *Aerospace Industries Association – Organization and Functions,* a booklet available from the Aerospace Industries Association.

17. See, for example, James J. Haggerty, *Spinoff 1984* (Washington, D.C.: National Aeronautics and Space Administration, July 1984 [available from the U.S. Government Printing Office]).

18. See, for example, "AIA Report Urges Policies to Encourage Exports," *Aerospace Daily,* January 27, 1984, p. 151; "NASA's FY'85 Aeronautical Budget Inadequate: AIA," *Aerospace Daily,* April 3, 1984, p. 191; "Technology Export List is Too Complex, AIA's Harr says," *Aerospace Daily,* May 10, 1984, p. 64; and "DOD Position No Perle of Wisdom," *Aerospace America,* July 1984, pp. 16-18.

19. Interview with John Loosbrock and James J. Haggerty, October 20, 1983.

20. Interview with David Wilkinson, October 18, 1983.

21. Interview with Harold Volkmer, December 21, 1983.

22. Interview with Victor Reis, March 8, 1984.

23. Interview with Darrell Branscome, December 19, 1983.

24. Interview with Frank Moss, January 27, 1984.

25. "Aerospace PACs Target Three Hill Defense Figures in Tough Races," *Aerospace Daily,* July 12, 1984, p. 57.

26. "Big Spenders," *Aerospace Daily,* October 29, 1984, p. 305.

27. "Rockwell Tops Procurement List 11th Straight Year," *Space Commerce Bulletin,* February 1, 1985, p. 6.

28. "Aviation/Space Writers Association – Who We Are, What We Do, Why You Should Join Us," booklet published by the Aviation/Space Writers Association.

29. This section is based on *AIAA at 50* (New York: American Institute of Aeronautics and Astronautics, 1981) and on an interview with Jerry Grey, November 22, 1983.

30. Interview with Jerry Grey, November 22, 1983.

31. Ibid.

32. *AIAA at 50,* p. 32.

33. Ibid.

34. Interview with Jerry Grey.

35. *Space Policy – An AIAA View* (New York: American Institute of Aeronautics and Astronautics, 1982).

36. *AIAA Bulletin,* July 1985, p. B4; *AIAA Bulletin,* June 1985, p. B7.

37. *AIAA Bulletin,* October 1983, p. B5.

38. "Who's Where," *Aviation Week and Space Technology,* October 29, 1984, p. 13.

39. Interview with Theodore R. Simpson, January 4, 1984.

40. Interview with Philip M. Hamilton, January 26, 1984.

41. Interview with William Winpisinger, May 29, 1984.

42. Allen L. Hammond, "Exploring the Solar System (IV): What Future for Space Science," *Science* 186 (1974):1011-13.

43. John Newbauer, "Away Space Station," *Aerospace America*, March 1984, p. 13.

44. Based on an interview with Elizabeth Young, March 7, 1984, and on published materials provided by the Public Service Satellite Consortium.

45. From an undated document entitled "1983 Board of Governors," provided by the GEOSAT Committee.

46. See Frederick B. Henderson, "Not Defaulting on Land Remote Sensing," *Aerospace America*, June 1985, pp. 28, 92.

CHAPTER 3

1. Stephen M. Cobaugh, as quoted in "Speaking of Space: Dynamic Presentations on the New Frontier," an undated leaflet published by the United States Space Education Association.

2. Interview with G. Harry Stine, April 5, 1984.

3. Homer E. Newell, *Beyond the Atmosphere: Early Years of Space Science* (Washington, D.C.: National Aeronautics and Space Administration, 1980), p. 295.

4. Interview with G. Harry Stine.

5. As quoted in Tom Buckley, "Caribbean Cruise Attempts to Seek Meaning of Apollo," *New York Times*, December 12, 1972.

6. Interview with Jesco von Puttkamer, February 28, 1984.

7. See "Teachers Aim High at Space Camp," *New York Times*, October 29, 1984. The figure is from an undated (1984) letter from Lee Sentell of the ASRC.

8. Barbara Marx Hubbard, *The Hunger of Eve* (Harrisburg, Pa.: Stackpole Books, 1976).

9. Ibid., p. 53.

10. Ibid., p. 109.

11. Ibid., p. 123.

12. Earl Hubbard, *The Search is On* (Los Angeles: Pace, 1969).

13. Hubbard, *Hunger of Eve*, p. 139.

14. Ibid., p. 140.

15. Edward S. Cornish, "A Quest for the Meaning of Life (a review of *The Hunger of Eve*)," *The Futurist*, December 1976, pp. 336-45, 340.

16. Committee for the Future Fact Sheet 71-722-203 on "Harvest Moon."

17. Hubbard, *Hunger of Eve*, p. 144.

18. Ibid., p. 150.

19. Interview with Alan Ladwig, March 1, 1984.

20. There is a diagram of the SYNCON wheel in William S. Bainbridge, *The Spaceflight Revolution* (New York: John Wiley & Sons, 1976), p. 173.

21. Interview with Alan Ladwig, March 1, 1984.

22. Earl Hubbard, *Our Need for New Worlds* (New York: Interbook, 1976).

23. Interview with Alan Ladwig.

24. Cornish, "Quest for Meaning of Life," p. 344.

25. Theodore Taylor, "Strategies for the Future," *Saturday Review World,* December 14, 1974, pp. 56-59, 56.

26. Hubbard, *Hunger of Eve,* p. 129.

27. Ibid., p. 216.

28. A copy of this letter is in the files of the Space Coalition, kindly loaned by Leigh Ratiner.

29. Barbara Marx Hubbard, *The Evolutionary Journey: A Personal Guide to a Positive Future* (Berkeley, Calif.: Mindbody Press, 1985).

30. Interview with Alan Ladwig.

31. Bainbridge, *Spaceflight Revolution,* p. 197.

32. Much of this section is based on interviews with Alan Ladwig, March 1, 1984; Thomas A. Heppenheimer, February 10, 1984; and Leonard W. David, January 23, 1984.

33. *University of Michigan News,* March 10, 1972.

34. Letter to Senator Walter F. Mondale, February 1, 1972.

35. *FASST News,* January-February 1976, p. 2.

36. Ibid., pp. 6-7.

37. United States Space Education Association release, July 11, 1983, on the organization's tenth anniversary.

38. Undated leaflet entitled "United States Space Education Association."

39. "USSEA Milestones," *Space Age Times* 10 (July/August 1983):8.

40. Much of this section is based on interviews with Harry S. Dawson, Jr., November 18, 1973, and Neil Ruzic, January 30, 1984.

41. The chronology is from an undated fact sheet entitled "History of the National Space Institute," kindly provided by Harry S. Dawson, Jr.

42. Interview with Harry S. Dawson, Jr.

43. Interview with Frederick C. Durant III, February 29, 1984.

44. *National Space Institute Newsletter,* November/December, 1975, pp. 2-4.

45. Ibid., p. 3.

46. Interview with Harry S. Dawson, Jr.

47. Interview with William Winpisinger, May 29, 1984.

48. Interview with Neil Ruzic.

49. See "840,000 Dialed Columbia's Phone,"*Washington Post,* December 10, 1983.

50. Interview with Benjamin Bova.

51. Interview with Mark R. Chartrand III, October 11, 1983.

52. Undated letter from Mark R. Chartrand.

53. National Space Institute press release, June 20, 1984.

54. Interview with James Muncy, November 18, 1983.

55. Interview with Louis D. Friedman, February 6, 1984.

56. Interview with Robert L. Forward, February 6, 1984.

57. Interview with Benjamin Bova.

58. Interview with Dennis Stone, December 17, 1983.

59. Krafft A. Ehricke, "The Anthropology of Astronautics," *Astronautics* 2 (November 1957):26-29, 65-68.

60. Krafft A. Ehricke, "The Extraterrestrial Imperative," *Bulletin of the Atomic Scientists,* November 1971, pp. 18-26. An expanded version appeared in the *Journal of the British Interplanetary Society* 32 (1979):311-17, 410-18.

61. Interview with Gerald W. Driggers, February 13, 1984.

62. Dandridge M. Cole and Donald W. Cox, *Islands in Space* (New York, Chilton, 1964). See also Dandridge M. Cole, "Extraterrestrial Colonies," *Navigation* 7 (Summer-Autumn 1960):86.

63. See news note in *Astronomy,* February 1977, p. 56.

64. G. Harry Stine, "The Third Industrial Revolution," *Analog,* January 1973, pp. 30-45, and *Analog,* February 1973, pp. 94-115. See also G. Harry Stine, "The Third Industrial Revolution: The Exploitation of the Space Environment," *Spaceflight,* September 1974, pp. 327-34.

65. G. Harry Stine, *The Third Industrial Revolution* (New York: G. Putnam's Sons, 1975).

66. *Outlook for Space: Report to the NASA Administrator by the Outlook for Space Study Group* (Washington, D.C.: National Aeronautics and Space Administration, 1976).

67. Interview with Jesco von Puttkamer.

68. *Space Industrialization Study Final Report,* Science Applications Incorporated, April 15, 1978; *Space Industrialization Study Final Report,* Rockwell International Space Division, April 1978.

CHAPTER 4

1. Arnold J. Toynbee, *A Study of History,* volume III (New York: Oxford University Press, 1934), p. 89.

2. Arthur Kantrowitz, "The Relevance of Space," *Bulletin of the Atomic Scientists,* April 1971, pp. 32-33.

3. As quoted in *Omni,* November, 1983, p. 153.

4. Dennis L. Meadows, et al., *The Limits to Growth* (New York: Universe, 1972).

5. Interview with Carolyn Meinel, December 19, 1983.

6. Interview with H. Keith Henson, April 3, 1984.

7. Nigel Calder, *Spaceships of the Mind* (New York: Viking Press, 1978), pp. 7-8.

8. The idea was well treated in a fictional guise in Arthur C. Clarke, *The Fountains of Paradise* (New York: Harcourt, Brace, Jovanovich, 1978). Another popular treatment is Robert L. Forward and Hans P. Moravec, "High Wire Act," *Omni,* July 1981, pp. 44-47.

9. Brian T. O'Leary, *The Making of an Ex-Astronaut* (Boston: Houghton-Mifflin, 1970).

10. Brian T. O'Leary, *The Fertile Stars* (New York: Everest House, 1981) and *Project Space Station* (Harrisburg, Pa.: Stackpole Books, 1983).

11. Interview with Gerard K. O'Neill, May 28, 1984.

12. This story is recounted in several places, including Gerard K. O'Neill, *The High Frontier: Human Colonies in Space* (New York: William Morrow, 1976), pp. 233-63.

13. Interview with Brian T. O'Leary, February 7, 1984.

14. Interview with George Hazelrigg, February 29, 1984.

15. See Arthur C. Clarke, "Electromagnetic Launching as a Major Contributor to Space Flight," *Journal of the British Interplanetary Society* 9 (November 1950):260-67. Dandridge M. Cole and Donald W. Cox noted the linear accelerator idea in *Islands in Space* (New York: Chilton, 1964), p. 118.

16. A summary of the Proceedings is in Jerry Grey, ed., *Space Manufacturing Facilities (Space Colonies)* (New York: American Institute of Aeronautics and Astronautics, 1977), Appendix A.

17. Ibid., p. A-6.

18. Walter Sullivan, "Proposal for Human Colonies in Space is Hailed by Scientists as Feasible Now," *New York Times*, May 13, 1974.

19. Richard Reis, "Colonization of Space," *Mercury*, July/August 1974, pp. 4-10. See also Gerard K. O'Neill, "Space Colonies are Far Out – But So Was a Man on the Moon," *Los Angeles Times*, July 30, 1974.

20. Barbara Marx Hubbard, *The Hunger of Eve* (Harrisburg, Pa.: Stackpole Books, 1976), pp. 207-8.

21. Gerard K. O'Neill, "The Colonization of Space," *Physics Today*, September 1974, pp. 32-40.

22. Konstantin Tsiolkowsky, *Beyond the Planet Earth*, trans. Kenneth Syers (New York: Pergamon Press, 1960), pp. 83-84, 89.

23. Ibid., pp. 93-94.

24. Konstantin Tsiolkowsky, quoted by Nicholas Daniloff, *The Kremlin and the Cosmos* (New York: Knopf, 1972), p. 20.

25. J. D. Bernal, *The World, the Flesh, and the Devil* (London: Methuen, 1929); a more modern edition was published by the Indiana University Press in 1969.

26. J. N. Leonard, *Flight into Space* (New York: Signet, 1954), pp. 162-65.

27. Arthur C. Clarke, *Islands in the Sky* (New York: Holt, Rinehart, & Winston, 1954).

28. See Jerry Grey, *Beachheads in Space: A Blueprint for the Future* (New York: Macmillan, 1983), pp. 10-11.

29. Cole and Cox, *Islands in Space*, pp. 142-50.

30. Interview with James E. Oberg, March 13, 1984.

31. Krafft A. Ehricke wrote about "space cities" in "Extraterrestrial Imperative," *Bulletin of the Atomic Scientists*, November 1971, pp. 18-26. The Lockheed design appeared in Joseph Newman, *1994: The World of Tomorrow* (Washington, D.C.: U.S. News and World Report, 1973), pp. 48-49.

32. O'Neill, "The Colonization of Space," p. 36.

33. Interview with Keith Henson.

34. Interview with Jesco von Puttkamer.

35. Interview with Gerard K. O'Neill. Mark confirmed his early support for O'Neill in an interview on August 21, 1984. See also Hans Mark, "The Space Station – Mankind's Permanent Presence in Space," *Aviation, Space, and Environmental Medicine*, October 1984, pp. 948-56, 952.

36. Peter Glaser, "Energy from the Sun – Its Future," *Science* 162 (1968):857-60. For a detailed proposal, see Peter E. Glaser, "Solar Power via Satellite," *Astronautics and Aeronautics*, August 1973, pp. 60-68. A useful overview of the idea

was presented in "An Orbiting Solar Power Station," *Sky and Telescope,* April 1975, pp. 224-28.

37. Interviews with Gerard K. O'Neill and Jesco von Puttkamer.

38. Gerard K. O'Neill, "Space Colonies and Energy Supply to the Earth," *Science* 190 (1975):943-47.

39. Interview with Brian T. O'Leary.

40. Interview with Konrad Dannenberg, February 14, 1984.

41. Interview with Gerald W. Driggers.

42. The proceedings are in Grey, *Space Manufacturing Facilities.*

43. Interview with Gerald W. Driggers.

44. O'Neill's testimony can be found in the fall 1975 issue of *Coevolution Quarterly,* pp. 10-19, and in Stewart Brand, ed., *Space Colonies* (New York: Penguin, 1977).

45. O'Neill's testimony can be found in Gerard K. O'Neill, *The High Frontier,* pp. 264-75.

46. Undated press release entitled "NASA/Ames-Stanford ASEE 1975 Summer Study of Space Colonization." See also Richard D. Johnson and Charles Holbrow, eds., *Space Settlements: A Design Study* (Washington, D.C.: National Aeronautics and Space Administration, 1977).

47. For example, Mike Dunstan, "Space City Envisioned," *Washington Post,* August 23, 1975, and Marvin Miles, "Experts Foresee Space City of 10,000 Inhabitants," *Los Angeles Times,* August 23, 1975.

48. Thomas O. Paine, "Humanity Unlimited," *Newsweek,* August 25, 1975, p. 11.

49. Isaac Asimov, "The Next Frontier," *National Geographic,* July 1976, pp. 76-89; Gerard K. O'Neill, "Colonies in Orbit," *New York Times Magazine,* January 18, 1976, pp. 10-11, 25-29; Isaac Asimov, "Colonizing the Heavens," *Saturday Review,* June 28, 1975, pp. 12-17; Gwyneth Cravens, "The Garden of Feasibility," *Harpers,* August 1975, pp. 66-75.

50. Ron Chernow, "Colonies in Space Might Turn Out to be Nice Places to Live," *Smithsonian,* February 1976, pp. 62-68.

51. Interview with G. Harry Stine.

52. See Gerard K. O'Neill, "Engineering a Space Manufacturing Center," *Astronautics and Aeronautics,* October 1976, pp. 20-28, and "Space-Based Manufacturing from Non-Terrestrial Materials: The NASA/Ames Research Center Study," NASA technical memorandum 73,265 (Moffett Field, Calif.: Ames Research Center, August 1977).

53. Brian T. O'Leary, "Mining the Apollo and Amor Asteroids," *Science* 197 (1977):363-66.

54. See, for example, Thomas Heppenheimer and Mark M. Hopkins, "Initial Space Colonization: R & D Aims," *Astronautics and Aeronautics,* March 1976, pp. 58-64, 72.

55. O'Neill, *The High Frontier,* p. 232.

56. As quoted by O'Neill in his column in *SSI Update,* September 1981, p. 1.

57. See John Billingham, William Gilbreath, and Brian T. O'Leary, eds. *Space Resources and Space Settlements* (Washington, D.C.: National Aeronautics and Space Administration, 1979). Meanwhile, a summer 1977 workshop on Near Earth

Resources had supported the mining of the Moon and the asteroids. See John Noble Wilford, "Scientists Call on NASA to Prospect for Mining on Moon and Asteroids," *New York Times,* August 15, 1977.

58. Interview with Brian T. O'Leary.

59. Interview with Gerard K. O'Neill.

60. Louis J. Halle, "A Hopeful Future for Mankind," *Foreign Affairs,* Summer 1980, pp. 1129-36.

61. See the fall 1975 issue of *Coevolution Quarterly* and Brand, ed., *Space Colonies.*

62. Interview with Stewart Brand, April 16, 1984.

63. Interview with Thomas A. Heppenheimer.

64. Carl Sagan, quoted in "Interest in Space Colonies Rises," *The Futurist,* February 1976, p. 32. Intrestingly, Sagan had written in 1973 of the need for experimental societies. See the chapter "Experiments in Utopias," in Sagan's book *The Cosmic Connection,* pp. 35-39.

65. Interview with Stewart Brand.

66. Interview with Brian T. O'Leary.

67. See Magoroh Maruyama, "Diversity, Survival Value, and Enrichment: Design Principles for Extraterrestrial Communities," in Grey, *Space Manufacturing Facilities,* pp. 159-74, and Magoroh Maruyama, "Social and Political Interactions Aong Extraterrestrial Human Communities: Contrasting Models," *Technological Forecasting and Social Change* 9 (1976):349-60.

68. J. Peter Vajk, "The Impact of Space Colonization on World Dynamics," Lawrence Livermore Laboratory, October 15, 1975.

69. J. Peter Vajk, *Doomsday Has Been Cancelled* (Palo Alto, Calif.: Peace Press, 1978).

70. Interview with Gerard K. O'Neill.

71. Lewis Mumford, letter in Brand, ed. *Space Colonies,* p. 34.

72. Paul L. Csonka, "Space Colonization: An Invitation to Disaster," *The Futurist,* October 1977, pp. 285-90.

73. "Colonies in Space," *Newsweek,* November 27,1978, pp. 95-101.

73. Interview with Brian T. O'Leary.

75. Interview with T. Stephen Cheston, November 25, 1983.

76. Interview with Stewart Nozette, February 10, 1984.

77. Interview with Thomas A. Heppenheimer.

78. White House Fact Sheet on U.S. Civil Space Policy, October 11, 1978,

79. Interview with Brian T. O'Leary.

80. Memorandum from Harrell Graham dated November 16, 1977.

81. Circular letter from Barbara Marx Hubbard, January 3, 1978.

82. See Bob Rankin, "Giant Solar Powered Electric Plants in the Sky . . . Are the Aerospace Industry's Newest Dream," *Congressionl Quarterly,* April 22, 1978, pp. 964-65.

83. Interview with Leigh Ratiner, January 24, 1984.

84. A debate on the issue took place in the letters pages of *Astronomy* after the publication of "Solar Power Satellites: A Threat to Astronomy," March 1979, p. 60.

85. Interview with Leonard W. David, January 23, 1984.

86. Letter from Gael Baudino to *Astronomy,* November 1979, p. 58.

87. Interview with George Hazelrigg.

88. Gerard K. O'Neill, *SSI Subscribers Newsletter*, December 1978, p. 3. See also Luther J. Carter, "House Gives a Nod to Solar Power Satellite," *Science* 206 (1979):1052-54.

89. *Program Assessment Report Staement of Findings: Satellite Power Systems Concept Development and Evaluation Program* (Washington, D.C.: U.S. Department of Energy, November 1980); "Sunny Outlook for Sunsats," ITime, December 5, 1980, pp. 46-47; "Solar Power Satellite Deemed Possible," *Aviation Week and Space Technology*, December 15, 1980, p. 95.

90. *Electric Power from Orbit: A Critique of a Satellite Power System* (Washington, D.C.: National Academy of Sciences, 1981).

91. *Solar Power Satellites* (Washington, D.C.: Office of Technology Assessment, 1981).

92. Letter to the author dated November 13, 1984.

93. Interview with George Hazelrigg.

94. Interview with Leigh Ratiner.

95. Interview with Todd C. Hawley, November 10, 1983.

96. Gerard K. O'Neill, *Space Studies Institute Subscribers Newsletter*, Fall 1979, p. 3.

97. Gerard K. O'Neill, *Space Studies Institute Update*, May/June 1984, p. 1.

98. Undated brochure on the Institute for the Social Science Study of Space.

99. *The Space Humanization Series* (Washingotn, D.C.: Institute for the Social Science Study of Space, 1979).

100. *Social Sciences and Space Exploration: New Directions for University Education* (Washington, D.C.: National Aeronautics and Space Administration, 1984).

101. Interview with T. Stephen Cheston.

102. Interview with Irwin Pikus.

103. Gerard K. O'Neill, *2081: A Hopeful View of the Future* (New York: Simon and Schuster, 1981), pp. 62, 71.

104. Interview with G. Harry Stine.

105. Interview with Irwin Pikus.

106. Interview with Courtney Stadd.

107. Interview with T. Stephen Cheston.

CHAPTER 5

1. K. Eric Drexler, "Have We Changed Our Goals?" *L-5 News*, October 1983, pp. 1-2, 2.

2. The words are reproduced in the front of Thomas A. Heppenheimer, *Colonies in Space* (New York: Warner Books, 1977). Reprinted with the permission of the author.

3. Thomas A. Heppenheimer, *Toward Distant Suns* (Harrisburg, Pa.: Stackpole Books, 1979); Thomas A. Heppenheimer, *The Man-Made Sun* (Boston: Little, Brown, 1984). Much of the amterial in these paragraphs is based on an interview with Thomas A. Heppenheimer.

4. Much of this is based on an interview with Mark M. Hopkins, February 6,

1984, and on personal communciations from Mr. Hopkins to the author.

5. K. Eric Drexler, *Engines of Creation* (New York: Doubleday, forthcoming). Some of this paragraph is based on a private communication from Mr. Drexler to the author.

6. Interview with Mark M. Hopkins.

7. Interview with H. Keith Henson.

8. Interview with Carolyn Meinel, December 19, 1983.

9. Interview with H. Keith Henson.

10. Carolyn Henson, "A Message from the Ex-President," *L-5 News,* December 1979, p. 3.

11. Interviews with H. Keith Henson, Carolyn Meinel, and Gerald W. Driggers. The Hensons' paper is in Jerry Grey, *Space Manufacturing Facilities (Space Colonies),* (New York: American Institute of Aeronautics and Astronautics, 1977), pp. 105-14.

12. Interview with Gerald W. Driggers. His paper is in Jerry Grey, *Space Manufacturing Facilities,* pp. 33-50.

13. Interviews with Carolyn Meinel, H. Keith Henson, and Gerald W. Driggers.

14. Interviews with H. Keith Henson and Carolyn Meinel.

15. Interviews with H. Keith Henson and Carolyn Meinel. The Udall letter is reproduced on the front page of the September 1975 *L-5 News.*

16. "The L-5 Society," *L-5 News,* September 1975, p. 2.

17. Interview with H. Keith Henson.

18. "The L-5 Society," *L-5 News,* September 1975, p. 2.

19. Interview with Randall Clamons, April 4, 1984.

20. "Woes of Success – Words of Caution," *L-5 News,* May 1976, supplement.

21. Interview with Carolyn Meinel.

22. Carolyn Henson, "A Letter from the Ex-President," p. 3.

23. Interview with Randall Clamons. The first Starseed seminar reportedly was in August 1976. See Robert Anton Wilson, "Starseed Seminar Aims for Mutation, Migration, Rejuvenation," *L-5 News,* October 1976, pp. 11-12.

24. Tim Leary on Snake Oil, Liberals, Amino-Uganda," *L-5 News,* August 1976, pp. 4-5.

25. Interview with J. Peter Vajk, April 21, 1984. See "Vajk Addresses Club of Rome," *L-5 News,* May 1976, p. 1.

26. Norie Huddle, *Surviving: The Best Game on Earth* (New York: Schocken Books, 1984).

27. Interview with Gerald W. Driggers.

28. Interview with Gerard K. O'Neill.

29. Interview with H. Keith Henson.

30. Interview with Jerry E. Pournelle, February 8, 1984.

31. Interview with Randall Clamons.

32. "Woes of Success – Words of Caution."

33. Interview with B. J. Bluth, January 27, 1984.

34. Undated circular letter from the L-5 Society.

35. See Kenneth McCormick, "So You Want to Lobby?" *L-5 News,* December 1977, p. 18.

36. Robert A. Freitas Jr., *Lobbying for Space* (santa Clara, Calif.: Space Initiative, 1978).

37. Interview with Randall Clamons.

38. See the December 1979 *L-5 News.*

39. See "Responses to Space Mines," *L-5 News,* June 1978, pp. 10-13.

40. Arthur Dula, "Free Enterprise and the Proposed Moon Treaty," *L-5 News,* October 1979, pp. 1-2.

41. H. Keith Henson, "Bulletin from the Moon Treaty Front," *L-5 News,* January 1980, p. 16.

42. Interview with H. Keith Henson.

43. Interviews with H. Keith Henson and Carolyn Meinel.

44. Interview with Leigh Ratiner. For a general account, see Helen Dewar, "Would-Be Space Colonists Lead Fight Against Moon Treaty," *Washington Post,* October 30, 1979.

45. The text of the letter is reproduced in the December 1979 *L-5 News,* p. 5.

47. "Salin and Pergamit: A Philosophy of Space Entrepreneurship," *L-5 News,* April 1984, pp. 7-8, 17.

48. Interview with Thomas A. Heppenheimer.

49. Interview with Gerald W. Driggers.

50. Gerald W. Driggers, "The President's Column," *L-5 News,* July 1980, p. 18.

51. Interview with Jerry E. Pournelle.

52. Interview with James E. Oberg.

53. Interview with Jerry E. Pournelle.

54. See Gerald W. Driggers and Rebecca Wright, "The Space Coalition: Bringing the Pro-Space Message to Government," *L-5 News,* July 1980, pp. 2-7.

55. Most of this is based on an interview with David Brandt-Erichsen and Sandra Adamson, April 4, 1984.

56. Interview with Mark M. Hopkins.

57. Interview with Gerald W. Driggers.

58. "Who is Gerald Driggers?" *L-5 News,* February 1980, pp. 2-3.

59. Interview with Randall Clamons.

60. Interview with Mark M. Hopkins.

61. Ken McCormick, "High Frontier Politics," *L-5 News,* June 1979, pp. 11-13.

62. Interviews with Jerry E. Pournelle and Mark M. Hopkins.

63. Interview with Mark M. Hopkins.

64. Interview with T. Stephen Cheston.

65. Interview with Mark M. Hopkins.

66. Interview with Sandra Adamson and David Brand-Erichsen.

67. The first L-5 chapter to testify (on March 1, 1984) was Washington State Citizens for Space, before the Senate Subcommittee on Science, Technology, and Space. The subcommittee's chairman, Senator Slade Gorton, is from Washington state.

68. Interview with Jerry E. Pournelle.

69. Interview with Mark M. Hopkins.

70. Interview with Jerry E. Pournelle.

71. Ibid.

72. Interview with Sandra Adamson and David Brandt-Erichsen.

73. Interview with James E. Oberg.

74. Stephen I. Thompson, "The L-5 Society as a Revitalization Movement,"

paper presented at the 81st Annual Meeting of the American Anthropological Association, Washington, D.C., December 3-9, 1982.

75. Interview with Sandra Adamson and David Brandt-Erichsen.

76. Ibid.

77. Ibid.

78. "Space Colonization Organizations," *L-5 News,* April 1976, pp. 6-7.

79. Interview with Randall Clamons.

80. Interview with Keith Morton, January 5, 1984.

81. Much of this is based on an interview with Todd C. Hawley, November 10, 1983.

82. "Students Exploring . . . Developing . . . Space," undated pamphlet from SEDS.

83. Interview with Todd C. Hawley.

84. Undated fact sheet from the Space Settlement Studies Project, Sociology Department, Niagara University, New York.

85. Interview with Howard and Janelle Gluckman, February 7, 1984.

CHAPTER 6

1. Ronald Reagan, in a speech to the graduating class at the U.S. Air Force Academy, May 30, 1984. See *Weekly Compilation of Presidential Documents,* Volume 20, Number 22, June 4, 1984, pp. 782-86. Also see "Washington Roundup," *Aviation Week and Space Technology,* June 4, 1984, p. 15.

2. The slogan was created in response to a 1984 *USA Today* poll that showed that 45 percent of Americans wanted to fly on the space shtutle. See note 7.

3. *Civilian Space Policy and Applications* (Washington, D.C.: Office of Technology Assessment, 1982), p. 136.

4. Robert D. McWilliams, "The Improving Socio-Political Situation of the American Space Program in the Early 1980s," in Jerry Grey and Lawrence A. Hamdan, eds., *Space Manufacturing: Proceedings of the Fifth Princeton/AIAA Conference, May 18-21, 1981* (New York: American Institute of Aeronautics and Astronautics, 1981), pp. 251-60.

5. "Poll Finds that Most Americans Support Nation's Space Program," *New York Times,* August 19, 1981, and "Public Majority Support Space Effort," *Insight* (National Space Institute), September/October 1981, p. 6. See also *Civilian Space Policy and Applications,* p. 141.

6. See "Poll Indicates Growing Public Support for Space," *Insight,* July/August 1981, p. 3. Louis Harris, *The Harris Survey* (Orlando, Fla.: Tribune Media Services, June 1, 1981). Reprinted with permission. See also "Poll Finds 63% of Surveyed Back Full Spending on Shuttle," *Aviation Week and Space Technology,* June 15, 1981, p. 33, and *Civilian Space Policy and Applications,* p. 141.

7. "45% of Us Would Be Astronauts," *USA Today,* April 19, 1984.

8. Jon D. Miller, "The Information Needs of the Public Concerning Space Exploration: A Special Report for the National Aeronautics and Space Administration," Public Opinion Laboratory, Northern Illinois University, July 20, 1982.

9. National Space Institute News Release, March 2, 1984.

10. McWilliams, "Improving Socio-Political Situation," p. 252.

11. Ibid., p. 254.

12. Ibid., p. 255.

13. Ibid.

14. Ibid., p. 253.

15. Ibid., p. 254.

16. Interview with Johan Benson, October 24, 1983. Trudy E. Bell made a similar observation in her unpublished 1982 paper "From Little Acorns . . . : American Space Interest Groups, 1980-1982."

17. *Civilian Space Policy and Applications*, pp. 138-39.

18. Miller, "Information Needs of Public," pp. 16-18.

19. McWilliams, "Improving Socio-Political Situation," p. 257.

20. Ibid., p. 256.

21. Ibid., p. 258.

22. Ibid., p. 259.

23. Brochure entitled "National Space Institute," dated October 1983.

24. "Report from the Executive Director," *Insight*, May 1985, pp. 2a-3a.

25. "Preliminary Results from Most Recent Survey," unpublished document kindly provided by Randall Clamons.

26. "Demographic Survey Results," *SSI Update*, Second Quarter 1983, p. 6.

27. "You Tell Us – The Planetary Society Members' Survey," *The Planetary Report*, May/June 1985, p. 13.

28. Tenth Anniversary Press Kit, United States Space Education Association, July 11, 1983.

29. It should be noted that these are not necessarily final figures.

30. Testimony of George M. Low, Acting NASA Administrator, *1972 NASA Authorization*, Hearings before the Committee on Science and Astronautics, U.S> House of Representatives, March 2, 1971 (Washington, D.C.: U.S. Government Printing Office, 1971), p. 76.

31. Undated document kindly supplied by Randall Clamons.

32. "Crowds Gather to View Launch, Landing," *Aviation Week and Space Technology*, April 20, 1981, p. 26.

33. Speech by James Beggs at Georgetown University seminar on education for international cooperation in science and technology, January 29, 1985.

34. "840,000 Dialed Columbia's Phone," *Washington Post*, December 10, 1983, and conversation with Bonny Lee Michelson, January 27, 1984. The National Space Institute puts out a press release each time this service is available.

35. "Air and Space Museum: Four Million in Five Months," *Astronautics and Aeronautics*, January 1977, p. 16.

36. "News Digest," *Aviation Week and Space Technology*, June 4, 1984, p. 27.

37. "From Pate to Hot Dogs, Tourism Rises," *New York Times*, July 28, 1985.

38. "NASA Names 17 Astronaut Candidates," *Aviation Week and Space Technology*, May 28, 1984, p. 29.

39. "Industry Observer," *Aviation Week and Space Technology*, June 4, 1984.

40. Miller, "Information Needs of Public," pp. 53-54.

41. Ibid., p. 54.

42. Ibid., pp. 54-56.

43. "Preliminary Results from Most Recent Survey."

44. See, for example, "Here Come the Baby Boomers," *U.S. News and World Report*, November 5, 1984, pp. 68-73; "A Party in Search of Itself," *Time*, July 16, 1984, pp. 12-20; "Tomorrow," *U.S. News and World Report*, May 21, 1984, pp. 19-20.

45. See "Texas, Florida Areas Burgeoning," *Washington Post*, June 5, 1985.

46. "Tomorrow," *U.S. News and World Report*, June 24, 1985, p. 16.

47. Miller, "Information Needs of Public," pp. 27, 77.

48. McWilliams, "Improving Socio-Political Situation," p. 259.

49. Interview with John Loosbrock and James J. Haggerty.

CHAPTER 7

1. James Gunn, *Alternate Worlds: The Illustrated History of Science Fiction* (New York: Visual Library, 1975), p. 48.

2. William S. Bainbridge, *The Spaceflight Revolution: A Sociological Study* (New York: John Wiley & Sons, 1976), p. 234.

3. Lucian of Samosata, *A True History* (New York: Murray, Scribner, & Wedford, 1880).

4. H. Bruce Franklin, as quoted by Curt Suplee in "In the Strange Land of Robert Heinlein," *Washington Post*, September 5, 1984.

5. William S. Bainbridge, *Spaceflight Revolution* and William S. Bainbridge, "The Impact of Science Fiction on Attitudes Toward Technology," in Euguene M. Emme, ed., *Science Fiction and Space Futures* (San Diego: American Astronautical Society [Univelt], 1982), pp. 121-35, 121.

6. Interview with Gerard K. O'Neill.

7. Isaac Asimov, "The Truth Isn't Always Stranger Than Science Fiction, Just Slower," *New York Times*, February 12, 1984.

8. Peter Nicholls, ed., *The Science Fiction Encyclopedia* (Garden City, N.Y.: Doubleday, 1979), p. 473.

9. Peter Nicholls, ed., *The Science in Science Fiction* (New York: Knopf, 1983).

10. Jesco von Puttkamer, "Reflections on a Crystal Ball: Science Fact vs. Science Fiction," in Emme, *Science Fiction and Space Futures*, pp. 137-50, 137.

11. Nicholls, *The Science in Science Fiction*, p. 6.

12. Interview with Jesco von Puttkamer. Similar views were suggested by Peter S. Prescott, "Science Fiction: The Great Escape," *Time*, December 22, 1975, pp. 68-74.

13. Ursula K. LeGuin, *The Language of the Night* (New York: G. P. Putnam's Sons, 1979), p. 40.

14. Interview with B. J. Bluth.

15. Nicholls, *The Science in Science Fiction*, p. 201.

16. Bainbridge, "The Impact of Science Fiction," p. 124.

17. Ibid., p. 130.

18. Ibid., p. 133.

19. Ibid.

20. Ibid., pp. 1251-26.

21. William S. Bainbridge and Murray Dalziel, "The Shape of Science Fiction as Perceived by the Fans," in *Science Fiction Studies* 5 (1978):165-71, 166-67.

22. Bainbridge, "The Impact of Science Fiction on Attitudes Toward Technology," p. 135.

23. "Red Shift – Leftist Science Fiction Writers Association Formed," *Space for All People* 3 (April 1982):3.

24. Interview with G. Harry Stine.

25. Stanley Schmidt, "A Portrait of You," *Analog Science Fiction/Science Fact* C1 (April 27, 1981): 5-14.

26. Bainbridge, "The Impact of Science Fiction," p. 133.

27. Rita Kempey, "Enterprise Keeps on Trekking, *Washington Post*, June 8, 1984.

28. Interview with Jesco von Puttkamer.

29. See Mark Starr, "Cosmic Cult: Fans of Star Trek Are, Well, Spaced Out," *Wall Street Journal*, September 4, 1975; "The Trekkie Fad," *Time*, September 8, 1975, p. 70; "Star Trek Lives," *Newsweek*, September 11, 1975.

30. Interview with Jesco von Puttkamer.

31. Conversation with Richard Preston, 1984.

32. See, for example, "Enterprise Undergoes Sound Suppression Test," *Aviation Week and Space Technology*, May 6, 1985, p. 101.

33. Some of this section is based on an interview with Bjo Trimble, February 7, 1984. She is the author of *The Star Trek Concordance* (New York: Ballantine Books, 1976) and *On the Good Ship Enterprise: My 15 Years with Star Trek* (Norfolk/Virginia Beach, Va.: Donning, 1982).

34. Nathan C. Goldman and Michael Fulda, "The Outer Space Lobby and the 1980 Electronics," in Paul Amaejionu, Nathan C. Goldman, and Philip J. Meeks, eds., *Space and Society: Challenges and Choices* (San Diego: American Astronautical Society [Univelt], 1984), pp. 15-28.

35. See John Culhane, "George Lucas: Skywalker Supreme," *Readers Digest*, September 1982, pp. 66-70 (condensed from *Families*, April 1982).

36. Lawrence Meyer and Joel Garreau, "Making Movies with the Same Message as John Wayne and Buddha" (interview with Gary Kurtz), *Washington Post*, December 30, 1984.

37. Rita Kempey, "A Joyous Ride in The Last Starfighter," *Washington Post*, July 13, 1984.

38. Robert D. McWilliams, "An Analysis of the Socio-Political Status of Efforts Toward the Development of Space Manufacturing Facilities," in *Space Manufacturing III: Proceedings of the Fourth Princeton/AIAA Conference, May 14-17, 1979* (New York: American Institute of Aeronautics and Astronautics, October 31, 1979), pp. 173-82, 174.

39. See Ron Miller, "The Astronomical Visions of Lucien Rudaux," *Sky and Telescope*, October 1984, pp. 293-95.

40. Willy Ley and Chesley Bonestell, *The Conquest of Space* (New York: Viking Press, 1949).

41. A small reproduction appeared in *Discover*, November 1984, p. 100.

42. Interview with Kathy Keeton, May 25, 1984.
43. Ibid.
44. Ibid.
45. See Ross Parker Simons, "Science Magazines," *Astronomy*, June 1982, pp. 24-28.
46. See, for example, Carolyn Meinel Henson, "Alternate Space," *Future Life*, November 1981, pp. 22-23.
47. Carl Sagan, *Cosmos* (New York: Random House, 1980).
48. See Pat Glossop, "Sky's No Longer the Limit," *Millimeter*, October 1984, pp. 201-13.

CHAPTER 8

1. Trudy E. Bell, "Space Activists on Rise," *Insight* (National Space Institute), August/September 1980, pp. 1, 3, 10.
2. Testimony of T. Stephen Cheston before the Subcommittee on Science, Technology, and Space of the Committee on Commerce, Science, and Transportation, U.S. Senate, March 17, 1977. Reproduced in *L-5 News*, April 1977, pp. 2-3, 12.
3. Bell, "Space Activists on Rise."
4. Nathan C. Goldman and Michael Fulda, "Outer Space Groups: Galaxies of Interests, Galaxies of Groups," paper to be published in volume 2 of the *Space Humanization Series* (Institute for the Social Science Study of Space).
5. This is based on an interview with Mickey Farrance, April 21, 1984, and on materials provided by the Hypatia Cluster.
6. This is based on an interview with Dennis Stone and on materials provided by Spaceweek headquarters.
7. On this early period, see Ron Jones, "UAL to Form Civilian Space Corps," *L-5 News*, July 1981, p. 11.
8. See David C. Koch, "Chairman's Report: Circling the Wagons," *ASAP Update*, April 1984, p. 1.
9. "1984 Space and Aviation Camp a Success," *ASAP Update*, June/July/August 1984, p. 5, and "ASAP Activity Sparks Vitality, Growth in Oregon," *Lawyers Title News*, July/August 1984, pp. 4-7, 17.
10. David C. Koch, "Creating Tomorrow," *Lawyers Title News*, July/August 1984, pp. 8-10.
11. David C. Koch, "Chairman's Report: Happy Birthday ASAP," *ASAP Update*, September 1984, p. 1.
12. David C. Koch, "Free Enterprise in Space," *ASAP Update*, September 1984, p. 2.
13. See Larry van Dyne, "Diamonds in the Sky," *Washingtonian*, November 1983, pp. 164-71, 194-97. This section is based on conversations with Polly Rash, December 12, 1983, and Elizabeth Harrington, January 9, 1984, and on materials provided by the Society of Satellite Professionals.
14. This is based on an interview with Stewart Nozette, February 10, 1984, and on a brochure entitled "California Space Institute."
15. See Leonard C. David, "Personality Profile: Stan Kent," *Insight* (National

Space Institute), October/November 1980, p. 6; "Mars Outlook," *Aviation Week and Space Technology*, January 12, 1981, p. 15; Van R. Kane, "Come Explore a Comet, *Astronomy*, July 1981, pp. 24-26; "Halley Fund Announced at Shuttle Liftoff," *Astronomy*, July 1981, pp. 56-57; letter from Van R. Kane, *Astronomy*, May 1981, p. 36; Juana E. Doty, "Launching Private Space Fund Drive," *Los Angeles Times*, July 20, 1982.

16. Interview with Nancy Wood, March 13, 1984; also materials provided by The Space Foundation. On Arthur Dula, see Ron Bitto, "Cosmic Counselor," *Omni*, August 1981, pp. 48-51, 94 and Arthur Dula, "Getting Involved With the Future," *Analog Science Fiction/Science Fact*, July 1979, pp. 8-13.

17. Based on materials provided by the Radio Amateur Satellite Corporation and on "Satellites on a Shoestring," paper presented by Roger Soderman at the Second Annual Conference on Private Sector Space Research and Exploration, May 27, 1984.

18. Based on an interview with Jesse Eichenlaub, May 27, 1984, and on materials from the Independent Space Research Group, including its newsletter *Focal Point*. See also "Amateur Space Telescope," *Astronomy*, October 1980, pp. 59-60; Robert J. Sawyer, "An Amateur Space Telescope," *Sky and Telescope*, August 1982, pp. 127-29; "Amateur Space Scope," *Astronomy*, September 1982, p. 64; "Amateur Space Telescope Nears Completion," *Astronomy*, March 1984, p. 60; "AST Camera Fund," *Sky and Telescope*, April 1985, p. 312.

19. See Chauncey Uphoff, "The First Solar Sail," *L-5 News*, July 1981, p. 10.

20. Based on an interview with Robert L. Staehle, February 6, 1984, and on materials from the World Space Foundation.

21. Save the Apollo Launch Tower was described in "Help Save the Past," *Insight* (National Space Institute), December 1983, p. 2a.

22. Interview with Jim Heaphy, April 20, 1984.

23. See "Big is Beautiful, Too," *Time*, August 22, 1977, pp. 25-26 and "California Plans Space Program Symposium," *Aviation Week and Space Technology*, July 25, 1977, p. 20.

24. See "Jimmy C. Will Increase NASA's Budget if Good Friend Jerry B. Will Take Permanent Space Walk," *L-5 News*, April 1978, p. 2.

25. As quoted in Stewart Brand, ed., *Space Colonies* (New York: Penguin), pp. 146-48.

26. Press release from the April Coalition, February 25, 1978.

27. Interview with Jim Heaphy.

28. Interview with Tim Kyger, November 28, 1984.

29. See Jay Miller, "L-5ers Develop a State Space Program," *L-5 News*, April 1982, pp. 2-3 and "Golden State Ready for Liftoff," *Insight* (National Space Institute), December 1983, p. 3a.

30. See Nathan C. Goldman, "Working for Unified Space Lobbying," *L-5 News*, March 1981, p. 4.

31. Interview with Tim Kyger.

32. Much of this section is based on interviews with Mark R. Chartrand III, October 11, 1983 and with David C. Webb, October 13, 1983; also Bell's surveys and materials provided by the National Space Institute. For an action by the ad hoc committee, see "Solar Power Possible," *Aviation Week and Space Technology*, December 15, 1980, p. 95.

33. Letter to President Reagan on National Space Institute letterhead, September 11, 1981.

34. Interview with Victor Reis, March 8, 1984.

35. The author was present at the meeting.

36. Trudy E. Bell, "From Little Acorns . . . : American Space Interest Groups, 1980-82," unpublished paper.

37. This section is based in part on interviews with Gerald W. Driggers and Leigh Ratiner.

38. "Explanatory Note" from the files of The Space Coalition, kindly loaned by Leigh Ratiner.

39. Letter from Gerald W. Driggers to Hugh Downs, June 3, 1980, from the files of the Space Coalition.

40. Gerald W. Driggers and Rebecca Wright, "The Space Coalition: Bringing the Pro-Space Message to Washington," *L-5 News,* July 1980, pp. 2-7.

41. Interview with Leigh Ratiner.

42. Ibid.

43. Letterhead of Space Coalition letter dated April 7, 1981.

44. Interview with Leigh Ratiner.

45. Trudy E. Bell, "From Little Acorns . . . :" unpublished paper.

46. John Kenneth Galbraith, *The Anatomy of Power* (Boston: Houghton, Mifflin, 1983), p. 150.

47. Letter from James Logan, February 10, 1980. Interestingly, Logan's letter proposed a National Space Coalition "dedicated to educating the public and Washington to the virtually infinite possibilities and benefits of a strong national presence in space."

48. Mark R. Chartrand III, "From the Executive Director," *Insight* (National Space Institute), November 1983, p. 1a.

49. See "Sixties People," *Omni,* March 1984, p. 41.

50. Interview with Charles Chafer, October 6, 1983.

CHAPTER 9

1. Thomas J. Frieling, undated Campaign for Space circular, 1984.

2. As quoted in Trudy E. Bell, "Space Activists on Rise," *Insight* (National Space Institute), August/September 1980, p. 10.

3. Attributed to the American Space Foundation.

4. Interview with Trudy E. Bell, November 22, 1983.

5. Letter from William Blair, *Astronomy,* August 1981, p. 41.

6. Nathan C. Goldman and Michael Fulda, "The Outer Space Lobby and the 1980 Election," in Paul Amaejionu, Nathan C. Goldman, and Philip J. Meeks, eds., *Space and Society: Challenges and Choices* (San Diego: Univelt, 1984), pp. 15-28, 21.

7. Much of this is based on materials kindly provided by Trudy E. Bell and on an interview with her.

8. "Campaign for Space," draft document dated March 16, 1980, kindly provided by Trudy E. Bell.

9. Bell, "Space Activists on Rise," p. 10, and interview with Trudy E. Bell.

10. Undated press release from Campaign for Space.

11. Trudy E. Bell, "From Little Acorns . . . : American Space Interest Groups, 1980-82," unpublished paper.

12. Undated Campaign for Space letter from Thomas J. Frieling, late 1984.

13. Clay F. Richards, "PACs Donate a Record $113 Million," *Washington Post,* December 1, 1985.

14. "Senator Glenn Receives Contribution," *Campaign for Space Update,* January/February 1984, p. 1.

15. "President Reagan Endorsed by Campaign for Space," *Campaign for Space Update,* September/October 1984, p. 1.

16. Undated circular from Campaign for Space, 1984.

17. Interview with Charles Chafer.

18. "1984 Races Become Battleground for Competing Interest Groups," *Congressional Quarterly,* September 1, 1984, pp. 2147-52, 2150.

19. Much of the section on the Anderson campaign is based on materials kindly loaned by Michael Fulda. See also Goldman and Fulda, "The Outer Space Lobby and the 1980 Elections," pp. 22-24 and Goldman and Fulda, "Outer Space Groups: Galaxies of Interests, Galaxies of Groups," paper to be published in volume 2 of the *Space Humanization Series.*

20. Goldman and Fulda, "The Outer Space Lobby and the 1980 Elections," p. 24.

21. Copies are in the files of the Space Coalition.

22. See "Flash L-5 Launches Nationwide Telegraph Campaign," *L-5 News,* March 1981, p. 5.

23. Letter from Barbara Marx Hubbard to Vice-President Bush dated January 20, 1981 (copy in the files of the Space Coalition).

24. See Citizens Advisory Council on National Space Policy, *Space: The Crucial Frontier, Spring 1981,* published by the L-5 Society, 1981.

25. Ibid., p. vii.

26. Ibid., p. 1.

27. Ibid.

28. Ibid., p. 9.

29. Citizens Advisory Council on National Space Policy, *Space and Assured Survival, Executive Summary Report, September 28, 1983,* published by the L-5 Society, 1983, p. 7.

30. Interview with Jerry E. Pournelle.

31. Ibid.

32. Undated letter from Mark M. Hopkins.

33. Interview with Victor Reis.

34. Much of this section is based on interviews with Franklin L. Lavin, October 31, 1983; Fred Whiting, October 13, 1983; William F. Norton, May 7, 1985 and May 14, 1985; and on materials from the American Space Foundation. Also see Jonathan Z. Agronsky, "Return of The Right Stuff," *Regardies,* June/July 1983, pp. 45-48.

35. Interview with Frank Lavin.

36. Ibid.

37. Interview with James Muncy, November 18, 1983.

38. Interview with Frank Lavin.

39. Interview with Mark M. Hopkins.

40. Undated letter from Edward Gibson, 1983.

41. Interview with Frank Lavin.

42. "ASF Members Make Impact on Capitol Hill," *ASF News,* Spring 1984, p. 5.

43. "Issue Brief: Why the Space Station Should be Manned," Washington, D.C.: American Space Foundation, May 13, 1984.

44. "Space Movement Happenings," *ASF News,* Fall 1983, p. 6.

45. "ASF Sets Goals for 1984 – and Beyond," *ASF News,* Spring 1984, p. 7.

46. "Space Happenings," *ASF News,* Spring 1984, p. 6.

47. This section draws heavily on interviews with James Muncy, Diana Hoyt, Karl Pflock, and William G. Norton.

48. Interview with James Muncy, March 8, 1984.

49. Ibid.

50. Interview with James Muncy, November 18, 1983. The bill was H.R. 4286, The National Space and Aeronautics Act, July 28, 1981.

51. Interview with James Muncy, March 8, 1984.

52. Interview with Karl Pflock, December 9, 1983.

53. Interview with James Muncy, March 8, 1984.

54. Interview with Diana Hoyt, October 17, 1983.

55. Letter dated November 20, 1981, signed by Daniel Akaka, Newt Gingrich, Tom Bevill, Wayne Grisham, Timothy E. Wirth, Joe Skeen, Norman Y. Mineta, and Ken Kramer.

56. Dear Colleague letter from Daniel Akaka and Newt Gingrich, March 9, 1982.

57. Dear Colleague letter from Daniel Akaka and Newt Gingrich, June 8, 1983.

58. Dear Colleague letter from Daniel Akaka, June 16, 1983.

59. Interview with William G. Norton, May 14, 1985.

60. Undated letter from Mark M. Hopkins, 1982. See also Mark M. Hopkins, "Hopkins Announces: Birth of L-5 Spacepac," *L-5 News,* May 1982, p. 1.

61. A longer list is contained in Mark M. Hopkins' Spacepac letter of January 6, 1984.

62. Duncan Forbes, ed., *The Space Activist's Handbook* (Santa Monica, Calif.: Spacepac, 1984).

63. Undated Spacepac brochure.

64. Interview with Mark M. Hopkins.

65. Undated letter from Mark M. Hopkins, 1983.

66. Letter from Mark M. Hopkins, January 6, 1984.

67. Ibid.

68. Scott Pace, "Space Politics: The Role of Spacepac," *L-5 News,* August 1984, p. 9.

69. Interview with Gary Paiste.

70. Scott Pace, "Spacepac and the 1984 U.S. Election," *L-5 News,* April 1985, p. 12.

71. See "Glenn Stumps for Expanded Space Station," *Astronautics and Aeronautics,* October 1983, p. 14.

72. "Presidential Endorsement on Hold," *Campaign for Space Update,*

January/February 1984, p. 1.

73. See "Text of 1984 Democratic Party Platform," *Congressional Quarterly,* July 21, 1984, pp. 1747-80, 1774.

74. See "Text of 1984 Republican Party Platform," *Congressional Quarterly,* August 25, 1984, pp. 2096-2117, 2104, 2117.

75. See Lois Romano, "Newt Gingrich, Maverick on the Hill," *Washington Post,* January 3, 1985.

76. Interview with Diana Hoyt.

77. See letter from Newt Gingrich, *Omni,* December 1981, p. 14.

78. Interview with Newt Gingrich, March 9, 1984.

79. Interview with James Muncy, November 18, 1983.

80. See James Muncy, "Needed: A Bold Leap Into Space," *Washington Times,* December 9, 1983.

CHAPTER 10

1. Horace Freeland Judson, "Century of the Sciences," *Science 84,* November 1984, pp. 41-43, 43.

2. Walter A. McDougall, "Technocracy and Statecraft in the Space Age – Toward the History of a Saltation," *American Historical Review,* 87 (1982):1010-40, 1028.

3. From remarks by Congressman George E. Brown on "The Fledgling Science Lobby," *Congressional Record,* June 18, 1983, p. E3227.

4. "50 and 25 Years Ago," *Sky and Telescope,* March 1985, p. 215.

5. William J. Broad, "Golden Age of Astronomy Peers to Edge of Universe," *New York Times,* May 8, 1984.

6. Bruce Murray, "Space Exploration – Is it Worth the Cost?" *The Planetary Report,* December 1980/January 1981, pp. 1, 15.

7. Interview with James E. Wilson.

8. Interview with Frank Moss.

9. Interview with Stephen H. Flajser, January 11, 1984.

10. Interview with Marc Rosenberg, December 15, 1983.

11. Interview with Marc Rosenberg.

12. Much of the descriptive material in this section is based on an interview with NASA official Nathaniel Cohen, March 12, 1984.

13. Interview with Peter Boyce, December 21, 1983.

14. The section on the USRA is based on an interview with USRA Executive Director David Cummings, November 1, 1983, and on materials supplied by the USRA.

15. James A. Van Allen, "U.S. Space Science and Technology," *Science* 214 (1981):495, reprinted in *The Planetary Report,* January/February 1982, p. 3. See also Craig Covault, "Shuttle Costs Threatening Science Programs," *Aviation Week and Space Technology,* July 6, 1981, pp. 16-18.

16. Interview with Carl Sagan, March 12, 1984.

17. From "The Universities and NASA Space Sciences – Initial Report of the NASA/University Relations Study Group, July 1983," appendix 2, p. 1. The same chart appeared later in Hans Mark, "The Space Station," *The Planetary Report,*

July/August 1984, p. 7.

18. See Clayton R. Koppes, *JPL and the American Space Program: A History of the Jet Propulsion Laboratory* (New Haven: Yale University Press, 1982).

19. Interview with John Naugle, March 28, 1984.

20. Interview with Noel Hinners, March 2, 1984. See also Thomas O'Toole, "3 Space Projects Threatened by $3 Billion Budget Limit," *Washington Post*, January 4, 1972.

21. Interview with Carl Sagan.

22. "Of Space Ships and Tall Ships," *Science* 194 (1976):39.

23. See *Ground-Based Astronomy: A Ten-Year Program* (Washington, D.C.: National Academy of Sciences, 1964); *Astronomy and Astrophysics for the 1970s,* volume 1, Report of the Astronomy Survey Committee (Washington, D.C.: National Academy of Sciences, 1972); *Astronomy and Astrophysics for the 1980s,* volume 1, Report of the Astronomy Survey Committee (Washington, D.C.: National Academy Press, 1982).

24. *Astronomy and Astrophysics for the 1980s,* p. 8. For a description of how the Space Telescope was then envisioned, see C. R. O'Dell, "The Large Space Telescope Program," *Sky and Telescope,* December 1972, pp. 369-71.

25. For a description of the 1974 effort, see Paul A. Hanle, "Astronomers, Congress, and the Large Space Telescope," *Sky and Telescope,* April 1985, pp. 300-5. See also George Alexander, "House Vote to Cancel Orbiting Telescope May Peril Shuttle," *Los Angeles Times,* July 14, 1974.

26. Space Science Board, *Opportunities and Choices in Space Science, 1974* (Washington, D.C.: National Academy of Sciences, 1975), p. 17; Space Science Board, *Report on Space Science, 1975* (Washington, D.C.: National Academy of Sciences, 1976), p. 13.

27. See John N. Bahcall, "Galaxies, Quasars, and Beyond – The Space Telescope," in John H. McElroy and E. Larry Heacock, eds., *Space Applications at the Crossroads* (San Diego: American Astronautical Society [Univelt], 1983), pp. 177-88. See also John Walsh, "Large Space Telescope: Astronomers Go into Orbit," *Science* 191 (1976):544-45.

28. Interview with James E. Wilson.

29. Bahcall, "Galaxies, Quasars, and Beyond," p. 180.

30. Allen L. Hammond, "Pioneer-Venus: Did Astronomers Undercut Planetary Science?" *Science* 189 (1975):270-71; "Pioneer Venus Probe Suffers Large Cutback," *Astronomy,* November 1975, p. 60.

31. Bahcall, "Galaxies, Quasars, and Beyond," p. 181.

32. M. Mitchell Waldrop, "Astronomy and the Realities of the Budget," *Science* 237 (1985):283-85.

33. See M. Mitchell Waldrop, "Space Telescope in Trouble," *Science* 220 (1983):172-74; M. Mitchell Waldrop, "New Worries About Space Telescope," *Science* 224 (1984):1077-78; J. Kelley Beatty, "HST: Astronomy's Greatest Gambit," *Sky and Telescope,* May 1985, pp. 409-14.

34. In Space Science Board, *Report on Space Science 1975.*

35. Committee on Planetary and Lunar Exploration, *Strategy for Exploration of the Inner Planets: 1977-1987* (Washington, D.C.: National Academy of Sciences, 1978).

36. Van R. Kane, "The Planets or Bust," *Astronomy*, May 1982, pp. 6-17.

37. Richard Berry, "A Gathering of Astronomers," *Astronomy*, February 1980, pp. 24-26.

38. Brigette Rouson, "Little Lobbying to Stop Space, Science Cuts," *Congressional Quarterly*, April 12, 1980, p. 956. By contrast, see "Scientists Beginning to Lobby for More Research Funding," *Congressional Quarterly*, March 19, 1983, pp. 555-59.

39. "JPL Scientists Formulate Future Mission Proposals," *Astronomy*, November 1976, p. 6.

40. Nathan C. Goldman and Michael Fulda, "The Outer Space Lobby and the 1980 Elections," in Paul Amaejionu, Nathan C. Goldman, and Philip J. Meeks, eds., *Space and Society: Challenges and Choices* (San Diego: Univelt, 1984), pp. 15-28, 21; interview with Carolyn Meinel; interview with Clark Chapman, April 4, 1984. See also John Noble Wilford, "Two Major Space Projects Facing a Money Roadblock in Congress," *New York Times*, June 10, 1977; and Clark R. Chapman, *Planets of Rock and Ice* (New York: Charles Scribner's Sons, 1982), pp. 199-206.

41. "Galileo Mission in Trouble," *Astronomy*, December 1979, pp. 62-63.

42. See "Galileo Cut," *Aviation Week and Space Technology*, December 17, 1981, p. 17 and Craig Covault, "Galileo Reinstated in Budget," *Aviation Week and Space Technology*, December 28, 1981, p. 10.

43. "Galileo Mission Saved – Just Barely," *Astronomy*, April 1982, pp. 78-79.

44. "Halley's Comet Mission Jeopardized," *Aviation Week and Space Technology*, January 21, 1980, p. 51; "News Flash: Comet Mission in Trouble," *Astronomy*, February 1980, p. 23; Richard A. Kerr, "Planetary Science on the Brink Again," *Science* 206 (1979):1288-89.

45. Mark Washburn, "In Pursuit of Halley's Comet," *Sky and Telescope*, February 1981, pp. 111-13, 113.

46. "Spacelab, Solar-Polar Curtailed," *Aviation Week and Space Technology*, February 23, 1981, p. 18.

47. Van R. Kane, "Bruce Murray Interview," *Astronomy*, September 1982, pp. 24-28.

48. Covault, "Galileo Reinstated in Budget."

49. Louis Friedman, "Washington Report," *The Planetary Report*, December 1979/January 1980, p. 11.

50. Richard F. Hirsh, "The Future of Space Astronomy," *Astronomy*, January 1981, pp. 24-28, 28.

51. Richard Berry, "Astronomers Meet in Tucson," *Astronomy*, March 1981, pp. 24-28, 24.

52. Some of this section is based on interviews with John Naugle and Noel Hinners, the first chairmen of the SSEC.

53. See Jesse W. Moore, "Effective Planetary Exploration at Low Cost," *Astronautics and Aeronautics*, October 1982, pp. 28-38, 52.

54. *Planetary Exploration Through the Year 2000 – A Core Program* (Part 1 of a Report by the Solar System Exploration Committee of the NASA Advisory Council) (Washington, D.C.: National Aeronautics and Space Administration, 1983).

55. Interview with Harold Volkmer, December 21, 1983.

56. Interview with Victor Reis.

57. See John Noble Wilford, "Plans to Explore Planets Revived," *New York Times*, February 20, 1983.

58. Louis D. Friedman, "Washington Watch," *The Planetary Report*, April/May 1981, p. 10.

59. "Bruce Murray on the Future of America in Space," *Astronomy*, July 1981, pp. 26-28, 26.

60. Craig Covault, "Shuttle Costs Threatening Science Programs," *Aviation Week and Space Technology*, July 6, 1981, pp. 16-18, 16.

61. Earl Lane, "The Politics of Planetary Science," *Omni*, July 1982, pp. 20, 118; M. Mitchell Waldrop, "Planetary Science in Extremis," *Science* 214 (1981):1322-24, 1322.

62. "Galileo Cut," p. 17 and William H. Gregory, "Bean-Counting the Solar System," *Aviation Week and Space Technology*, December 14, 1981, p. 11.

63. Interview with Darrell Branscome.

64. William H. Gregory, "Bean-Counting the Solar System," p. 11.

65. "Clouds over the Cosmos," *Time*, October 26, 1981, p. 49.

66. Interview with Clark Chapman.

67. Richard Berry, "Planetary Scientists Fight Funding Cuts," *Astronomy*, January 1982, pp. 24-32. The letter to Edwin Meese is reproduced on p. 26.

68. Interview with Clark Chapman.

69. Ibid.

70. Interview with Peter Boyce.

71. Van R. Kane, "Congressional Hearings on Space," *Astronomy*, July 1982, pp. 24-28, 24.

72. M. Mitchell Waldrop, "Planetary Science: Up from the Ashes," *Science* 218 (1982):665-66.

73. Interview with Geraldine Shannon, October 28, 1983. See also "University Researchers Lobby for Space Science," *Science* 216 (1982):157.

74. Interview with Stephen H. Flajser.

75. Interview with Carl Sagan.

76. Interview with Louis D. Friedman, February 6, 1984.

77. "Society News," *The Planetary Report*, December 1980/January 1981, p. 15.

78. Carl Sagan, "The Adventure of the Planets," *The Planetary Report*, December 1980/January 1981, p. 3.

79. Undated Planetary Society brochure.

80. Interview with Louis D. Friedman.

81. See "Hailing Frequencies Open: Gene Roddenberry Looks at the Planetary Society," *The Planetary Report*, April/May 1981, p. 3.

82. Undated Planetary Society brochure.

83. Ibid.

84. Interview with Leigh Ratiner.

85. Interview with Carl Sagan.

86. Ibid.

87. Interview with Peter Boyce.

88. Interview with Dennis Stone.

89. Interview with Carl Sagan.

CHAPTER 11

1. Leonard W. David, "Space: The Next Battleground?" *L-5 News*, June 1976, p. 2.

2. Interview with Benjamin Bova.

3. Arthur C. Clarke, *The Promise of Space* (New York: Harper & Row, 1968), p. 313.

4. William S. Bainbridge, *The Spaceflight Revolution* (New York: John Wiley & Sons, 1976), p. 7.

5. J. N. Leonard, *Flight into Space* (New York: Signet, 1954), p. 177. See also Curtis Peebles, *Battle for Space* (New York: Beaufort Books, 1983), pp. 47-49.

6. Dandridge M. Cole and Donald W. Cox, *Islands in Space* (New York: Chilton, 1964), pp. xiii, 155-59.

7. Robert Salkeld, *War and Space* (Englewood Cliffs, N.J.: Prentice-Hall, 1970).

8. For discussions of this issue, see "The Great Frontier: Military Space Doctrine: The Final Report from the United States Air Force Academy Military Space Doctrine Symposium, 1-3 April, 1981," distributed by the Defense Technical Information Center, Defense Logistics Agency, Cameron Station, Alexandria, Va. 22304. See also Lieutenant Colonel David Lupton (USAF, Ret.), "Space Doctrines," *Strategic Review*, Fall 1983, pp. 36-47 and Lieutenant Colonel Dino A. Lorenzini, USAF and Major Charles L. Fox, USAF, "2001: A U.S. Space Force," *Naval War College Review*, March/April 1982, pp. 48-67.

9. See Donald F. Robertson, "The Rise and Fall of the X-20 Dyna-Soar," *Astronomy*, June 1985, pp. 26-28; Philip P. Chandler, Leonard W. David, and Courtland S. Lewis, "MOL, Skylab, and Salyut," in Theodore R. Simpson, ed., *The Space Station: An Idea Whose Time Has Come* (New York: IEEE Press, 1985), pp. 32-37; and Jack Manno, *Arming the Heavens* (New York: Dodd, Mead, 1984), pp. 48ff.

10. See "Trans-Atmospheric Craft," *Popular Science*, October 1984, p. 12; "Industry Observer," *Aviation Week and Space Technology*, September 9, 1985, p. 15; and "Air Force Weighs Transatmospheric Vehicle," *Aviation Week and Space Technology*, October 28, 1985, p. 24.

11. Thomas Karas, *The New High Ground: Systems and Weapons of Space Age War* (New York: Simon & Schuster, 1983), pp. 9-13.

12. "Reagan Approves Unified Space Command," *Aerospace Daily*, December 3, 1984, p. 156; "Reagan Sees Consolidation," *Aerospace Daily*, December 4, 1984, p. 162; "Space Command to Locate in Colorado Springs," *Aviation Week and Space Technology*, May 6, 1985, p. 21; "Roger, Houston . . . Er, Colorado," *Time*, May 13, 1985, p. 30; and General James V. Hartinger, "The Air Force Space Command: An Update," paper presented to the 30th Annual Conference of the American Astronautical Society, October 3, 1983.

13. See, for example, "Tomorrow," *U.S. News and World Report*, May 31, 1982, p. 16; James E. Oberg, "Is NASA Being Militarized?" *Astronomy*, February 1985, pp. 24-26; and James E. Oberg, "The Myth of Militarization," *Space World*, June 1985, pp. 21-24.

14. *Congressional Quarterly*, July 24, 1982, pp. 1764-65; "Implications of Joint NASA/DOD Participation in Space Shuttle Operations," Washington, D.C., General Accounting Office, November 7, 1983; and R. Jeffrey Smith, "A New Image for the

Space Shuttle," *Science* 227 (1985):276-77.

15. John Noble Wilford, "Military Mission of Space Shuttle to be Kept Secret," *New York Times,* December 18, 1984; Walter Pincus and Mary Thornton, "U.S. to Orbit Sigint Craft from Shuttle," *Washington Post,* December 19, 1984; John Noble Wilford, "Shuttle Can't be Kept Fully Secret but Aides Say It is Worth the Try," *New York Times,* December 19, 1984; "The Space Shuttle's Uneasy Rider," *Washington Post,* December 23, 1984; and John Noble Wilford, "Shuttle Launched on Secret Mission," *New York Times,* January 25, 1985.

16. Bruce A. Smith, "NASA, Air Force Decide to Delay Initial Vandenberg Shuttle Launch," *Aviation Week and Space Technology,* January 14, 1985, pp. 23-24, and Robert Lindsey, "Getting Vandenberg Off the Ground," *New York Times,* March 24, 1985.

17. See Paul B. Stares, *The Militarization of Space: U.S. Policy, 1945-1984* (Ithaca, N.Y.: Cornell University Press, 1985), p. 95; John W. Finney, "Pentagon Fearful of Soviet Effort to Develop Hunter-Killer Satellites," *New York Times,* November 24, 1976; and *Aviation Week and Space Technology,* September 2, 1964, p. 21. For a good general history of ASAT programs, see Peebles, *Battle for Space,* pp. 77-123.

18. See Philip J. Klass, *Secret Sentries in Space* (New York: Random House, 1971), p. 221; Lawrence Freedman, "The Soviet Union and Anti-Space Defense," *Survival,* January/February 1977, pp. 50-56; Richard D. Lyons, "Soviet Satellite Destroyer in Orbit," *New York Times,* February 6, 1970; Joseph Alsop, "Blinders on U.S. Eyes," *Washington Post,* November 22, 1971; and "War's Fourth Dimension," *Newsweek,* November 29, 1976, pp. 46-48, 47.

19. See "2 Magazines Say Soviet Lasers Destroyed a U.S. Space Satellite," *New York Times,* November 23, 1976 and "Pentagon Denies Soviet Anti-Satellite Laser Use," *Defense/Space Daily,* November 24, 1976, p. 137.

20. Philip J. Klass, "Progress Made on High Energy Laser," *Aviation Week and Space Technology,* March 7, 1977, pp. 16-17; Jeff Smith, "U.S. Plans Laser Satellite Killer," *Atlanta Constitution,* November 5, 1977; "Killers Acknowledged," *Aviation Week and Space Technology,* November 14, 1977, p. 23; and "Anti-satellite Laser Weapons Planned," *Aviation Week and Space Technology,* June 16, 1980, p. 244.

21. Deborah Shapley, "Soviet Killer Satellites: U.S. Ponders A Response," *Science* 193 (1977):865-66, and "War's Fourth Dimension," *Newsweek,* November 29, 1976, pp. 46-48.

22. News Conference by Secretary of Defense Harold Brown, October 4, 1977; George C. Wilson, "Brown Says Some Satellites are Vulnerable to Soviet Hunters," *Washington Post,* October 5, 1977; and "Targeting a Hunter-Killer," *Time,* October 7, 1977, p. 10.

23. See Stares, *Militarization of Space,* pp. 180-200; "Industry Observer," *Aviation Week and Space Technology,* July 4, 1977, p. 11; Donald L. Hafner, "Averting a Brobdignagian Skeet Shoot," *International Security,* 5 (Winter 1980/81):41-60, 45; and R. Jeffrey Smith, "Reagan Announces a New ASAT Test," *Science* 229 (1985):946-47.

24. As quoted in "U.S., Soviet ASAT Agreement Seen Vital," *Aerospace Daily,* April 30, 1984, p. 341.

25. *Seventeenth Strategy for Peace Conference Report, October 7-10, 1976*

(Muscatine, Iowa: The Stanley Foundation, 1977); *Eighteenth Strategy for Peace Conference Report, October 13-16, 1977* (Muscatine, Iowa: The Stanley Foundation, 1978); and *Nineteenth Strategy for Peace Conference Report, October 5-8, 1978* (Muscatine, Iowa: The Stanley Foundation, 1979).

26. One of the few examples was "The Militarization of Space," *Insight* (National Space Institute), September 1978, pp. 4-5.

27. Benson D. Adams, *Ballistic Missile Defense* (New York: American Elsevier, 1971), p. 239.

28. *Newsweek*, July 17, 1944, as quoted in G. Edward Pendray, *The Coming Age of Rocket Power* (New York: Harper & Brothers, 1945), p. 223.

29. Ibid., p. 221.

30. Adams, *Ballistic Missile Defense*, p. 239.

31. See, for example, Curtis Peebles, *Battle for Space*, pp. 82-94.

32. Eugen Sanger, *Spaceflight – A Technical Way of Overcoming War* (in German) (Hamburg, Germany: Rowohlt, 1958), p. 90.

33. Arthur Kantrowitz, "Some Military Potentialities of Space," *Astronautics and Aeronautics*, November 1965, pp. 36-38.

34. Stefan T. Possony and Jerry E. Pournelle, *The Strategy of Technology: Winning the Decisive War* (Cambridge, Mass.: Dunellen, 1970), p. 131.

35. Maxwell W. Hunter II, *Thrust into Space* (Holt, Rinehart, & Winston, 1966), p. 210.

36. Interview with Maxwell W. Hunter II, April 19, 1984.

37. Ibid.

38. Maxwell W. Hunter II, "Strategic Dynamics and Space-Laser Weaponry," unpublished paper dated October 31, 1977.

39. "Arms Race in Space," *Newsweek*, February 13, 1978, pp. 53, 55.

40. "Max Hunter: The Force Behind Reagan's Star Wars Strategy," *Business Week*, June 20, 1983, p. 54, and interview with Maxwell W. Hunter II.

41. Malcolm Wallop, "Opportunities and Imperatives of Ballistic Missile Defense," *Strategic Review*, Fall 1979, pp. 13-21.

42. William H. Gregory, "Laurels for 1980," *Aviation Week and Space Technology*, January 5, 1981, p. 7.

43. Interview with John D. Rather, March 7, 1984.

44. John D. Rather, "Space Transportation, Solar Power from Space, and Space Industrialization: A Better Way," unpublished paper dated December 1976.

45. "Laser Technology – Development and Applications," Hearings before the Subcommittee on Science, Technology, and Space of the Committee on Commerce, Science, and Transportation, December 12 and 14, 1979, January 8 and 12, 1980 (Washington, D.C.: U.S. Government Printing Office, 1980).

46. H. Keith Henson, "Military Aspects of SSPS Power," *L-5 News*, May 1976, pp. 2-3.

47. Letter from Michael Mautner, *L-5 News*, July 1976, p. 5.

48. H. Keith Henson, "Space Forts, or Where Are You Obi Wan Kenobi?" *L-5 News*, June 1979, pp. 1-3.

49. Carolyn Meinel, "Fighting MAD," *Technology Review*, April 1984, pp. 31-37, 50-51.

50. Stewart Nozette, ed., "Defense Applications of Near-Earth Resources,"

California Space Institute, August 1983; revised October 31, 1983.

51. "The Laser Whammy," *Newsweek*, January 22, 1976, p. 39. See also David Binder, "U.S. and Soviet Reported Trying to Build Missile-Destroying Beam," *New York Times*, February 5, 1977.

52. Clarence A. Robinson, Jr., "Soviets Push for Beam Weapon," *Aviation Week and Space Technology*, May 2, 1977, pp. 16-23.

53. Clarence A. Robinson, Jr., "U.S. Pushes Development of Beam Weapons," *Aviation Week and Space Technology*, October 2, 1978, pp. 14-22.

54. "Technology Eyed to Defend ICBMs, Spacecraft," *Aviation Week and Space Technology*, July 28, 1980, pp. 32-63, 61.

55. "Senate Directs Air Force to Formulate Laser Plans," *Aviation Week and Space Technology*, May 25, 1981, pp. 52-53, 52.

56. *Space: The Crucial Frontier*, p. 25.

57. See William R. Van Cleave and W. Scott Thompson, eds., *Strategic Options for the Early Eighties: What Can Be Done?* (New York: National Strategy Information Center, 1979).

58. Interview with Brigadier General Robert C. Richardson III (Ret.), October 31, 1983.

59. See Tom Nugent, "Daniel Graham: Sheriff of the High Frontier," *The Washington Times*, November 1, 1983, and George Lardner, Jr., "Gen. Graham's Star Wars," *Washington Post Magazine*, November 17, 1985, pp. 14-15, 26, 29.

60. Lieutenant General Daniel O. Graham (USA, Ret.), "Toward a New U.S. Strategy: Bold Strokes Rather than Increments," *Strategic Review*, Spring 1981, pp. 9-16.

61. Colin Norman, "Keyworth to Review Space Program," *Science* 213 (1981):519.

62. "Interview: Dr. Lowell Wood, Lawrence Livermore National Laboratory," *Defense Science and Electronics*, January 1983, pp. 57-63, 61.

63. "Remarks and a Question-and-Answer Session with Reporters on the Announcement of the United States Strategic Weapons Program, October 2, 1981," in *Public Papers of the Presidents: Ronald Reagan, 1981*, pp. 878-80, 879.

64. Lieutenant General Daniel O. Graham, *High Frontier: A New National Strategy* (Washington, D.C.: High Frontier, 1982), p. xi.

65. John Bosma, *High Frontier: Supplemental Report* (Washington, D.C.: High Frontier, undated).

66. Manno, *Arming the Heavens*, p. 165.

67. Interview with Brigadier General Robert Richardson.

68. Jonathan Schell, *The Fate of the Earth* (New York: Knopf, 1982).

69. Brent Scowcroft, "Strategic System Development and New Technology: Where Should We Go," in *New Technology and Western Policy*, Adelphi Paper 197, Part I (London: International Institute of Strategic Studies, 1985), p. 9.

70. See, for example, David Hoffman, "Reagan Seized Idea Shelved in '80 Race," *Washington Post*, March 3, 1985; William J. Broad, "Reagan's Star Wars Bid: Many Ideas Converging," *New York Times*, March 4, 1984; and "How Reagan Became a Believer," *Time*, March 11, 1985, p. 16.

71. "Speech to the Nation on Defense and National Security," March 23, 1983, in *Public Papers of the Presidents: Ronald Reagan, 1983, I*, pp. 437-43, 442-43.

72. "Defense Against Ballistic Missiles: An Assessment of Technologies and Policy Implications" (Washington, D.C.: Department of Defense, April 1984). See also James C. Fletcher, "Technologies for Strategic Defense," *Issues in Science and Technology 1* (1985):15-29, and R. Jeffrey Smith, "Star Wars Plan Gets a Green Light," *Science* 222 (1983):901-2.

73. "President to Pick Ballistic Missile Defense Option," *Aerospace Daily,* November 30, 1983, p. 145; Patrick E. Tyler, "Reagan Set to Grapple with Space-Based Missile Defense Decision," *Washington Post,* November 27, 1983; Francis X. Clines, "Reagan Reported to Agree on Plan to Repel Missiles," *New York Times,* December 1, 1983.

74. Michael Getler, "Reagan Sets Missile-Defense Research in Motion," *Washington Post,* January 26, 1984.

75. Charles Mohr, "General to Head Missile Program," *New York Times,* March 28, 1984.

76. Caspar Weinberger, in an introduction to "Defense Against Ballistic Missiles: An Assessment of Technologies and Policy Implications" (Washington, D.C.: Department of Defense, April 1984).

77. Interview with Brigadier General Robert C. Richardson; Trudy E. Bell, *Upward: Status Report and Directory of the American Space Interest Movement, 1984-1985,* available from author. A more current figure was supplied in Tom Nugent, "Daniel Graham."

78. For example, the Committee on the Present Danger's press releases of April 19 and April 26, 1984; "A Star is Born: Strategic Defense Has Unconditional Support," *Policy Review,* Summer 1985, pp. 94-96; and "The Public Wants Defenses," *High Frontier Newsletter,* September 1985, pp, 1, 6. Some other polls show the reverse, such as George Lardner, Jr., "Most Polled Disapprove of SDI," *Washington Post,* August 14, 1985.

79. Circular letter from Daniel O. Graham, December 1984.

80. Interview with Robert Dornan, March 22, 1984.

81. Interview with Carolyn Meinel.

82. A partial list of the invitees was given in "A PAC for Star Wars," *Science* 222 (1983):304-05. See also *Executive Intelligence Review,* October 11, 1983, p. 63.

83. Interview with Robert Dornan.

84. Undated circular from the American Space Frontier Committee, 1984.

85. Newt Gingrich, with David Drake and Marianne Gingrich, *Window of Opportunity: A Blueprint for the Future* (New York: Tom Doherty & Associates, 1984).

86. Citizens Advisory Council on National Space Policy, *Space and Assured Survival,* p. 7.

87. As quoted in "House Panel Cuts $317 Million From Ballistic Missile Defense," *Aviation Week and Space Technology,* May 7, 1984, pp. 22-23. Dr. Rather has confirmed the accuracy of the quotation.

88. Daniel O. Graham, "Space to Provide for the Common Defense," *L-5 News,* December 1983, p. 12; David C. Webb, "Looking Before We Leap," *L-5 News,* January 1984, pp. 7-9; and Jerry E. Pournelle, "Truths, Pleasant and Unpleasant, about Military Space," *L-5 News,* February 1984, pp. 11, 13.

89. Fred Hiatt, "U.S. Tests Satellite Destroyer," *Washington Post,* January 22, 1984; "ASAT Firing is First of 12, Air Force Says," *Aerospace Daily,* January 24, 1984, p. 122.

90. Charles Mohr, "Army Test Missile is Said to Destroy A Dummy Warhead," *New York Times,* June 12, 1984 and Rick Atkinson and Walter Pincus, "Test of Missile Interceptor Called Success by Pentagon," *Washington Post,* June 12, 1984.

91. "Disarmament Groups Seeking Rallying Point After Failure on Nuclear Freeze," *New York Times,* August 18, 1985 and Emmett Tyrell, Jr., "Star Wars vs. the Freeze," *Washington Post,* March 4, 1985.

92. See, for example, Mark C. Hallam, *Star Wars Revisited: A Look at the Arms Race in Outer Space* (Washington, D.C.: SANE, 1978).

93. See "News Briefs," *L-5 News,* June 1981, p. 12; John Parmentola and Kosta Tsipis, "Particle Beam Weapons," *Scientific American,* April 1979, pp. 54-65; and Kosta Tsipis, "Laser Weapons," *Scientific American,* December 1981, pp. 51-57.

94. Interview with Jim Heaphy, April 20, 1984.

95. Jim Heaphy, "SD – Bridge Between Two Movements," *Space for All People,* June 1980, p. 2.

96. Interview with Jim Heaphy.

97. Undated Progressive Space Forum draft program.

98. Undated letter from Jim Heaphy.

99. Interview with Jim Heaphy.

100. This is based on interviews with Carol S. Rosin and Robert M. Bowman, and on the account in Jack Manno, *Arming the Heavens,* pp. 172-86.

101. Robert M. Bowman and Carol S. Rosin, "The Socio-Economic Benefits of an International Communications and Observation Platform Program," paper presented to the XXXII Congress of the International Astronautical Federation, Rome, Italy, September 6-12, 1981.

102. This is based on *Unispace 82: A Context for International Cooperation and Competition – A Technical Memorandum* (Washington, D.C.: Office of Technology Assessment, March 1983); David Dickson, "U.N. Space Conference Ends in Compromise," *Science* 217 (1982):915-16; Jerry Grey, "UNISPACE Passes up an Opportunity," *Astronautics and Aeronautics,* October 1982, pp. 12-14; and interview with David C. Webb.

103. Richard Halloran, "U.S. Plans Weapon Against Satellites" *New York Times,* June 7, 1982.

104. White House Fact Sheet, July 4, 1982.

105. See note 101.

106. Interview with James McGovern, December 2, 1983.

107. See especially *Eighteenth Strategy for Peace Conference Report, October 13-16, 1977,* pp. 27-28. The text of the resolution was printed in *Space for All People,* December 1982, p. 3.

108. "To Prevent the Militarization of the Solar System," *The Planetary Report,* January/February 1983, p. 13.

109. Interview with James McGovern.

110. Daniel Deudney, *Space: The High Frontier in Perspective,* Worldwatch Paper 50 (Washington, D.C.: Worldwatch Institute, August 1982), p. 52.

111. *Anti-Satellite Weapons: Arms Control or Arms Race* (Cambridge, Mass.: The

Union of Concerned Scientists, 1983). The text of the February 1983 petition for a ban on space weaponry is in Appendix A to *Space-Based Missile Defense* (Cambridge, Mass.: Union of Concerned Scientists, 1984).

112. Some of this history is based on Joseph Moakley, "Space Weapons and Congress," *Arms Control Today*, December 1983, pp. 8-9.

113. John F. Burns, "Andropov Issues a Promise on Antisatellite Weapons," *New York Times*, August 19, 1983.

114. Interview with John Pike, October 7, 1983.

115. R. Jeffrey Smith, "Aerospace Experts Challenge ASAT Decision," *Science* 224 (1984):693-96. Senator Malcolm Wallop had written in his 1979 *Strategic Review* article that "such integration of the elements as is being undertaken is occurring under the rubric of anti-satellite warfare." Wallop, "Opportunities and Imperatives," p. 20.

116. See Moakley, "Space Weapons and Congress"; "House Panel Keeps Big-Ticket Defense Items," *Congressional Quarterly*, October 22, 1983, pp. 2169-70; R. Jeffrey Smith, "Administration Resists Demand for ASAT Ban," *Science* 222 (1983):394-96; and Charles Mohr, "Conferees Debate Effort for Antisatellite Arms," *New York Times*, November 17, 1983.

117. "Reagan Report to Congress on ASAT Arms Control," *Congressional Quarterly*, April 7, 1984, pp. 797-800; Francis X. Clines, "Reagan Dismisses Satellite Arms Control," *New York Times*, April 3, 1984; and Walter Pincus, "Reagan Considering an Anti-Satellite Pact," *Washington Post*, April 3, 1984.

118. Interview with John Pike.

119. Press releases by Congressmen Brown, Moakley, Stark, Schroeder, Bedell, Gore, and Levine and by Senator Tsongas, March 28, 1984.

120. Union of Concerned Scientists, *Space-Based Missile Defense* (Cambridge, Mass.: Union of Concerned Scientists, 1984); Jerry Grey, "A Soviet View of Star Wars," *Aerospace America*, August 1984, pp. 26, 28, 30.

121. "Chernenko Asks U.S. to Give Up Space Arms," *New York Times*, May 20, 1984.

122. "Air Force Tests Antisatellite Payload," *Aviation Week and Space Technology*, November 19, 1984, p. 28; "2nd U.S. Anti-Satellite Rocket is Fired in Test by Air Force," *New York Times*, November 14, 1984.

123. *Congressional Quarterly*, November 3, 1984, p. 2861; "Rep. Brown Sees Sharp SDI Cut Next Year," *Aerospace Daily*, December 4, 1984, pp. 163-64; "ASAT Testing Slips Further," *Aerospace Daily*, October 26, 1984, pp. 298-99; R. Jeffrey Smith, "Congress Approves Nuclear Weapons Buildup," *Science* (1984): 422-23.

124. *Congressional Quarterly*, November 3, 1984, p. 2857.

125. Clyde W. Farnsworth, "U.S. and Russians Agree to Resume Armaments Talks," *New York Times*, November 23, 1984; Don Oberdorfer, "U.S., Soviets to Resume Arms Talks," *Washington Post*, January 9, 1985.

126. See, for example, Dusko Doder, "New Start Said to be Result of Reassessment by Soviets," *Washington Post*, January 9, 1985.

127. Interview with Charles Sheffield.

128. Matsunaga's bill was S.J. Res. 236, introduced February 9, 1984. See also articles by Spark Matsunaga, "With Moscow to Mars," *Washington Post*, September 9, 1984, and "First Word," *Omni*, December 1984, p. 6.

129. Conversation with Harvey Meyerson, April 17, 1984.

130. "Cooperative East-West Ventures in Space," in *Presidential Documents, Administration of Ronald Reagan, 1984*, p. 1691.

131. John L. McLucas, "Could Space Thaw the Freeze?," *Aerospace America*, January 1985, p. B32.

132. Louis D. Friedman, "New Era of Global Security: Reach for the Stars," *Aerospace America*, August 1984, p. 4.

133. Howard Kurtz, "The Collapse of U.S. Global Strategy," *Military Review*, May 1969, pp. 43-52. For background on the Kurtz enterprise, see Norie Huddle, *Surviving: The Best Game on Earth* (New York: Schocken Books, 1984), pp. 165-75.

134. M. Callaham and K. Tsipis, *Crisis Management Satellite: Progress Report of Study Group* (Cambridge, Mass.: Department of Physics, Massachusetts Institute of Technology, 1978.

135. *The Implications of Establishing an International Satellite Monitoring Agency*, Disarmament Study Series 9 (New York, United Nations, 1983).

136. On Clarke's views, see Robert T. Brodbeck, "Self-Interest and Peace," *L-5 News*, September 1983, p. 5. On Daniel Deudney, see his *Space: The High Frontier in Perspective*, Worldwatch Paper 50 (1982), and *Whole Earth Security: A Geopolitics of Peace*, Worldwatch Paper 55 (Washington, D.C.: Worldwatch Institute, 1983).

137. John L. McLucas, "Whither LANDSAT," *Aerospace America*, January 1985, p. 6. See also John L. McLucas, "Open Skies A Fresh Challenge," *Aerospace America*, April 1985, p. 6.

138. See Carol S. Rosin, "Cooperation in Space: An Alternative to Star Wars," *Breakthrough* (Global Education Associates), Winter 1985, p. 1.

139. Arthur C. Clarke, "The Rocket and the Future of Warfare," in Arthur C. Clarke, ed., *Ascent to Orbit: A Scientific Autobiography – The Technical Writings of Arthur C. Clarke* (New York: John Wiley & Sons, 1984), pp. 69-79.

140. Bainbridge, *The Spaceflight Revolution*, pp. 241-43.

141. *Preventing Nuclear War* (San Francisco: World Security Council, 1983).

142. Richard H. Ullman, "U.N.-doing Missiles," *New York Times*, April 28, 1983.

143. Benjamin Bova, *Assured Survival: Putting the Star Wars Defense in Perspective* (Boston: Houghton Mifflin, 1984), p. 277.

144. William J. Broad, "Science Fiction Writers Choose Sides in Star Wars," *New York Times*, February 26, 1985.

145. H. Keith Henson, "Space Foxholes or Beetle Bailey in Oribt," *L-5 News*, October 1979, pp. 3-4.

146. H. Keith Henson, "L-5 and the Military in Space," *L-5 News*, November 1983, pp. 5-8.

147. H. Keith Henson, "Weapons for Peace," *L-5 News*, July 1984, pp. 8-11. See also K. Eric Drexler, "Can Space Weapons Serve Peace?" *L-5 News*, July 1983, pp. 1-2.

148. Letter from Jon Alexandr, *Astronomy*, March 1983, pp. 32-33.

149. Clark R. Chapman, "News and Reviews," *The Planetary Report*, January/February, 1982, p. 6.

150. See, for example, Dr. Andrew G. Adelman, "On the Strategic Defense Initiative," *American Astronautical Society Newsletter*, September 1984, p. 1.

151. Broad, "Science Fiction Writers Choose Sides in Star Wars."

152. Trudy E. Bell, "From Little Acorns . . . : American Space Interest Groups, 1980-1982," unpublished paper.

153. Letter from Jim Klann, *Astronomy*, September 1981, p. 36.

154. Interview with Norman Cousins, February 7, 1984.

155. Bernard Weinraub, "Mondale Asks Ban on Arms in Space," *New York Times*, April 25, 1984.

CHAPTER 12

1. E. P. Wheaton and Maxwell Hunter, "Space Commerce," paper presented to the Fourth Goddard Memorial Symposium of the American Astronautical Society, Washington, D.C., March 15-16, 1966.

2. As quoted in Ned Scharff, "Too Crowded Here? Why Not Fly Into Space," *Washington Star*, November 3, 1977.

3. Interview with Klaus Heiss, November 22, 1983.

4. Interview with Thomas O. Paine.

5. See Bernard Gwertzman, "U.S. Expected to Seek Competition to INTELSAT, *New York Times*, May 2, 1984; Michael Schrage, "Private Global Satellites Win Backing," *Washington Post*, November 29, 1984; Jay C. Lowndes, "FCC Considers Policy Favoring Competition with INTELSAT," *Aviation Week and Space Technology*, January 7, 1985, pp. 24-26; and Jay C. Lowndes, "FCC Authorizes New Spacecraft, Sanctions Competition with INTELSAT," *Aviation Week and Space Technology*, August 26, 1985, pp. 63-66.

6. David C. Webb, *A Current Perspective on Space Commercialization* (Washington, D.C.: Aerospace Industries Association, 1985), p. 12.

7. Interview with Elizabeth Harrington.

8. For a brief overview, see the articles of Barbara Luxenberg, "Preliminary OK for Direct Broadcast Satellites," *Astronautics and Aeronautics*, September 1982, pp. 20-21, and "Hat in the Ring for Direct Broadcast Satellites," *Astronautics and Aeronautics*, September 1982, pp. 32, 86.

9. See Jay C. Lowndes, "Fourteen Seek Direct Broadcast Rights," *Aviation Week and Space Technology*, August 10, 1981, pp. 60-61.

10. Michael Schrage, "COMSAT Drops Plans for DBS Network," *Washington Post*, December 1, 1984 and Andrew Pollack, "Plan for TV by Satellite Falls Apart Over Risks," *New York Times*, December 1, 1984.

11. White House Fact Sheet, June 20, 1978.

12. White House Fact Sheet On U.S. Civil Space Policy, October 11, 1978.

13. See, for example, the books of Charles Sheffield: *Earth Watch* (New York: Macmillan, 1981) and *Man on Earth* (New York: Macmillan), 1983.

14. For reviews of these events, see Craig Covault, "Two-Step Operational LANDSAT Plan Set," *Aviation Week and Space Technology*, July 14, 1980, pp. 108-15; William H. Gregory, "Free Enterprise and LANDSAT," *Aviation Week and Space Technology*, July 14, 1980; Benjamin M. Elson, "Concern over LANDSAT Budget Cutbacks," *Aviation Week and Space Technology*, July 27, 1981, pp. 26-27; "LANDSAT Choices," *Aviation Week and Space Technology*, April 26, 1982, p. 17;

M. Mitchell Waldrop, "Imaging the Earth (I): The Troubled First Decade of LANDSAT," *Science* 215 (1982):1600-03; M. Mitchell Waldrop, "Imaging the Earth (2): The Politics of LANDSAT," *Science* 216 (1982):40-41; and Joel S. Greenberg, "Die Cast for Commercial Land Remote Sensing," *Aerospace America*, January 1984, pp. 33-38.

15. Interview with Charles Sheffield.

16. Reports concerning this period include "Congress Firming Position on Satellite Sale Proposal," *Aviation Week and Space Technology*, April 18, 1983, p. 30; Philip J. Hilts, "U.S. House Wants a Say on Satellites," *Washington Post*, April 28, 1983; "Senate Resolution Opposes Satellite Sale," *Aviation Week and Space Technology*, October 17, 1983, p. 19; "Hill Gets Report on LANDSAT Commercialization," *Aerospace Daily*, October 2, 1984, p. 168; "News Digest," *Aviation Week and Space Technology*, October 15, 1984, p. 29; M. Mitchell Waldrop, "OMB Move Threatens LANDSAT," *Science* 226 (1984):151; "LANDSAT Transfer Down to the Wire," *Space Business News*, November 19, 1984, p. 5; and "LANDSAT Funding Impasse," *Aerospace Daily*, November 26, 1984, p. 114.

17. See "SPARX to Replace Earthstar," *Space Business News*, December 5, 1983, p. 24; "SPARX Seeks ELV Launch After NASA Veto," *Space Business News*, June 18, 1984, p. 2; and David Dickson, "SPARX Fly over U.S.-German Space Venture," *Science* 227 (1985):617-18.

18. On OTRAG, see Robert R. Ropelevski, "Low-Cost Satellite Launcher Developed," *Aviation Week and Space Technology*, September 12, 1977, pp. 44-47; "OTRAG Considering Launch Sites in Brazil and Asia," *Wall Street Journal*, June 30, 1978; John Dornberg, "Bargain Basement Rocket," *Popular Science*, March 1978, pp. 76-80, 186-88; "OTRAG News," *L-5 News*, April 1978, pp. 12-13; Carolyn Henson, "OTRAG: Progress in The Face of Adversity," *L-5 News*, August 1978, pp. 4-7; "Zaire Terminates Rocket Launch Site Pact," *Aviation Week and Space Technology*, May 7, 1979, p. 18; "German Company Testing Launch Vehicles in Libya," *Aviation Week and Space Technology*, March 23, 1981, p. 25; Theo Pirard, "OTRAG Update" and James E. Oberg, "The Rise, Fall, and Rise of OTRAG," *L-5 News*, December 1980, pp. 8-11; Carolyn Henson, "The Third World Space Powers," *Future Life*, September 1981, p. 33; "OTRAG Ends Libyan Launch Work," *Aviation Week and Space Technology*, December 14, 1981, p. 22; Jeffrey M., Lenorovitz, "OTRAG to Market Sounding Rockets," *Aviation Week and Space Technology*, October 4, 1982; "OTRAG Prepares for Full Launch Service," *Aviation Week and Space Technology*, September 12, 1983, pp. 77-79; OTRAG Plans Spring Launch," *Space Business News*, April 23, 1984, p. 7; and "OTRAG Wants U.S. Partner," *Space Business News*, June 4, 1984, p. 1.

19. Much of this is based on an interview with Robert Truax, April 19, 1984. See Also Carolyn Henson, "Robert Truax: Crackpot or Pioneer?" *L-5 News*, September 1979, p. 1.

20. "No Joy Riding Allowed," *Newsweek*, June 13, 1983, p. 26.

21. Interview with Robert Truax. For the early years of the Truax venture, see Carolyn Henson, "First Private Enterprise Astronaut Selected," *L-5 News*, September 1979, p. 2; "Right Stuff of Captain Truax," *Newsweek*, July 7, 1980, p. 23; "Volksrocket Engine Test, *L-5 News*, August 1980, p. 5; Theo Pirard, "First Private Astronaut," *Space World*, November 1980, pp. 9-12; "Space Barnstorming," *Popular Mechanics*, March, 1981, p. 146; and "Space Venture in Need of Capital," *Newsweek*,

May 4, 1981, p. 8. See also "Truax Succeeds in Ground Test of Private Manned Rocket," *Space Commerce Bulletin*, July 20, 1984, pp. 6-7.

22. Much of this is based on an interview with Gary C. Hudson, April 19, 1984. William S. Bainbridge briefly treated Hudson's early years in *The Spaceflight Revolution* (New York: John Wiley & Sons, 1976), pp. 227-28. See also *L-5 News*, October 1976, p. 11.

23. Siegler was interviewed for Vernon Louviere, "Space: Industry's Newest Frontier," *Nation's Business*, February 1978, pp. 30-41. His article "Marketing Space" appeared in *L-5 News*, March 1980, pp. 14-17. The *Earth/Space* Newsletter was noted in the *Futurist*, February 1976, p. 33.

24. Interviews with Mark Frazier, October 20, 1983, and July 16, 1984, and interview with Gary C. Hudson. The Space Freeport Project put out a press release on February 10, 1977, and the Sabre Foundation put out an *Earthport Bulletin* during 1978. See also "International Launch Center Proposed,"*L-5 News*, June 1977, pp. 1, 6, and Carolyn Henson, "Gateway to the Stars: Earthport," *L-5 News*, April 1979, p. 8.

25. See undated "Earthport" brochure put out by the Sabre Foundation, 317 C Street, N.E., Washington, D.C. 20003.

26. Richard House, "Brazil Pursues Dream in Space," *Washington Post*, December 13, 1984.

27. This apparently was Ron Chernow, "Colonies in Space Might Turn Out to Be Nice Places to Live," *Smithsonian*, February 1976.

28. John W. Wilson, "Space," *Omni*, December 1981, pp. 14, 24, 161.

29. Much of this section is based on interviews with Gary C. Hudson, April 19, 1984 and Nancy Wood, March 13, 1984; James Bennett, April 19, 1984; and Philip Salin, August 2, 1984 and September 17, 1984.

30. See Robin Snelson, "Industrialization of Space Conference Postscripts," *L-5 News*, December 1977, p. 3.

31. See *Aviation Week and Space Technology*, August 10, 1981, p. 26; *Washington Post*, August 6, 1981, and Randy Clamons, "Percheron Tests: Private Enterprise in Space," *L-5 News*, September 1981, p. 14. For an overview of the whole venture, see Tom Richman, "The Wrong Stuff," *INC.*, July 1982, pp. 64-70.

32. White House Fact Sheet on National Space Policy, July 4, 1982.

33. Interview with T. Stephen Cheston. See also Dan Balz, "Private U.S. Investors Are Ready to Attempt Again to Launch Free Enterprise into Space," *Washington Post*, September 3, 1982.

34. See *Aviation Week and Space Technology*, September 20, 1982, p. 17, and "Outer Space Entrepreneurs," *Time*, September 20, 1982, p. 40.

35. Chafer presented a thoughtful paper on space commercialization in 1982. See Paul Amaejionu, Nathan C. Goldman, and Philip J. Meeks, eds., *Space and Society: Challenges and Choices* (San Diego, Ca.: Univelt, 1984), pp. 29-40.

36. See "Design for Conestoga II Shaping Up," *Space World*, January 1984, pp. 12-13; "Conestoga III in Works," *Space Business News*, March 26, 1984, p. 3.

37. See *Space Commerce Bulletin*, January 18, 1984, p. 4.

38. Interview with James Bennett. This section also draws on interviews with Philip Salin.

39. See "Starstruck Launches Prototype Dolphin Rocket in First Flight," *Aviation Week and Space Technology*, August 13, 1984, pp. 20-21, and "Starstruck

Launches Booster from Pacific," *Aviation Week and Space Technology*, August 27, 1984, pp. 58-59.

40. See "Booster Tests, Workforce Cut by Starstruck," *Aviation Week and Space Technology*, September 24, 1984, p. 25, and "Starstruck Management is Reorganized," *Aviation Week and Space Technology*, October 22, 1984, p. 30.

41. Speech by Congressman Daniel Akaka to the Federal Bar Association, November 3, 1983.

42. White House Fact Sheet, May 16, 1983.

43. The President's statement is in *Presidential Documents*, November 5, 1984, p. 1692. Diana Hoyt told the tortuous history of the bill in an unpublished paper entitled "The Dance of Legislation."

44. See "TCI Signs Delta Agreement," *Space Business News*, April 9, 1984, p. 8; Edmund L. Andrews, "The Rocket Man," *Regardies*, October 1984, pp. 103-13; "Martin Marietta Seeks Commercial Opportunities for Titan," *Space Commerce Bulletin*, March 29, 1985, p. 4.

45. Materials distributed by William A. Good at a Symposium on the Moon treaty, Center for Strategic and International Studies, Georgetown University, February 1980. See also Randall Clamons, "Why on Earth . . . Why Not in Space," *L-5 News*, December 1980, p. 12.

46. This is based primarily on an interview with Klaus Heiss. See also "Space Transportation Firm Agrees to Marketing of Titan," *Aviation Week and Space Technology*, December 6, 1982, pp. 26-27, and "Federal Express to Acquire Launcher Rights," *Aviation Week and Space Technology*, May 23, 1983, p. 24.

47. "Cyprus Mulls Fifth Orbiter," *Space Business News*, March 12, 1984, p. 1; "Cyprus/Astrotech Going Big into Space," *Space Business News*, June 4, 1984, p. 7.

48. *Space Business News*, June 4, 1984, p. 7.

49. See "Space Van," *Omni*, December 1981, p. 49; "The Amazing Mini-Shuttle," *Science Digest*, August 1982, pp. 80-81; "Transpace Reports Space Van Plans," *Space Business News*, April 23, 1984, p. 4; "Third Millenium to be New Name of Transpace," *Aerospace Daily*, May 25, 1984, p. 152; and "Space Van Tickets Sold," *Space Business News*, October 8, 1984.

50. See "Private Firm Planning Upper Stage," *Aviation Week and Space Technology*, December 27, 1982, pp. 8-9.

51. *Space Business News*, November 18, 1984, p. 5.

52. See *Materials Processing in Space* (Washington, D.C.: National Academy of Sciences, 1978) and "MPS Oversold, Panel Says," *Space Business News*, May 7, 1984, p. 2.

53. See "Microscopic Beads Made in Space Go on Sale," *New York Times*, July 18, 1985.

54. See Craig Covault, "Payload Tied to Commercial Drug Goal," *Aviation Week and Space Technology*, May 31, 1982, pp. 51-57; "Prototype Plan Follows Space Processing Test," July 19, 1982, pp. 26-27; "Drug Made in Space is Contaminated," *New York Times*, November 4, 1984, p. 18; "Space Hormone Found Contaminated, Unusable," *Aviation Week and Space Technology*, November 12, 1984, p. 18..

55. "3M, NASA Detail Ten-Year Agreement," *Aerospace Daily*, October 4, 1984, p. 184; "Shuttle Crystal-Growing Experiment Unqualified Success," *Aerospace Daily*, December 5, 1984, p. 174; "3M Plans Ambitious 10 Year Commercial Space Research Project," *Space Commerce Bulletin*, October 12, 1984, pp. 9-10; "3M

Unveils Ambitious Space Research Plan," *Space Business News*, October 8, 1984, p. 1. See also "Microgravity Research Raises $1.7 million with Limited R and D Partnership," *Space Commerce Bulletin*, August 17, 1984, p. 2.

56. See "New Space Venture: Micro-Gravity Factories," *Popular Science*, October 1984, p. 80.

57. Interview with Max Faget, March 13, 1984. See also "Industrial Space Facility," *Commercial Space*, Fall 1985, pp. 40-42.

58. *Encouraging Business Ventures in Space Technologies* (Washington, D.C.: National Academy of Public Administration, 1983).

59. Craig Covault, "Reagan Briefed on Space Station," *Aviation Week and Space Technology*, August 8, 1983, pp. 16-18.

60. M. Mitchell Waldrop, "The Commercialization of Space," *Science* 221 (1983):1353-54, 1354.

61. "Space Program: Address to the Nation, January 28, 1984," in *Presidential Documents*, February 3, 1984, pp. 113-14.

62. Carole A. Shifrin, "Reagan Backs Space Commerce," *Aviation Week and Space Technology*, July 30, 1984, pp. 16-17; "Reagan Issues Commercial Space Policy," *Aerospace Daily*, July 23, 1984, p. 114; Pat Jefferson, "President Pushes for Space Commercialization," *Aerospace America*, November 1984, pp. 15-16, 20.

63. "Import Duties Blocked on Space Products," *Aviation Week and Space Technology*, October 29, 1984, p. 66.

64. "NASA Opens Space to Entrepreneurs," *New York Times*, November 21, 1984; "NASA Unveils Commercial Space Policy; New Program Office Defined," *Aerospace Daily*, November 21, 1984, p. 106; Craig Covault, "NASA Formulates Policy to Spur Private Investment," Aviation Week and Space Technology, November 26, 1984, pp. 18-1965. As quoted in "Space Industry," *L-5 News*, August 1984, p. 14.

66. "Space Entrepreneurs Directory," *Space Calendar*, August 20-26, 1984, p. 5.

67. In *Space*, a publication by Shearson Lehman/American Express, 1985, p. 4.

68. John M. Logsdon, "Space Commercialization: How Soon the Profits?," *Futures* 16 (1984):71-78.

69. Interviews with Jerry Grey and Charles Sheffield.

70. Speech by Gregg Fawkes to a luncheon meeting of the American Astronautical Society and the Congressional Staff Space Group, January 27, 1984.

71. Interview with Robert L. Staehle.

CHAPTER 13

1. Willy Ley, *Space Stations* (Poughkeepsie, N.Y.: Guild Press, 1958), p. 44.

2. Ronald Reagan, State of the Union Address, January 24, 1984, in *Presidential Documents*, January 24, 1984, pp. 87-94.

3. Interview with Thomas A. Heppenheimer.

4. See E. E. Hale, "The Brick Moon," *Atlantic Monthly*, volume XXIV (October, November, December 1869), and Konstantin Tsiolkowsky, *Dreams of Earth and Heaven* (Moscow, 1895).

5. Hermann Oberth, *The Rocket into Interplanetary Space* (Munich: R. Oldenbourg, 1923).

6. Hermann Noordung, *The Problems of Spaceflight* (Berlin: Schmidt & Company, 1929). The von Braun quote is from Egon Eis, *The Forts of Folly* (London: Oswald Wolff, 1959), p. 248.

7. See Sylvia Doughty Fines, "Function, Form, and Technology: The Evolution of Space Station in NASA," paper presented at the Congress of the International Astronautical Federation, Stockholm, October 1985.

8. Walter B. Olstad, "Targeting Space Station Technologies," *Astronautics and Aeronautics*, March 1983, pp. 28-32. See also the special section on Skylab in the June 1971 issue of *Astronautics and Aeronautics*.

9. Interview with Hans Mark and James E. Oberg, *The New Race for Space* (Harrisburg, Pa.: Stackpole Books, 1984). See also "Window for Space Detente," *Aerospace America*, November 1984, pp. 86-87.

10. *Space Station* (Washington, D.C.: National Aeronautics and Space Administration, 1970).

11. Philip E. Culbertson, "Current NASA Space Station Planning," *Astronautics and Aeronautics*, September 1982, pp. 36-43, 38, 59.

12. *Manned Orbital Facility: A User's Guide* (Washington, D.C.: National Aeronautics and Space Administration, 1975).

13. Donald W. Patterson, John W. Gurr, and George V. Butler, "Earth Orbiting Stations," *Astronautics and Aeronautics*, September 1975, pp. 22-29.

14. "Expanded Utilization of Shuttle Studied," *Aviation Week and Space Technology*, November 8, 1976, p. 134. See also "NASA Weighing Space Station Approach," *Aviation Week and Space Technology*, April 18, 1977, pp. 42-43.

15. "U.S. and Russia Announce Talks on Operating Space Station in 80's," *New York Times*, May 18, 1977.

16. See Craig Covault, "Soviets Developing 12-Man Space Station," *Aviation Week and Space Technology*, June 16, 1980, pp. 26-29. See also *Salyut: Soviet Steps Toward Permanent Human Presence in Space* (Washington, D.C.: Office of Technology Assessment, 1983).

17. "Floating in Space," *Aviation Week and Space Technology*, August 4, 1980, p. 11.

18. Craig Covault, "NASA Termed Underutilized," *Aviation Week and Space Technology*, October 13, 1980, pp. 16-17.

19. Brian T. O'Leary, *Project Space Station* (Harrisburg, Pa.: Stackpole Books, 1983), p. 14.

20. "Industry Pessimism," *Aviation Week and Space Technology*, September 8, 1980, p. 15.

21. "A Space Base for the 80s?" *L-5 News*, October 1976, pp. 1-6.

22. Gerald W. Driggers, "Space Station – Pathway into the Universe," *L-5 News*, April 1980, pp. 4-5.

23. "To Everyone Who Wants an Expanded Space Program," *L-5 News*, January 1981, p. 1.

24. *Space: The Crucial Frontier*, p. 15.

25. "The Washington Scene," *L-5 News*, June 1981, p. 14.

26. Craig Covault, "Planners Set Long-Term Space Goals," *Aviation Week and Space Technology*, March 9, 1981, pp. 75-78, 75.

27. "NASA Endorses Space Station Development," *Aviation Week and Space*

Technology, May 25, 1981, p. 16.

28. "NASA Nominees Back Expanded Goals," *Aviation Week and Space Technology,* June 29, 1981, p. 56.

29. "Citizens Council Discusses Space Future," *L-5 News,* November 1981, p. 1.

30. "Space: America's Frontier for Growth, Leadership, and Freedom," Rockwell International, October 2, 1981.

31. "Interview: James M. Beggs," *Omni,* December 1981, pp. 93-94, 148-51, 94.

32. "Official Doubts Space Station Need," *Aviation Week and Space Technology,* November 30, 1981, p. 20.

33. See letter from Newt Gingrich, *Omni,* December 1981, p. 14.

34. Craig Covault, "Consensus Nearing on Orbital Facilities," *Aviation Week and Space Technology,* February 15, 1982, pp. 119-25, 119.

35. Craig Covault, "NASA Mulls International Effort on Space Station," *Aviation Week and Space Technology,* March 1, 1982, pp. 20-22 and "News Digest," *Aviation Week and Space Technology,* March 22, 1982, p. 28.

36. Letter from Louis D. Friedman, *Aviation Week and Space Technology,* January 4, 1981, p. 68. A rebuttal from Leonard W. David appeared in *Aviation Week and Space Technology,* March 22, 1982, p. 80.

37. "Space Policy: An AIAA View," *Astronautics and Aeronautics,* May 1982, pp. 12-14.

38. "Industry Observer," *Aviation Week and Space Technology,* September 6, 1982, p. 39; Culbertson, "Current NASA Space Station Planning."

39. White House Fact Sheet, July 4, 1982.

40. Remarks by President Reagan at Edwards Air Force Base, July 4, 1982.

41. R. Jeffrey Smith, "Squabbling Over Space Policy," *Science* 217 (1982):331-33.

42. Richard G. O'Lone, "Keyworth Urges Definition of Space Station Objective," *Aviation Week and Space Technology,* July 4, 1983, pp. 25-26, 25.

43. Interview with James Muncy, March 8, 1984.

44. R. Jeffrey Smith, "Squabbling Over Space Policy."

45. Philip E. Culbertson, "Current NASA Space Station Planning," pp. 36-43, 59.

46. M. Mitchell Waldrop, "NASA Wants a Space Station," *Science* 217 (1982):1018-21, 1021.

47. "What's In a Step," *Aviation Week and Space Technology,* February 28, 1983, p. 15.

48. *Astronautics and Aeronautics,* March 1983, pp. 28-68, and David Dooling, "Space Station Faces Difficult Birth," *Astronautics and Aeronautics,* March 1983, pp. 14-18.

49. Craig Covault, "Space Station Pivotal in NASA Future," *Aviation Week and Space Technology,* March 14, 1983, pp. 83-89.

50. Letter from Louis D. Friedman, *Aviation Week and Space Technology,* May 2, 1983, p. 110.

51. "Keyworth Calls for Bold Push in Space," *Science* 221 (1983):132.

52. Craig Covault, "NASA Chief Foresees Space Station Approval," *Aviation Week and Space Technology,* July 25, 1983, pp. 18-21, 18.

53. "AIAA/NASA Space Station Symposium," *L-5 News,* November 1983,

pp. 3-4, 3.
54. Undated letter from Spacepac, Summer 1983.
55. M. Mitchell Waldrop, "The Selling of the Space Station," *Science* 223 (1984):793-94, 794.
56. Interview with James Muncy, March 8, 1984.
57. The author was present.
58. M. Mitchell Waldrop, "The Selling of the Space Station."
59. "Interview: John Glenn,"*Omni*, October 1983, pp. 127-32, 190.
60. See M. Mitchell Waldrop, "Spacelab: Science on the Shuttle," *Science* 222 (1983):405-7.
61. "Space Station View," *Aviation Week and Space Technology*, August 15, 1983, p. 17.
62. "No Consensus Yet on a Space Station," *Science* 222 (1983):34.
63. Johan Benson, "Science Board Urges Hold on Space Station," *Astronautics and Aeronautics*, November 1983, p. 77 and "No Consensus Yet on Space Station."
64. T. M. Donahue, "Science and a Space Station," *The Planetary Report*, July/August 1984, p. 7.
65. Space Applications Board, *Practical Applications of a Space Station* (Washington, D.C.: National Academy Press, 1984). See also "Space Applications Board Takes Supportive Stand on Space Station," *Aerospace Daily*, December 21, 1983, p. 269.
66. Interview with James Muncy, March 8, 1984.
67. Speech by President Reagan, October 19, 1983, in *Weekly Compilation of Presidential Documents*, October 24, 1983, pp. 1460-63, 1462-63. See also "Reagan Urges NASA to be More Visionary," *Aviation Week and Space Technology*, October 24, 1983, p. 25.
68. See Robert C. Toth and Sara Fritz, "Reagan Likely to Back Permanent Station in Space," *Los Angeles Times*, November 17, 1983.
69. "Soviet Threat," *Aerospace Daily*, November 21, 1983, p. 105.
70. "Reagan to Get Space Station Options Soon, NSC Staffer Says," *Aerospace Daily*, November 29, 1983, pp. 139-40.
71. The author was present at the meeting.
72. See "Space Station," *Aviation Week and Space Technology*, November 28, 1983, p. 17; "President Hears Space Station Options," *Aerospace Daily*, December 2, 1982, p. 164; "Washington Roundup," *Aviation Week and Space Technology*, December 5, 1983, p. 15.
73. M. Mitchell Waldrop, "The Selling of the Space Station."
74. "President Hears Space Station Options," *Aerospace Daily*, December 2, 1983, p. 164.
75. See "Station Decision Overrode Strong Opposition," *Aviation Week and Space Technology*, January 30, 1984, p. 16.
76. "The Attorney General's Space Shot," *Newsweek*, January 30, 1984, p. 11.
77. M. Mitchell Waldrop, "The Selling of the Space Station."
78. "Space Station," *Aerospace Daily*, December 12, 1983, p. 210.
79. "ESA Pursuing Space Station Role," *Aviation Week and Space Technology*, December 5, 1983, pp. 16-20; "Japanese Firms Study U.S. Space Station," *Aerospace Daily*, December 13, 1983, p. 223.

80. Philip M. Boffey, "President Seems Near Commitment on Space Station," *New York Times,* December 14, 1983.

81. "Cooper Predicts OSD Backing of Space Station if Reagan Supports It," *Aerospace Daily,* December 15, 1983, p. 235.

82. See "Senator Gorton Urges Presidential Go-Ahead on Space Station Proposal," *Aerospace Daily,* January 9, 1984, p. 37.

83. Letter from Mark M. Hopkins, December 9, 1983.

84. Postcard from Campaign for Space, postmarked December 10, 1983.

85. Undated letter from Benjamin Bova.

86. Memorandum from Gregg Maryniak dated January 12, 1984.

87. L-5 Society release dated January 24, 1984.

88. The author was present at the meeting.

89. Press release from the National Coordinating Committee on Space, January 24, 1984.

90. "The State of the Union," in *Presidential Documents,* volume 20 (January 30, 1984), pp. 87-94, 90.

91. John Newbauer, "Away Space Station," *Aerospace America,* March 1984, pp. 12-13.

92. Letter from Mark M. Hopkins.

93. Letter from Hans Mark, April 4, 1984.

94. Circular letter from Mark M. Hopkins and David Brandt-Erichsen, April 16, 1984.

95. "New Task Force to Involve Scientists in Space Station Use Study," *Aerospace Daily,* January 20, 1984, p. 107.

96. Press release from Scientists for a Manned Space Station, April 30, 1984.

97. Kenneth J. Frost and Frank B. MacDonald, "Space Research in the Era of the Space Station," *Science* 226 (1984):1381-85.

98. Speech to National Space Club Luncheon, November 14, 1984.

99. *Civilian Space Stations and the U.S. Future in Space* (Washington, D.C.: Office of Technology Assessment, 1984).

100. Peter E. Glaser, "Space Station – A Boon to Industry," *Aerospace America,* November 1984, p. 4.

101. James M. Beggs, "Space Station: The Next Logical Step," *Aerospace America,* September 1984, pp. 47-52, 48.

102. Interview with Brian Duff.

103. Interviews with Victor Reis and James Muncy.

CHAPTER 14

1. Richard Hutton, *The Cosmic Chase* (New York: New American Library, 1981), p. 181.

2. Ralph Waldo Emerson, *Essays: First Series (1841): Circles,* as quoted in John Bartlett, *Familiar Quotations,* 15th ed. (Boston: Little, Brown, 1980), p. 497.

3. Interview with Klaus Heiss.

4. William S. Bainbridge, *The Spaceflight Revolution* (New York: John Wiley & Sons, 1976), p. 3.

5. Harrison Schmitt, "A Milennium Project: Mars 2000," *ASF News,* Summer 1985, pp. 6-7, 7.

6. As quoted in Joel Kotchin, "Reaching for the Last Frontier," *Washington Post,* August 3, 1980.

7. Burdett A. Loomis and Allan J. Cigler, "Introduction: The Changing Nature of Interest Group Politics," in Allan J. Cigler and Burdett A. Loomis, eds., *Interest Group Politics* (Washington, D.C.: Congressional Quarterly Press, 1983), pp. 1-28.

8. "Who and What is ASAP?" *ASAP Update,* May 1985, p. 1.

9. According to a letter from astronomer Yaron Sheffer in *Sky and Telescope,* May 1985, p. 388.

10. Undated Planetary Society circular letter from Louis D. Friedman.

11. *The Space Shuttle Passenger Project: A Design Study* (Santa Clara, Calif.: Space Age Review Foundation, 1983). Durst's complaint was in "America's Space Shuttle: For What? For Whom?" *Space Age Review,* January 1979, p. 3.

12. Letter from Maxwell Hunter II to the author, August 14, 1984.

13. Remarks by President Reagan to a joint session of Congress, April 28, 1981, in *Public Papers of the Presidents of the United States: Ronald Reagan, 1981* (Washington, D.C.: U.S. Government Printing Office, 1982), p. 394.

14. See Eliot Marshall, "Public Attitudes to Technological Progress," *Science* 205 (1979): 281-85.

15. Howard P. Segal, *Technological Utopianism in American Culture* (Chicago: University of Chicago Press, 1985), p. 152.

16. Joseph J. Corn, *The Winged Gospel: America's Romance with Aviation* (New York: Oxford University Press, 1983).

17. Ibid., pp. 143-44. See also the excerpt from Corn's book in "Seeking Salvation in the Stars," *Aerospace America,* March 1984, pp. 98-99.

18. "The Exploration of Space," *Spaceflight* 27 (1985):241.

19. See Richard Hutton, *Cosmic Chase,* p. 179 and Freeman J. Dyson, "Pilgrims, Saints, and Spacemen," *L-5 News,* August 1979, pp. 1-4.

20. Joel Garreau, "The Solid West Is More Than Votes: It's a State of Mind," *Washington Post,* February 9, 1984.

21. Dennis Overbye, "The Last Frontier," *Discover,* April 1984, pp. 30-31.

22. Daniel Deudney, "Space: The High Frontier in Perspective," p. 52. Richard Slotkin has written that the frontier was the source of a myth that men could regenerate themselves by conquering virgin lands and destroying savage Indians. See his book *The Myth of the Frontier in the Age of Industrialization, 1800-1890* (New York: Atheneum, 1985).

23. Interview with Carolyn Meinel.

24. See Curt Suplee, "In the Strange Land of Robert Heinlein," *Washington Post,* September 5, 1984.

25. Letter from Jon F. Snyder, *Aerospace America,* June 1984, p. 9.

26. Gerard K. O'Neill, "President's Column," *SSI Update,* September/October 1984, p. 1.

27. Brochure on Freeland II, obtained at the L-5 Society's Space Development Conference, San Francisco, April 1984.

28. Gary C. Hudson, "Some Commercial Possibilities in Space," in Larry Geis and Fabrice Florin, eds., *Moving into Space* (New York: Perennial Library, 1978), pp.

96-104, 102.

29. Jerome C. Glenn and George S. Robinson, *Space Trek: The Endless Migration* (Harrisburg, Pa.: Stackpole Books, 1978), p. 203. For a related argument, see Michael A. G. Michaud, "Spaceflight, Colonization, and Independence: A Synthesis," *Journal of the British Interplanetary Society* 30 (1977):83-95, 203-12, 323-31.

30. Remarks by James Muncy during a panel discussion at the Space Development Conference, Washington, D.C., April 27, 1985.

31. Ibid.

32. Back cover of the *L-5 News*, October 1985.

33. William H. Gregory, "The Next Four Years," *Aviation Week and Space Technology*, November 19, 1984, p. 11.

34. *Space: The Crucial Frontier*, p. 1.

35. See the Special Project Issue entitled "American Renewal," February 23, 1981. New York Times/CBS polls showed American optimism rising after 1981. See also Howell Raines, "Optimism in Nation is Increasing," *New York Times*, January 21, 1985. A similar upturn between 1979 and 1984 is reported in "What Makes America Keep on Smiling," *U.S. News and World Report*, January 13, 1986, p. 77.

36. Bainbridge, *The Spaceflight Revolution*, p. 197.

37. Hutton, *The Cosmic Chase*, p. 166.

38. H. Keith Henson, "Memes, L-5, and the Religion of the Space Colonies," *L-5 News*, September 1985, pp. 5-8.

39. Interview with Dennis Stone, December 17, 1983.

40. Trudy E. Bell, *Upward: Status Report and Directory of the American Space Interest Movement, 1984-1985*, available from the author.

41. Interview with Jerry Grey.

42. Interview with Mark R. Chartrand III.

43. Interview with Gary Oleson.

44. See Walter A McDougall, "Technocracy and Statecraft in the Space Age," p. 1013.

45. Trudy E. Bell, "From Little Acorns . . . : American Space Interest Groups, 1980-1982," unpublished paper.

46. Interview with James E. Oberg.

47. Ralph Turner and Lewis Killian, *Collective Behavior*, 2nd ed. (Englewood Cliffs, N.J.: Prentice-Hall, 1972), p. 246.

48. Robert D. McWilliams, "An Analysis of the Socio-Political Status of Efforts Toward the Development of Space Manufacturing Facilities," in Jerry Grey and Christine Krop, eds., *Space Manufacturing III: Proceedings of the Fourth Princeton/AIAA Conference, May 14-17, 1979* (New York: American Institute of Aeronautics and Astronautics, 1979), p. 178.

49. Trudy E. Bell, "Space Activists on Rise," p. 10.

50. McWilliams, "Analysis of Socio-Political Status," p. 178.

51. Trudy E. Bell, "Space Activists on Rise," p. 10.

52. Trudy E. Bell, "From Little Acorns. . ."

53. Trudy E. Bell, "Know Thyself: The Myth of the Space Movement," *Space World*, June 1985, p. 2.

54. Hutton, *Cosmic Chase*, p. 177.

55. A classic statement is in Mancur Olson, Jr., *The Logic of Collective Action*

(Cambridge, Mass.: Harvard University Press, 1965).

56. Terry M. Moe, *The Organization of Interests: Incentives and the Dynamics of Political Interest Groups* (Chicago: University of Chicago Press, 1980).

57. Nathan C. Goldman and Michael Fulda, "Space Interest Groups: Galaxies of Interests, Galaxies of Groups," unpublished paper, to be published in Volume 2 of the *Space Humanization Series.*

58. Norman J. Ornstein and Shirley Elder, *Interest Groups, Lobbying, and Policymaking* (Washington, D.C.: Congressional Quarterly Press, 1978).

59. Interview with Howard and Janelle Gluckman.

60. Interview with James Muncy, November 18, 1983.

61. Goldman and Fulda, "Space Interest Groups."

62. Loomis and Cigler, "Introduction," p. 1.

63. Trudy E. Bell, "From Little Acorns. . ."

64. Interview with James Wilson.

65. Interview with Joseph Allen.

66. See, for example, Ben Bova, *The High Road* (New York: Pocket Books, 1983), p. 272.

67. Interview with Gary Oleson.

68. Interview with Gary Paiste.

69. Interview with Darrell Branscome.

70. Interview with John Loosbrock and James J. Haggerty.

71. Interview with Patrick Templeton and Russell Ritchie, February 22, 1984.

72. John Kenneth Galbraith, *The Anatomy of Power* (Boston: Houghton Mifflin, 1983), p. 149.

73. Interview with Courtney Stadd.

74. Trudy E. Bell, "From Little Acorns . . ."

75. Interview with David Cummings.

76. Gregg Maryniak, "Return to the Moon," *SSI Update*, March/April 1985, p. 1.

77. Interview with Irwin Pikus.

78. Trudy E. Bell, "From Little Acorns . . ."

79. Ibid.

80. Bainbridge, *The Spaceflight Revolution*, p. 236.

81. Gerard K. O'Neill, "President's Column" *SSI Update*, second quarter 1982, p. 2.

82. In 1985 issues of the *L-5 News.*

83. Glenn and Robinson, *Space Trek*, p. 34. See also Jesco von Puttkamer, "Developing Space Occupancy: Perspectives on NASA Future Space Program Planning," *Journal of the British Interplanetary Society* 29 (1976):147-73.

84. Undated Campaign for Space circular letter from Thomas J. Frieling, 1984.

85. Remarks by Nathan C. Goldman during a panel discussion at the Space Development Conference, Washington, D.C., April 26, 1985.

86. Conversation with Nathan C. Goldman, April 26, 1985. Von R. Eshelman presented a space scientist's critique of these two sets of assertions in "Deciding on Means and Ends in Space," *Space World*, June 1984, p. 2.

87. As quoted by Scott Holton and Ed Naha in "The Shape of Things to Come," *Future*, May 1978, pp. 30-33, 33.

88. As quoted by Arthur C. Clarke in *The Promise of Space* (New York: Harper

& Row, 1968), p. 310.

89. Speech by James Beggs to the Society of Automotive Engineers, May 21, 1985, as quoted in the American Astronautical Society *News Letter,* September 1985.

90. See Louis D. Friedman, "Visions of 2010: Human Missions to Mars, the Moon, and the Asteroids," *The Planetary Report,* March/April 1985, pp. 4-6, 22.

91. Bainbridge, *Spaceflight Revolution,* chapter 2.

92. Hutton, *Cosmic Chase,* p. 166.

93. Leonard W. David, "Are You Pro-Space?" *Space World,* November 1982, p. 2.

CHAPTER 15

1. Joseph P. Allen, *Entering Space* (New York: Stewart, Tabori, & Chang, 1984), p. 218.

2. See "Roaming the High Frontier," *Time,* November 26, 1984, pp. 16-20. For a more critical view, see William J. Broad, "In Harsh Light of Reality, the Shuttle is Being Reevaluated," *New York Times,* May 14, 1985.

3. "Four More Years: Aerospace Wins Big," *Aerospace America,* January 1985, pp. 10-14. See also "Reagan Victory Margin Aids Defense, Space Objectives," *Aviation Week and Space Technology,* November 12, 1984, pp. 16-17.

4. Jeffrey M. Lenorovitz, "Joint Agreements Set Stage for Space Station Cooperation," *Aviation Week and Space Technology,* April 1, 1985, pp. 16-17; "Canada Accepts Space Station Invitation," *Aviation Week and Space Technology,* March 25, 1985, p. 23; "Japan Agrees to Participate in U.S. Space Station," *New York Times,* May 10, 1985; and "ESA/NASA Space Station Agreement Signed at Paris Air Show," *Aviation Week and Space Technology,* June 7, 1985.

5. See note 15, Chapter 11.

6. "Straphangers On," *Aviation Week and Space Technology,* September 16, 1985, p. 15.

7. Iver Peterson, "U.S. Activates Unit for Space Defense," *New York Times,* September 24, 1985; "Pentagon Christens U.S. Space Command," *Washington Post,* September 24, 1985.

8. See note 122, Chapter 11 and Walter Pincus and Michael Weiskopf, "ASAT Weapon Destroys Satellite in Difficult Test," *Washington Post,* September 14, 1985.

9. "Coalition for SDI Formed," *High Frontier Newsletter,* October 1985, p. 1.

10. "Congress Approves 1986 Defense Budget, Bans ASAT Tests in Space," *Aviation Week and Space Technology,* December 23, 1985, p. 19.

11. "Space-Processed Latex Spheres Sold," *Aviation Week and Space Technology,* July 22, 1985, p. 22.

12. "Flight Produces Contamination-Free Hormone," *Aviation Week and Space Technology,* April 22, 1985. See also "Commercial Space Prospects Brighten," *Aerospace America,* April 1985, p. 1 and Thomas O'Toole, "Out-of-this-World Experiments," *Washington Post,* April 9, 1985.

13. Thomas O'Toole, "NASA Agrees to Launch Private Chemical Factory," *Washington Post,* August 21, 1985; Craig Covault, "NASA Approves Fly-Now, Pay-Later Plan for Orbiting Industrial Facility," *Aviation Week and Space Technology,*

August 26, 1985, pp. 16-17; and William H. Gregory, "New Ideas in Space," *Aviation Week and Space Technology*, August 26, 1985, p. 9.

14. "New Study Finds Commercial Space Potential Uncertain, Expectations Too High," *Space Commerce Bulletin*, May 24, 1985, p. 7.

15. "New Pricing Policy," *Aviation Week and Space Technology*, August 5, 1985, p. 13. For some critical views, see Chris Peterson, "Shuttle Pricing and Space Development," *L-5 News*, January/February 1985, pp. 8-16.

16. See "Former NASA Official Urges Space Commerce Subsidies," *Aviation Week and Space Technology*, July 8, 1985, p. 86.

17. "Conestoga Booster Will Launch Human Ashes for Space Burial," *Aviation Week and Space Technology*, January 21, 1985, pp. 20-21; "Ashes to Ashes – to Orbit," *Science* 227 (1985):615; "Ashes of the Stars," *Sky and Telescope*, June 1985, p. 491.

18. See, among others, "LANDSAT Threatened Again," *Science* 228 (1985):308; "OMB Approves Funds to Shift LANDSAT to Private Sector," *Aviation Week and Space Technology*, May 27, 1985, p. 19; "LANDSAT Logjam Finally Cleared," *Aerospace America*, July 1985, p. 1.

19. Craig Covault, "NASA, Defense Dept. Drop Idea of Private Shuttle Management," *Aviation Week and Space Technology*, April 29, 1985, pp. 42-43.

20. "Teacher Picked for Space Trip," *Washington Post*, July 20, 1985; "Teacher Is Picked for Shuttle Trip," *New York Times*, July 20, 1985; Richard A. Methia, "Riding the Blackboard Shuttle," *Newsweek*, July 8, 1985, p. 8; Edwards Park, "Around the Mall and Beyond," *Smithsonian*, August 1985, pp. 20, 22.

21. "Garn, Head of Senate Space Panel, Is Chosen to Fly Aboard Shuttle," *New York Times*, November 8, 1984; "NASA Invites Rep. Nelson to be Next Legislator Aboard Shuttle," *Washington Post*, September 7, 1985.

22. "Competition Set to Select Reporter for Shuttle Flight," *Aviation Week and Space Technology*, October 25, 1985, p. 28; "Journalist to Fly on 1986 Shuttle Mission," *Aerospace Daily*, October 25, 1985, p. 300; "Journalists Race for Space," *Space Business News*, November 4, 1985, p. 4.

23. Letter from Al Stewart, *Aviation Week and Space Technology*, September 9, 1985, p. 120.

24. Thomas F. Rogers, "Homesteading the New Frontier," *Space World*, June 1985, pp. 4-6. He had presented a paper on this subject in May 1985 at the AIAA Annual Meeting in Washington, D.C. See "Station Role Proposed for Shuttle Tanks," *Aviation Week and Space Technology*, April 29, 1985, p. 145.

25. William E. Schmidt, "Young Space Enthusiasts Crowd Alabama Camp," *New York Times*, September 8, 1985; Lance Morrow, "In Alabama: The Right Stuff," *Time*, October 14, 1985, pp. 14, 20.

26. A Society Expeditions advertisement for "Project Space Voyage" appeared in the *New York Times* on May 12, 1985. See also Leonard W. David, "Commuting into the Cosmos," *Space World*, August 1985, p. 9; "Project Space Voyage," brochure from Society Expeditions, June 1985; and "Launcher Company, Travel Agency Reach Space Tour Pact," *Aviation Week and Space Technology*, September 30, 1985, p. 24.

27. David White, "Space Resort Site Being Sought in Ventura County," *Los Angeles Times*, March 25, 1985.

28. L-5 Society letter from Gordon R. Woodcock, June 17, 1985 (with enclosed survey); Ben Bova, "President's Message August 1985," *Insight* (National Space

Institute), August 1985, pp. 1a, 3a; "L-5/NSI Merger Plan Summary," undated L-5 Society circular, 1985.

29. Letter from H. Keith Henson, *L-5 News*, September 1985, p. 15.

30. Undated Spacepac letter from Ken Swezey, early 1985; undated National Space Institute Letter from Hugh Downs, January 1985; "The Space Station War: The Battle of Senate Authorizations," undated L-5 Society circular, August 1985.

31. Jack Anderson, "President Hopes to Spur Youths into Space Age," *Washington Post*, July 6, 1984.

32. National Space Institute Press Release, June 20, 1984. See Ellen I. Kelley, "The Young Astronauts," *Aerospace*, Summer 1985, pp. 14-16 and T. Wendell Butler, "The American Young Astronaut Program: A First Year Status Report," paper presented to the Congress of the International Astronautical Federation, Stockholm, Sweden, October 1985.

33. Trudy E. Bell, *Upward: Status Report and Directory of the American Space Interest Movement, 1984-1985*, pp. 69-72, available from author.

34. See, for example, the World Space Foundation advertisement in *Discover*, August 1985, p. 82.

35. See Louis D. Friedman, "Visions of 2010: Human Missions to Mars, the Moon, and the Asteroids," *The Planetary Report*, March/April 1985, pp. 4-5.

36. Undated Planetary Society letter from Louis D. Friedman, 1985. Spielberg flipped the switch to begin the operation of the Megachannel Extraterrestrial Assay on September 29, 1985.

37. Undated circular letter from the United States Space Foundation, ca. 1984; undated leaflet "The United States Space Foundation: Overview," ca. 1984; T. R. Reid, "Life Looks Up for City After a Cosmic Coup," *Washington Post*, December 3, 1984.

38. Craig Covault, "Soviets in Houston Reveal New Lunar, Mars, Asteroid Flights," *Aviation Week and Space Technology*, April 1, 1985, pp. 18-20; M. Mitchell Waldrop, "A Soviet Plan for Exploring the Planets," *Science* 228 (1985):698, 703; James E. Oberg, "Soviets Unveil Upcoming Deep-Space Probes," *Aerospace America*, June 1985, pp. 92-94; and "Soviets Planning 1988 Mission to Study Martian Moon Photos," *Aviation Week and Space Technology*, October 28, 1985, pp. 66-67.

39. "Soviet Strategic Defense Programs," a booklet released by the U.S. Department of Defense and the U.S. Department of State, October 1985.

40. Jeffrey M. Lenorovitz, "Europeans Developing Space Plans for the 1990s," *Aviation Week and Space Technology*, December 21, 1981, pp. 52-53; "ESA Defining Future in Space," *Aviation Week and Space Technology*, October 4, 1982, p. 22; Jeffrey M. Lenorovitz, "Germany, Italy Propose Space Station," *Aviation Week and Space Technology*, February 20, 1984, pp. 55-56; David Dickson, "French Take Steps for European Space Station," *Science* 224 (1984):1413; Jeffrey M. Lenorovitz, "ESA Approves New Ariane Launcher, U.S. Station Role," *Aviation Week and Space Technology*, February 4, 1985; Kenneth Owen, "Ariane 5 and Columbus Spearhead New ESA Program," *Aerospace America*, April 1985, p. 24; Pierre Langereux, "Hermes: France's Winged Space Messenger to Fly in 1995," *Aerospace America*, August 1985, pp. 13-18; "Japan Studies Small Shuttle Development," *Aviation Week and Space Technology*, October 4, 1982, p. 22; and "Japan Prepares for Next Space Phase," *Aviation Week and Space Technology*, March 12, 1984, pp. 129-30.

41. See John Noble Wilford, "European Probe is Aimed at Halley's Veiled Heart," *New York Times,* July 2, 1985.

42. James J. Harford, "Chinese Launch Vehicles for Sale," *Aerospace America,* July 1985, pp. 20, 90, 92.

43. "British Institutionalize Space," *Aerospace America,* May 1985, p. 1; "Italy Plans Cabinet-Level Space Agency," *Aviation Week and Space Technology,* September 24, 1984, p. 19.

44. See Johan Benson, "SDI Presages New Space Transportation," *Aerospace America,* May 1985, pp. 28-30.

45. James Beggs, "Space Station: The Next Logical Step," *Aerospace America,* September 1984, pp. 47-52, 48.

46. Walter Froelich, *Space Station: The Next Logical Step* (Washington, D.C.: National Aeronautics and Space Administration, 1985), p. 47.

47. See, for example, "Industry, Scientists Seek Approaches to Tap SDI's Commercial Potential," *Aviation Week and Space Technology,* November 25, 1985, p. 75.

48. John Noble Wilford, "Seminar Envisions US-Soviet Mars Venture," *New York Times,* July 17, 1985; Thomas O'Toole, "U.S.-Soviet Mars Mission Urged," *Washington Post,* July 17, 1985; "Humans on Mars? Why Not?" *Time,* July 29, 1985, p. 67.

49. See Jerry Grey, "Mars or Bust," *Aerospace America,* September 1981, pp. 24, 26, 102.

50. *U.S.-Soviet Cooperation in Space: A Technical Memorandum* (Washington, D.C.: Office of Technology Assessment, 1985). See also John Noble Wilford, "Study Sees Gains from US-Soviet Space Efforts," *New York Times,* July 18, 1985.

51. Arthur Kantrowitz, "The Ming Navy and the U.S. Space Program," *Astronautics and Aeronautics,* September 1981, pp. 44-46.

52. Walter A. McDougall, "Technocracy and Statecraft in the Space Age," *American Historical Review* 87 (1982):1010-40.

53. "Space Commercialization Group Includes Non-Aerospace Firms," *Aviation Week and Space Technology,* March 4, 1985, p. 20.

54. William S. Bainbridge, *The Spaceflight Revolution* (New York: John Wiley & Sons, 1976), pp. 247-52.

55. *1985 Long-Range Program Plan* (Washington, D.C.: National Aeronautics and Space Administration, 1984), p. II-2.

56. "Beggs Discusses Long-Term Space Plans," *Aerospace Daily,* March 27, 1984, p. 149 and "Space Colonies," *Aviation Week and Space Technology,* March 26, 1985, p. 15.

57. "Space Station Symbolizes Evolution of U.S. Space Activities: Beggs," *Aerospace Daily,* October 25, 1984, pp. 295-96, 196.

58. Thomas O'Toole, "NASA's Master Plan," *Omni,* December 1984, pp. 70-72, 148, 150.

59. "Space Station Is Only the Beginning, Keyworth Says," *Aerospace Daily,* July 3, 1984, p. 13.

60. "Paine Defines Space Goals for Next Century," *Aerospace Daily,* October 31, 1984, p. 323. Michaud had suggested the .1 percent goal in "Spaceflight, Colonization, and Independence," *Journal of the British Interplanetary Society* 30 (1977).

61. *Civilian Space Stations and the U.S. Future in Space* (Washington, D.C.: Office of Technology Assessment, 1984), p. 4.

62. As quoted in "Science and the Citizen," *Scientific American*, January 1985, p. 54.

63. *Civilian Space Stations and the U.S. Future in Space*, p. 113.

64. *Civilian Space Policy and Applications* (Washington, D.C.: Office of Technology Assessment, 1982), pp. 18, 276-77.

65. "Reagan Appoints Commission to Guide Civil Space Effort," *Aviation Week and Space Technology*, April 8, 1985, pp. 19-20. See also Philip J. Klass, "Commission Considers Joint Mars Exploration, Lunar Base Options," *Aviation Week and Space Technology*, July 29, 1985, pp. 47-48 and "Panel Will Report on Long-Term Space Plans," *Aviation Week and Space Technology*, July 29, 1985.

66. See David C. Webb's letter in *Aviation Week and Space Technology*, September 23, 1985, p. 168.

67. "Lunar Base," *Aviation Week and Space Technology*, December 19, 1983, p. 17 and Buzz Aldrin, "Let's Return to the Moon for Good," *Los Angeles Times*, July 22, 1984.

68. See Michael A. G. Michaud and Leonard W. David, "Return to the Moon," *Astronomy*, April 1980, pp. 6-22; Julian Loewe, "Lunar Habitats," *Omni*, December 1982, pp. 172-81; Andrew Chaikin, "Return to the Moon," *Sky and Telescope*, June 1983, p. 493; Wendell W. Mendell, "Return to the Moon," *Aviation/Space*, Fall 1983, pp. 28-29.

69. Craig Covault, "NASA Answers Planning Challenge," *Aviation Week and Space Technology*, October 31, 1983, pp. 21-23.

70. Michael B. Duke, Wendell W. Mendell, and Barney B. Roberts, "Toward a Lunar Base," *Aerospace America*, October 1984, pp. 70-73.

71. See M. Mitchell Waldrop, "Asking for the Moon," *Science* 226 (1984):948-49; Walter Sullivan, "Scientists Chart a Return to the Moon for New Exploits," *New York Times*, December 4, 1984; Mark Washburn, "The Moon – A Second Time Around?" *Sky and Telescope*, March 1985, pp. 209-11; "Beggs Sees Return to Moon Early in Next Century," *Aerospace Daily*, October 30, 1984, pp. 314-15; "Aerospace Perspectives," *Aerospace*, Winter 1984, inside front cover.

72. "Lunar Colony," *Aviation Week and Space Technology*, January 7, 1985, p. 15.

73. Letter from Milan Posposil, *Aerospace America*, February 1985, p. 9.

74. "Futurists See Growth in Robot Population," *New York Times*, December 27, 1984.

75. Wernher von Braun, *Mars Project* (Urbana: University of Illinois Press, 1953) and Wernher von Braun and Willy Ley, *The Exploration of Mars* (New York: Viking Press, 1956).

76. See William K. Hartmann and Odell Raper, *The New Mars: The Discoveries of Mariner 9* (Washington, D.C.: National Aeronautics and Space Administration, 1974).

77. Carl Sagan, "The Planet Venus," *Science* 133 (1961):849-58 and Carl Sagan, "Planetary Engineering on Mars," *Icarus* 20 (1973):513-14. For a general discussion of terraforming, see James E. Oberg, *New Earths* (Harrisburg, Pa.: Stackpole Books, 1981).

78. M. M. Averner and R. D. McElroy, eds., *On the Habitability of Mars: An Approach to Planetary Ecosynthesis,* prepared by the Ames Research Center (Washington, D.C.: National Aeronautics and Space Administration, 1976). See also Arthur L. Robinson, "Colonizing Mars: The Age of Planetary Engineering Begins," *Science* 195 (1977):668 and Carl Sagan's review of James Lovelock and Michael Allaby's *The Greening of Mars, New York Times Book Review,* January 6, 1985, p. 6.

79. See Penelope J. Boston, ed., *The Case for Mars,* volume 57, Science and Technology Series (San Diego: American Astronautical Society [Univelt], 1984). See also Christopher P. McKay, "A Focus on Mars," in Paul Anaejionu, Nathan C. Goldman, and Philip J. Meeks, *Space and Society: Challenges and Choices* (San Diego, Calif.: American Astronautical Society [Univelt], 1984), pp. 411-16.

80. See Andrew Chaikin, "Mars or Bust," *Discover,* September 1984, pp. 12-17, 12.

81. See Chaikin, "Mars or Bust"; Michael W. Carroll, "The First Colony on Mars," *Astronomy,* June 1985, pp. 6-21; Christopher P. McKay, ed., *The Case for Mars II* (San Diego: American Astronautical Society [Univelt], 1985). Carroll later wrote that the Planetary Society had made both Case for Mars conferences possible. Letter in *Astronomy,* September 1985, p. 39.

82. Senate resolution 18, introduced by Senators Spark Matsunaga, William Proxmire, and Paul Simon, January 21, 1985.

83. James E. Oberg, "Racing the Soviets to Mars," *Omni,* March 1985, pp. 45-46, 112, 115 and James E. Oberg, "Russians to Mars?" *Analog,* September 1985, pp. 52-61.

84. K. Eric Drexler, "Space Development: The Case Against Mars," *L-5 News,* October 1984, pp. 1-3.

85. "Goals of the American Space Foundation," paper dated February 25, 1985, copy kindly provided by William G. Norton.

86. Leslie R. Shepherd, "Interstellar Flight," *Journal of the British Interplanetary Society* 11 (1952):149-55.

87. Iosif Shklovskii and Carl Sagan, *Intelligent Life in the Universe* (New York: Delta, 1966), p. 449.

88. *Project Daedalus: The Final Report on the BIS Starship Study,* supplement to the *Journal of the British Interplanetary Society,* 1978.

89. Robert L. Forward, "A National Space Program for Interstellar Exploration," in *Future Space Programs 1975,* a compilation of papers prepared for the Subcommittee on Space Science and Applications of the Committee on Science and Technology, U.S. House of Representatives, Volume II (Washington, D.C.: U.S. Government Printing Office, 1975), pp. 279-326.

90. Eugene F. Mallove, Robert L. Forward, Zbigniew Paprotny, and Jurgen Lehmann, "Interstellar Travel and Communication: A Bibliography," *Journal of the British Interplanetary Society* 33 (1980):entire issue.

91. Interview with Robert L. Forward.

92. Eugene F. Mallove, "Mars: A Great Planet, But It Needs a Little Work," *Washington Post,* December 16, 1984.

93. "IRAS News: Tempel 2 and Vega," *Astronomy,* November 1983, pp. 62-63; Paul R. Weissman, "The Vega Particulate Shell: Comets or Asteroids?" *Science*

224 (1984):987-88; Thomas O'Toole, "Telescope Finds Third Solar System," *Washington Post*, December 17, 1983; Thomas O'Toole, "Mystery Heavenly Body Discovered," *Washington Post*, December 30, 1983; "NASA Scientist Finds Nascent-Planet Sites," *Washington Post*, June 13, 1984; and "Astronomers Locate Possible Solar System Around Star," *Aviation Week and Space Technology*, November 19, 1984, p. 99.

94. Bainbridge, *Spaceflight Revolution*, p. 235.

95. Interview with Benjamin Bova.

96. Robert H. Goddard, in a letter to H. G. Wells, 1932. Bainbridge noted Durkheim's observation that anomie can result from the loss of a ceiling to our aspirations. "The conquest of the universe," Bainbridge wrote, "is itself an anomic goal." *Spaceflight Revolution*, p. 193.

APPENDIX B

1. The history of the British Interplanetary Society is recounted by William S. Bainbridge in *The Spaceflight Revolution* (New York: John Wiley & Sons, 1976), pp. 145-57 and in Frank H. Winter, *Prelude to the Space Age: The Rocket Societies 1924-1940* (Washington, D.C.: Smithsonian Institution Press, 1983), pp. 87-97. Another source is *The British Interplanetary Society: A Descriptive Account, Regulations for Membership* (London: British Interplanetary Society, 1973).

2. Bainbridge, *Spaceflight Revolution*, p. 148.

3. Winter, *Prelude to Space Age*, p. 87.

4. Ibid., p. 93.

5. Bainbridge, *Spaceflight Revolution*, p. 150.

6. Winter, *Prelude to Space Age*, p. 90.

7. Ibid., pp. 95-97 and Bainbridge, *Spaceflight Revolution*, pp. 151-52.

8. Bainbridge, *Spaceflight Revolution*, 154.

9. Ibid., p. 155.

10. Ibid.

11. *British Interplanetary Society*, p. 12.

12. Interview with L. J. Carter, June 1, 1984.

13. "Britain Plans National Space Center," *Aviation Week and Space Technology*, February 4, 1985, p. 18.

14. *Project Daedalus* (London: British Interplanetary Society, 1978).

15. Bainbridge, *Spaceflight Revolution*, p. 156.

16. Interview with L. J. Carter.

17. Ibid.

18. Ibid.

19. "A Third Community in Space, *Spaceflight*, September/October 1984, p. 337.

20. Interview with L. J. Carter.

21. Ibid.

22. Ibid.

Bibliography

The literature on space exploration and spaceflight has grown to enormous proportions. In this bibliography I make no attempt to provide a comprehensive survey; this is a personal, highly selective listing of the works I have found most useful and interesting.

GENERAL AND HISTORICAL

Of the basic works that influenced my generation, the first was Willy Ley's *Rockets and Space Travel: The Future of Flight Beyond the Stratosphere* (New York: Viking, 1947), which is particularly useful for its descriptions of early German rocket experiments. Ley also collaborated with artist Chesley Bonestell on the memorable *The Conquest of Space* (New York: Viking, 1949), whose easily readable text and brilliant paintings inspired many space dreamers. Perhaps the best basic introductions to spaceflight of those years were Arthur C. Clarke's *Interplanetary Flight* (New York: Harper & Brothers, 1950) and *The Exploration of Space* (New York: Harper & Brothers, 1951); some of the same material was carried over into Clarke's excellent book *The Promise of Space* (New York: Harper & Row, 1968). A good, although now somewhat dated, introduction to space exploration and spaceflight was provided by Frederick I. Ordway, III, James Patrick Gardner, and Mitchell R. Sharpe, Jr., in *Basic Astronautics* (Englewood Cliffs, N.J.: Prentice-Hall, 1962) and by Frederick I. Ordway, III, James Patrick Gardner, Mitchell R. Sharpe, Jr., and Ronald C. Wakeford in *Applied Astronautics* (Englewood Cliffs, N.J.: Prentice-Hall, 1963). At about the same time, I. M. Levitt and Dandridge M. Cole published *Exploring the Secrets of Space: Astronautics for the Layman* (Englewood Cliffs, N.J.: Prentice-Hall, 1963), which included Cole's ideas about colonies in space. Wernher von Braun's visions of the human future in space appeared in *Collier's* magazine during 1952, in Cornelius Ryan, ed., *Across the Space Frontier* (New York: Viking, 1952), and in Wernher von Braun, *Space Frontier* (New York: Holt, Rinehart, & Winston, 1971).

In more recent years, Isaac Asimov and artist Robert McCall collaborated on the attractive large format book *Our World in Space* (Greenwich, Conn.: New York Graphic Society, 1974). NASA's *Outlook for Space* (Washington, D.C.: National Aeronautics and Space Administration, 1976), although it is a conservatively written official report, is a useful reference document. Periodically, the Congressional Research Service of the Library of Congress updates its massive published compilations of information on the United States and Soviet space programs. Richard Hutton produced a readable, informal survey of the space arena in *The Cosmic Chase* (New York: New American Library, 1981). The Office of Technology Assessment's *Civilian Space Policy and Applications* (Washington, D.C., 1982) provides a useful overview of U.S. space activities and the issues they raise. For the nonexpert reader, my current favorite among general books on spaceflight and space exploration is Kenneth W. Gatland, ed., *The Illustrated Encyclopedia of Space Technology* (New York: Harmony Books, 1981). Although it is not really either an encyclopedia or a comprehensive history, it is compactly written and superbly illustrated. Of the more personal works written by astronauts, I would recommend Michael Collins, *Carrying the Fire: An Astronaut's Journeys* (New York: Farrar, Straus, & Giroux, 1974) for the Apollo era. Joseph P. Allen's *Entering Space: An Astronaut's Odyssey* (New York: Stewart, Tabori, & Chang, 1984) conveys the Space Shuttle experience by text and magnificent photography, although the observations of the highly intelligent Allen appear to have been somewhat compromised by the "commercialization" of the text.

As for the periodical literature, I have found the most useful general publications to be the weekly *Aviation Week and Space Technology* (for current events), the British Interplanetary Society's monthly *Spaceflight,* and the American monthly *Space World,* now published in collaboration with the National Space Institute. The American Institute of Aeronautics and Astronautics' monthly magazine, which has been called *Aerospace America* since January 1984, has become a much more readable source for generalists under the direction of publisher Jerry Grey.

In my opinion, there is as yet no completely satisfactory history of the Space Age. Eugene M. Emme produced a straightforward history of the early years in *A History of Spaceflight* (New York: Holt, Rinehart, & Winston, 1965). Wernher von Braun and Frederick I. Ordway, III's

History of Rocketry and Space Travel, third revised edition (New York: Crowell, 1975) did a competent job of covering the history through the Skylab era, although much of it is about early rocketry. David Baker has compiled two massively detailed works on rocketry and spaceflight: *The Rocket: The History and Development of Rocket and Missile Technology* (New York: Crown, 1978) and *The History of Manned Space Flight* (New York: Crown, 1981). Although these are very useful as references, they are marred by strained syntax and are not easy reading. NASA has put out useful chronologies of individual years of the U.S. space program in its *Astronautics and Aeronautics* series, although as of this writing the most recent chronology published was for 1976. NASA also has published books on individual projects, such as Edgar M. Cortright, ed., *Apollo Expeditions to the Moon* (Washington, D.C.: National Aeronautics and Space Administration, 1975) and Leland F. Belew, ed., *Skylab, Our First Space Station* (Washington, D.C.: National Aeronautics and Space Administration, 1977), but these tend to be collections of papers without an overall point of view. Frederick I. Ordway, III, and Mitchell R. Sharpe tell an interesting story in *The Rocket Team* (New York: Crowell, 1979) about the German rocket experts who came to the United States. John M. Logsdon's *The Decision to Go to the Moon: Project Apollo and the National Interest* (Chicago: University of Chicago Press, 1970) has become a standard reference; as of this writing, Logsdon was completing a book on the Space Shuttle decision that should be equally useful. The American Astronautical Society has made helpful contributions through its AAS History Series, published by Univelt.

By far the most ambitious historical effort to date is Walter A. McDougall, . . . *The Heavens and the Earth: A Political History of the Space Age* (New York: Basic Books, 1985). This insightful, provocative, and sometimes controversial work is devoted primarily to the political history of the U.S. and Soviet space programs up to 1964, with more recent years covered in less depth. Although not everyone would agree with all of his conclusions, McDougall has shown that space history has come of age.

SPACE ADVOCATES AND SPACE GROUPS

For me, two works on this subject have been more influential than all others: William S. Bainbridge, *The Spaceflight Revolution: A*

Sociological Study (New York: Wiley, 1976) and Frank H. Winter, *Prelude to the Space Age – The Rocket Societies: 1924-1940* (Washington, D.C.: Smithsonian Institution Press, 1983). Arthur C. Clarke edited a useful collection of historical material about the space pioneers and the early years of spaceflight in *The Coming of the Space Age* (New York: Meredith Press, 1967). Some of the more established groups have published historical reminiscences, notably Eugene M. Emme, ed., *Twenty-Five Years of the American Astronautical Society, 1954-1979: Historical Reflections and Projections* (San Diego: Univelt, 1980 [AAS History Series, Volume 2]). For post-Apollo space interest groups, the standard references are Trudy E. Bell, "American Space Interest Groups," (*Star and Sky*, September 1980, pp. 53-60 [reprinted as "Space Activists on Rise," *Insight*, August/September 1980, pp. 1, 3, 10 and a revised version in "Space Activism," *Omni*, February 1981, pp. 50-54, 90-94]); "From Little Acorns . . . : American Space Interest Groups, 1980-1982" (unpublished paper available from the author); "Upward: Status Report and Directory of the American Space Interest Movement, 1984-1985" (published by the author in 1985 and available from her). There also have been a few other generalized articles on the space interest group phenomenon, notably Trudy E. Bell, "People for Space" (*Science 80*, September/October 1980, pp. 99-100); M. Mitchell Waldrop, "Citizens for Space" (*Science* 211 [1981], p. 152); Owen Davies, "Activist Update" (*Omni*, April 1982, pp. 20, 133, 134); Owen Davies, "Activist Update" (*Omni*, May 1983, p. 20); and Dennis Stone, "The Space Advocacy Movement" (*Sky and Telescope*, May 1983, pp. 452-53).

SPACE COLONIZATION AND SPACE INDUSTRIALIZATION

One can find the roots of these ideas in earlier works by Tsiolkowsky, Bernal, and several science fiction writers. For the modern era, however, the primary exponents are Krafft A. Ehricke (in numerous articles and papers), Dandridge M. Cole, and Gerard K. O'Neill. For Cole, I would recommend Dandridge M. Cole and Donald W. Cox, *Islands in Space* (New York: Chilton, 1964) and Dandridge M. Cole, *Beyond Tomorrow: The Next Fifty Years in Space* (Amherst, Wisc.: Amherst Press, 1965). For O'Neill, the basic references are Gerard K. O'Neill, "The Colonization of Space" (*Physics Today*, September 1974, pp. 32-40) and *The High Frontier: Human Colonies in Space* (New York:

William Morrow, 1977). A readable discussion of the broad concept is Thomas A. Heppenheimer, *Colonies in Space* (Harrisburg, Pa.: Stackpole Books, 1977). The proceedings of the biennial Princeton conferences on space manufacturing are published by the American Institute of Aeronautics and Astronautics. The results of the 1975 and 1977 summer workshops were published by NASA in Richard D. Johnson and Charles Holbrow, eds., *Space Settlements: A Design Study* (Washington, D.C.: National Aeronautics and Space Administration, 1977) and John Billingham, William Gilbreath, and Brian T. O'Leary, eds., *Space Resources and Space Settlements* (Washington, D.C.: National Aeronautics and Space Administration, 1979). The best introductory book on space manufacturing is G. Harry Stine, *The Third Industrial Revolution* (New York: G. P. Putnam's Sons, 1975).

DEMOGRAPHICS, SCIENCE FICTION, AND SCIENCE FACT

The only book I know of that deals extensively with the demographics of the space enterprise is Mary A. Holman, *The Political Economy of the Space Program* (Palo Alto, Calif.: Pacific Books, 1974). Jon D. Miller produced the most useful study of public opinion in "The Information Needs of the Public Concerning Space Exploration: A Special Report for the National Aeronautics and Space Administration" (De Kalb: Northern Illinois University, Public Opinion Laboratory, July 20, 1982). There is a useful section on this subject in *Civilian Space Policy and Applications* (Washington, D.C.: Office of Technology Assessment, 1982).

There are several good surveys of science fiction. My personal favorite is James Gunn, *Alternate Worlds: The Illustrated History of Science Fiction* (New York: Visual Library, 1975). A useful compendium of information, although in the dry format of an encyclopedia, is Peter Nicholls, ed., *The Science Fiction Encyclopedia* (Garden City, N.Y.: Doubleday, 1979). There are many periodicals in the science fiction field; *Analog* has been particularly space-oriented, and *Starlog* is good on space themes in visual media.

SPACE SCIENCE

Space science – especially planetary exploration – has been particularly well served by talented writers, spectacular photography, and imaginative art work. Over the years, NASA has published several fine books on the results of planetary missions. My personal favorite is William K. Hartmann and Odell Raper, *The New Mars: The Discoveries of Mariner 9* (Washington, D.C.: National Aeronautics and Space Administration, 1974). Others include Richard O. Fimmel, William Swindell, and Eric Burgess, *Pioneer Odyssey: Encounter with a Giant* (Washington, D.C.: National Aeronautics and Space Administration, 1974); David Morrison and Jane Samz, *Voyage to Jupiter* (Washington, D.C.: National Aeronautics and Space Administration, 1980); Richard O. Fimmel, James Van Allen, and Eric Burgess, *Pioneer: First to Jupiter, Saturn, and Beyond* (Washington, D.C.: National Aeronautics and Space Administration, 1980); and David Morrison, *Voyages to Saturn* (Washington, D.C.: National Aeronautics and Space Administration, 1982). A more generalized publication is Bevan M. French and Stephen P. Maran, *A Meeting with the Universe: Science Discoveries from the Space Program* (Washington, D.C.: National Aeronautics and Space Administration, 1981).

Of the commercially published books on planetary exploration, one of the best is Mark Washburn, *Planetary Encounters: The Exploration of Jupiter and Saturn* (New York: Harcourt, Brace, Jovanovich, 1983). Eric Burgess produced readable accounts in *To the Red Planet* (New York: Columbia University Press, 1978) and *By Jupiter: Odysseys to a Giant* (New York: Columbia University Press, 1982). An excellent survey of the solar system, drawing on the findings of planetary exploration, is J. Kelly Beatty, Brian T. O'Leary, and Andrew Chaikin, eds., *The New Solar System* (Cambridge, Mass.: Sky Publishing, 1981). Clark R. Chapman's well-written *Planets of Rock and Ice: From Mercury to the Moons of Saturn* (New York: Charles Scribner's Sons, 1982) presents a more personal and selective view. For those with a serious interest, Michael H. Carr's *The Surface of Mars* (New Haven, Conn.: Yale University Press, 1981) is a model of the clear, readable presentation of scientific findings.

Of more general books on space science and astronomy, Don Dixon's *Universe* (Boston: Houghton Mifflin, 1981) combines a clear explanatory text with photographs and spectacular paintings by the author. More philosophical and provocative, but also beautifully illustrated, is Carl

Sagan's *Cosmos* (New York: Random House, 1980). There are many excellent general textbooks on astronomy; my personal favorite is George O. Abell, *Exploration of the Universe,* fourth ed. (Philadelphia: Saunders College Publishing, 1982).

The decennial reports of the National Academy of Sciences on astronomy and astrophysics are basic references; each seems better written than its predecessor. They are *Ground-Based Astronomy: A Ten-Year Program* (Washington, D.C.: National Academy of Sciences, 1964); *Astronomy and Astrophysics for the 1970s* (Washington, D.C.: National Academy of Sciences, 1972); and *Astronomy and Astrophysics for the 1980s* (Washington, D.C.: National Academy of Sciences, 1982). For planetary exploration, the Solar System Exploration Committee's *Planetary Exploration Through the Year 2000 – A Core Program* (Washington, D.C.: National Aeronautics and Space Administration, 1983) is a helpful guide. As for the periodical literature, two magazines stand out: the long-established *Sky and Telescope* and the more popularized *Astronomy.* The *Lunar and Planetary Information Bulletin* put out by the Lunar and Planetary Institute of the Universities Space Research Association is good for current news and bibliography. For people with a serious scientific interest, I would recommend the journal *Icarus.*

MILITARY ACTIVITY IN SPACE

In 1982 and 1983, there was a sudden surge of books on the general subject of military activity in space. None is definitive, but several have good features: G. Harry Stine, *Confrontation in Space* (Englewood Cliffs, N.J.: Prentice-Hall, 1981); James Canaan, *War in Space* (New York: Harper & Row, 1982); David Baker, *The Shape of Wars to Come* (New York: Stein and Day, 1982); David Ritchie, *Spacewar* (New York: Atheneum, 1982); Thomas Karas, *The New High Ground: Systems and Weapons of Space Age War* (New York: Simon and Schuster, 1983); and Curtis Peebles, *Battle for Space* (New York: Beaufort Books, 1983). Uri Ra'anan and Robert L. Pfaltzgraff, Jr., edited *International Security Dimensions of Space* (Hamden, Conn.: Archon Books, 1984), based on a 1982 conference at which many speakers were sympathetic with strategic defense. General critiques of military activity in space include Bhupendra Jasani, ed., *Outer Space – A New Dimension of the Arms Race* (Cambridge, Mass.: Oelgeschlager,

Gunn, & Hain [for the Stockholm International Peace Research Institute], 1982) and William J. Durch, ed., *National Interests and the Military Use of Space* (Cambridge, Mass.: Ballinger, 1984). On U.S. policy, the definitive work is Paul B. Stares, *The Militarization of Space: U.S. Policy, 1945-1984* (Ithaca, N.Y.: Cornell University Press, 1985).

In my view, there is as yet no objective, detached, scholarly published work on space-based strategic defenses. As of 1985, the topic remained highly controversial; authors tended to be either very much for such systems or very much against them. Perhaps closest to an objective study is the Office of Technology Assessment's *Space-Based Missile Defenses* (Washington, D.C., 1985 [there is a parallel OTA study of anti-satellite weapons, also published in 1985]). A brief but relatively detached survey appeared in *Strategic Survey, 1984-1985* (London: International Institute of Strategic Studies, 1985, pp. 3-5 and 12-17). A set of pro and con articles appeared in *Issues in Science and Technology* (Fall 1984). Jeff Hecht produced a useful work, somewhat heavy on the technical aspects, in *Beam Weapons: The Next Arms Race* (New York: Plenum Press, 1984).

Basic official U.S. documents on the Strategic Defense Initiative include Fred S. Hoffman, study director, *Ballistic Missile Defenses and U.S. National Security: Summary Report* (Washington, D.C.: Department of Defense, October 1983); *Defense Against Ballistic Missiles: An Assessment of Technologies and Policy Implications* (Washington, D.C.: Department of Defense, April 1984); *The Strategic Defense Initiative: Defensive Technologies Study* (Washington, D.C.: Department of Defense, April 1984); *The President's Strategic Defense Initiative* (Washington, D.C.: Department of Defense, January 1985); *SDI: A Technical Progress Report Submitted to the Secretary of Defense by the Director of the Strategic Defense Initiative Organization* (Washington, D.C., June 1985); and *Report to the Congress on the Strategic Defense Initiative, 1985* (Washington, D.C.: Department of Defense, 1985).

As for private advocates of space-based missile defenses, Daniel O. Graham's *High Frontier: A New National Strategy* (Washington, D.C.: High Frontier, 1982) remains a political landmark, although it is not well written. Graham has followed this with other books, such as one done with Gregory A. Fossedal: *A Defense that Defends: Blocking Nuclear Attack* (Old Greenwich, Conn.: Devin-Adair, 1983); *The Non-Nuclear Defense of Cities: The High Frontier Space-Based Defense Against ICBM Attack* (Cambridge, Mass.: Abt Books, 1983); and the pamphletlike paperback *We Must Defend America: A New Strategy for National Survival* (Chicago: Regnery Gateway, 1983). Space scientist Robert Jastrow has been an aggressive advocate of the Strategic Defense Initiative in his 1985 book *How to Make Nuclear Weapons Obsolete* (Boston: Little, Brown) and in several articles and interviews. Prominent science fiction writers also have risen to the defense of the Strategic Defense Initiative: Benjamin Bova in *Assured Survival: Putting the Star Wars Defense in Perspective* (Boston: Houghton Mifflin, 1984) and Jerry E. Pournelle and Dean Ing in *Mutual Assured Survival* (New York: Baen Books, 1985). The work of Keith B. Payne, ed., *Laser Weapons in Space: Policy and Doctrine* (Boulder, Colo.: Westview Press, 1983) is a collection of essays that tends to be in support of the Strategic Defense Initiative. One of the authors, Colin S. Gray, supported space-based defenses and criticized space arms control in his sharply worded book *American Military Space Policy* (Cambridge, Mass.: Abt Books, 1983).

Among the critics of the Strategic Defense Initiative, the most active group has been the Union of Concerned Scientists, which published *The Fallacy of Star Wars* (New York: Vintage Books) in 1984. Generally critical essays can be found in Sidney D. Drell, Philip J. Farley, and David Holloway, *The Reagan Strategic Defense Initiative: A Technical, Political, and Arms Control Assessment* (Stanford, Calif.: Stanford University, July 1984) and in Jeffrey Boutwell, Donald Hafner, and F. A. Long, eds., *Weapons in Space: The Technology and Politics of Ballistic Missile Defense and Anti-Satellite Weapons,* (New York: W. W. Norton, 1985).

As for periodicals on military activity in space, the most useful to the generalist are *Aviation Week and Space Technology* and the Washington-based newsletter *Military Space*. The latter put out an expensive *Guide to the Strategic Defense Initiative* in 1985.

SPACE COMMERCIALIZATION

Most of the literature on space commercialization is in periodicals such as *Aviation Week and Space Technology, Commercial Space, Space Business News, Space Commerce Bulletin, Commercial Space Report,* and *International Space Business Review*. Particularly useful survey articles are Vernon Louviere, "Space: Industry's New Frontier" (*Nation's Business,* February 1978, pp. 25-41); Charles Chafer, "A Business Perspective on Space Policy," in Paul Amaejionu, Nathan C. Goldman, and Philip J. Meeks, eds., *Space and Society: Challenges and Choices* (San Diego, Calif.: American Astronautical Society [Univelt], 1984, pp. 29-40); and David Osborne, "Business in Space" (*The Atlantic Monthly,* May 1985, pp. 45-58). Jerry Grey provided some general coverage of this topic in his book *Beachheads in Space: A Blueprint for the Future* (New York: Macmillan, 1983). By 1985, larger studies had appeared, notably Edward Ridley Finch, Jr., and Amanda Lee Moore, *Astrobusiness: A Guide to the Commerce and Law of Outer Space* (New York: Praeger, 1985) and Nathan C. Goldman, *Space Commerce: Free Enterprise on the High Frontier* (Cambridge, Mass.: Ballinger Books, 1985). A useful survey of the issues can be found in David C. Webb, *A Current Perspective on Space Commercialization* (Washington, D.C.: Aerospace Industries Association of America, 1985). The communications satellite industry is well served with newsletters such as *Satellite Week* and *Satellite News.*

SPACE STATION

The general literature on space stations began to grow after President Regan's initiative of January 1984 but is still largely in the form of periodical articles. Jerry Grey's *Beachheads in Space* (New York: Macmillan, 1983) includes considerable material on space stations, and Brian T. O'Leary's *Project Space Station* (Harrisburg, Pa.: Stackpole Books, 1983) is almost entirely devoted to the subject; both are advocacy books. Theodore R. Simpson, ed., *The Space Station: An Idea Whose Time Has Come* (New York: IEEE Press, 1984) is a useful collection of papers and statements about the station. Hans Mark's "The Space Station – Mankind's Permanent Presence in Space" (*Aviation, Space, and Environmental Medicine,* October 1984, pp. 948-56) provides a helpful overview of the history of the American space station.

INTEREST GROUP POLITICS

Useful summaries of theories about interest group politics appear in Allan J. Cigler and Burdett A. Loomis, "Introduction: The Changing Nature of Interest Group Politics," in Allan J. Cigler and Burdett A. Loomis, eds., *Interest Group Politics* (Washington, D.C.: Congressional Quarterly Press, 1983) and in Section I of Norman J. Ornstein and

Shirley Elder, *Interest Groups, Lobbying, and Policymaking* (Washington, D.C.: Congressional Quarterly Press, 1978). Classic, influential works include Mancur Olson, Jr.'s economics-oriented studies, *The Logic of Collective Action* (Cambridge, Mass.: Harvard University Press, 1965) and *The Rise and Decline of Nations: Economic Growth, Stagflation, and Social Rigidities* (New Haven, Conn.: Yale University Press, 1982). Terry M. Moe's *The Organization of Interests: Incentives and the Internal Dynamics of Political Interest Groups* (Chicago: University of Chicago Press, 1980) is heavy reading but full of insights relevant to this book.

THE MOON, MARS, AND BEYOND

Of the many books that discuss Moon bases and colonies, my favorite is Neil P. Ruzic, *Where the Winds Sleep: Man's Future on the Moon* (Garden City, N.Y.: Doubleday, 1970). More than 50 papers from the 1984 lunar base symposium appear in Wendell W. Mendell, ed., *Lunar Bases and Space Activities of the 21st Century* (Houston: Lunar and Planetary Institute, 1985). For Mars, the classic work is Willy Ley and Wernher von Braun, *The Exploration of Mars* (New York: Viking, 1956). A good modern survey of the subject is James E. Oberg, *Mission to Mars: Plans and Concepts for the First Manned Landing* (Harrisburg, Pa.: Stackpole Books, 1982). Saul J. Adelman and Benjamin Adelman presented a fine, although slightly technical, discussion of solar system colonization and interstellar travel in *Bound for the Stars* (Englewood Cliffs, N.J.: Prentice-Hall, 1981).

Index

Abrahamson, James, 216-17, 230
Action Committee on Technology (ACT), 182-83
Adams, Benson D., 220
Adamson, Sandra, 94, 97, 98
Ad Hoc Coordinating Committee on Space, 154, 158
Advanced Communications Technology Satellite, 37
Advanced Propulsion Technologies, 258
Advanced X-Ray Astronomy Facility, 199
Aeronautical Chamber of Commerce of America, 29
Aerospace America, 33, 34
Aerospace and Electronic Systems Society, 35, 142
Aerospace Education Association (AEA), 31, 142, 153, 158
Aerospace Force (Proposed), 217
Aerospace Industries Association, 29-30, 31, 138, 139, 161, 307
Agnew, Spiro T., 15
Aircraft Industries Association, 29
Air Force, 216-17, 259, 264-65; Manned Orbiting Laboratory, 272; Space Command, 217, 278
Air Force Academy, 217
Air Line Pilots Association, 143
Akaka, Daniel, 173, 176, 177, 182, 261
Alabama Space and Rocket Center, 40
Alcorn, Fred, 265
Aldridge, Edward, 322
Aldrin, Edwin E. (Buzz), 336
Alford, Andrew, 171
Alfven, Hannes, 77
Alleman, John K., 47
Allen, Joseph P., 21, 308, 321
Allen, Richard, 170
Alvarez, Luis, 334
Amateur Space Telescope, 147
American Astronautical Society, 22, 100, 145, 153, 154, 164, 168, 177, 244, 247, 254, 256, 302, 324-25, 338
American Astronomical Society, 189,

202, 204-06
American Institute of Aeronautics and Astronautics, 11, 32, 56, 69, 104, 139, 153, 154, 227, 254-55, 279, 283, 285
American Interplanetary Society, 5, 124
American Interstellar Society, 268, 325
American Rocket Society (ARS), 5, 11, 12, 32
American Security Council Foundation, 227
American Society for Aerospace Education (ASAE), 31, 153, 325
American Society of Aerospace Pilots (ASAP), 143-44, 269, 294, 299, 305
American Space Foundation, 53, 160, 171-74, 175, 179, 243, 285, 325, 339; membership, 139, 153, 335
American Space Frontier Committee, 159, 231
American Space Political Action Committee, 171
American Telephone and Telegraph, 248
America on the Moon (Holmes), 10
Analog, 55, 125, 126-27
Analog Science Fiction/Science Fact, 126
Anderson, Charlene, 208
Anderson, Clinton, 19, 26
Anderson, Jack, 325
Anderson, John, 154, 167, 181
Anderson, Peter, 167, 168
Anderson, Poul, 169, 299
Anderson Supporters for Space Science and Technology, 167
Andropov, Yuri, 238
Applications Technology Satellites, 36, 37, 249
Apollo Society, 325
Apollo-Soyuz Test Project, 50, 329
Apollo spacecraft (*see* Project Apollo)
"April Coalition," 152
Arc Technologies, 260, 269
Ariane, 148
Arming the Heavens (Manno), 228
Armstrong, Neil, 75, 335

About the Author

Michael A. G. Michaud, a member of the Aviation/Space Writers Association, is the author of more than 40 published articles on spaceflight and related subjects. A Californian, he holds a master's degree in Political Science from the University of California at Los Angeles and has done further graduate work at UCLA and Georgetown University. Michaud has delivered papers on space-related subjects at the Princeton/AIAA Conference on Space Manufacturing, the Congress of the International Astronautical Federation, and the annual meeting of the American Anthropological Association. He also has spoken on such subjects at conferences of the American Association for the Advancement of Science and the American Astronautical Society. He is a member of the American Institute of Aeronautics and Astronautics and the American Astronautical Society and a Fellow of the British Interplanetary Society.

960